I0890856

Rebuilding and Restructuring the Tourism Industry:

Infusion of Happiness and Quality of Life

André Riani Costa Perinotto
Universidade Federal do Delta do Parnaíba, Brazil

Verônica Feder Mayer
Federal Fluminense University, Brazil

Jakson Renner Rodrigues Soares
Universidade da Coruña, Spain & Universidade Estadual do Ceará, Brazil

A volume in the Advances in Hospitality, Tourism, and the Services Industry (AHTSI) Book Series

IGI Global
PUBLISHER of TIMELY KNOWLEDGE

Published in the United States of America by
IGI Global
Business Science Reference (an imprint of IGI Global)
701 E. Chocolate Avenue
Hershey PA, USA 17033
Tel: 717-533-8845
Fax: 717-533-8661
E-mail: cust@igi-global.com
Web site: http://www.igi-global.com

Copyright © 2021 by IGI Global. All rights reserved. No part of this publication may be reproduced, stored or distributed in any form or by any means, electronic or mechanical, including photocopying, without written permission from the publisher.
Product or company names used in this set are for identification purposes only. Inclusion of the names of the products or companies does not indicate a claim of ownership by IGI Global of the trademark or registered trademark.

Library of Congress Cataloging-in-Publication Data

Names: Perinotto, André Riani Costa, 1981- editor. | Mayer, Verônica Feder, 1965- editor. | Soares, Jakson Renner Rodrigues, 1980- editor.
Title: Rebuilding and restructuring the tourism industry : infusion of happiness and quality of life / André Riani Costa Perinotto, Verônica Feder Mayer, and Jakson Renner Rodrigues Soares, Editors.
Description: Hershey, PA : Business Science Reference, [2021] | Includes bibliographical references and index. | Summary: "we want to understand from different angles, segments, and disciplines this fundamental relationship between tourism, happiness, and life quality. Despite the overall beneficial effects of travel on happiness levels, tourism is not an activity with solely positive implications on peoples' lives and therefore this book will help readers understand how happiness manifests through tourism and how it can be expanded to different tourism actors, such as tourists, workers, and communities"-- Provided by publisher.
Identifiers: LCCN 2021015974 (print) | LCCN 2021015975 (ebook) | ISBN 9781799872399 (hardcover) | ISBN 9781799872405 (paperback) | ISBN 9781799872412 (ebook)
Subjects: LCSH: Tourism--Psychological aspects. | Quality of life. | Tourism--Social aspects.
Classification: LCC G155.A1 R4132 2021 (print) | LCC G155.A1 (ebook) | DDC 306.4/819--dc23
LC record available at https://lccn.loc.gov/2021015974
LC ebook record available at https://lccn.loc.gov/2021015975

This book is published in the IGI Global book series Advances in Hospitality, Tourism, and the Services Industry (AHTSI) (ISSN: 2475-6547; eISSN: 2475-6555)

British Cataloguing in Publication Data
A Cataloguing in Publication record for this book is available from the British Library.

All work contributed to this book is new, previously-unpublished material.
The views expressed in this book are those of the authors, but not necessarily of the publisher.

For electronic access to this publication, please contact: eresources@igi-global.com.

Advances in Hospitality, Tourism, and the Services Industry (AHTSI) Book Series

ISSN:2475-6547
EISSN:2475-6555

Editor-in-Chief: Maximiliano Korstanje, University of Palermo, Argentina

MISSION

Globally, the hospitality, travel, tourism, and services industries generate a significant percentage of revenue and represent a large portion of the business world. Even in tough economic times, these industries thrive as individuals continue to spend on leisure and recreation activities as well as services.

The Advances in Hospitality, Tourism, and the Services Industry (AHTSI) book series offers diverse publications relating to the management, promotion, and profitability of the leisure, recreation, and services industries. Highlighting current research pertaining to various topics within the realm of hospitality, travel, tourism, and services management, the titles found within the AHTSI book series are pertinent to the research and professional needs of managers, business practitioners, researchers, and upper-level students studying in the field.

COVERAGE

- Cruise Marketing and Sales
- Health and Wellness Tourism
- Service Design
- Hotel Management
- Casino Management
- International Tourism
- Travel Agency Management
- Sustainable Tourism
- Destination Marketing and Management
- Food and Beverage Management

IGI Global is currently accepting manuscripts for publication within this series. To submit a proposal for a volume in this series, please contact our Acquisition Editors at Acquisitions@igi-global.com or visit: http://www.igi-global.com/publish/.

The Advances in Hospitality, Tourism, and the Services Industry (AHTSI) Book Series (ISSN 2475-6547) is published by IGI Global, 701 E. Chocolate Avenue, Hershey, PA 17033-1240, USA, www.igi-global.com. This series is composed of titles available for purchase individually; each title is edited to be contextually exclusive from any other title within the series. For pricing and ordering information please visit http://www.igi-global.com/book-series/advances-hospitality-tourism-services-industry/121014. Postmaster: Send all address changes to above address. Copyright © 2021 IGI Global. All rights, including translation in other languages reserved by the publisher. No part of this series may be reproduced or used in any form or by any means – graphics, electronic, or mechanical, including photocopying, recording, taping, or information and retrieval systems – without written permission from the publisher, except for non commercial, educational use, including classroom teaching purposes. The views expressed in this series are those of the authors, but not necessarily of IGI Global.

Titles in this Series

For a list of additional titles in this series, please visit:
http://www.igi-global.com/book-series/advances-hospitality-tourism-services-industry/121014

Socio-Economic Effects and Recovery Efforts for the Rental Industry Post-COVID-19 Strategies
Maximiliano Korstanje (University of Palermo, Argentina)
Business Science Reference • © 2021 • 281pp • H/C (ISBN: 9781799872870) • US $195.00

Risk, Crisis, and Disaster Management in Small and Medium-Sized Tourism Enterprises
Diego R. Toubes (University of Vigo, Spain) and Noelia Araújo-Vila (University of Vigo, Spain)
Business Science Reference • © 2021 • 330pp • H/C (ISBN: 9781799869962) • US $195.00

Resiliency Models and Addressing Future Risks for Family Firms in the Tourism Industry
Anita Zehrer (MCI The Entrepreneurial School, Austria) Gundula Glowka (MCI The Entrepreneurial School, Austria) Katrin Magdalena Schwaiger (MCI The Entrepreneurial School, Austria) and Victoria Ranacher-Lackner (MCI The Entrepreneurial School, Austria)
Business Science Reference • © 2021 • 361pp • H/C (ISBN: 9781799873525) • US $195.00

Strategic International Restaurant Development From Concept to Production
Angelo A. Camillo (Sonoma State University, USA)
Business Science Reference • © 2021 • 497pp • H/C (ISBN: 9781799843429) • US $195.00

ICT Tools and Applications for Accessible Tourism
Celeste Eusébio (University of Aveiro, Portugal) Leonor Teixeira (University of Aveiro, Portugal) and Maria João Carneiro (University of Aveiro, Portugal)
Business Science Reference • © 2021 • 356pp • H/C (ISBN: 9781799864288) • US $215.00

Impact of New Media in Tourism
Maria Gorete Dinis (GOVCOPP, Polytechnic Institute of Portalegre, Portugal) Luís Bonixe (ICNOVA, Polytechnic Institute of Portalegre, Portugal) Sónia Lamy (ICNOVA, Polytechnic Institute of Portalegre, Portugal) and Zélia Breda (GOVCOPP, University of Aveiro, Portugal)
Business Science Reference • © 2021 • 389pp • H/C (ISBN: 9781799870951) • US $195.00

For an entire list of titles in this series, please visit:
http://www.igi-global.com/book-series/advances-hospitality-tourism-services-industry/121014

IGI Global
PUBLISHER of TIMELY KNOWLEDGE

701 East Chocolate Avenue, Hershey, PA 17033, USA
Tel: 717-533-8845 x100 • Fax: 717-533-8661
E-Mail: cust@igi-global.com • www.igi-global.com

Editorial Advisory Board

Vicente de Paula Censi Borges, *Universidade Federal do Delta do Parnaíba (UFDPar), Brazil*
Jean Henrique Costa, *Universidade do Estado do Rio Grande do Norte (UERN), Brazil*
Christian Ehrlich, *Oxford Brookes University, UK*
Ricardo Jorge da Costa Guerra, *Polythecnic of Guarda, Portugal*
Sabrina Latusi, *University of Parma, Italy*
Osiris Ricardo Bezerra Marques, *Federal Fluminense University (UFF), Brazil*
Sarah Marroni Minasi, *Federal University of Rio Grande (FURG), Brazil*
Tiago Savi Mondo, *Federal Institute of Santa Catarina (IFSC), Brazil*
Keila Cristina Nicolau Mota, *Universidade Estadual do Ceará (UECE), Brazil*
José Pedro Da Ros, *Universidade Federal do Paraná (UFPR), Brazil*
Marinês da Conceição Walkowski, *Secretaria de Indústria, Comércio e Turismo de Urubici, Brazil*

Table of Contents

Detailed Table of Contents

This chapter presents happiness from the perspective of a public good, that is, of collective character understood as the basis for the constitution of public policies in tourism. Thus, the intimate relationship between tourism, happiness, and tourism development is recognized. The reflection undertaken on public policies in tourism is based on the analysis of the capacities to generate happiness in society and on the offer of a proposition that state interventions, whether of distributive, redistributive, regulatory, or constitutional nature, under pressure from collective needs, must attend to the tourism development process and sustained in a humanistic conception. Therefore, it is believed that tourism, in its various aspects, will only achieve the happiness of a tourist destination if it meets the premises of sustainable development to the extent that the citizen residing at the tourist destination feels part of the decision-making process and the formation of public policies.

This chapter is about residents' perceptions of the social impacts caused by tourism. The objective of the study is to analyze how the impacts of tourism influence the quality of life of residents of tourist destinations. The proposed methodology is based on a quali-quantitative approach with field research through the application of a questionnaire with residents of three coastal areas of Fortaleza: Grande Pirambu, Beira Mar, and Praia do Futuro (Northeast of Ceará-Brazil). The results show that tourism impacts positively on the quality of life, especially for residents of Beira-Mar and Praia do Futuro; however, residents of Grande Pirambu pointed to job generation as the greatest of the positive impacts in the area. About the negative impacts, the highest percentage was identified in Beira Mar, corroborating the need to plan tourism to expand benefits and minimize costs of tourism. It concludes that the impacts of tourism on the quality of life of residents in the studied areas are diverse, and that the most affected dimensions were related to economic aspects and citizenship.

In the past decade, we could see a myriad of efforts in fostering rural development. Two critical elements in travel, tourism, and leisure activities are the traveler's wellbeing and resident's welfare, which are inherent factors of rural tourism. Despite the importance of rural areas to tourism development, the focus of the studies on wellbeing usually is resident or employee oriented. There is a lack of inquiries trying to understand the relationship between rural tourism and welfare on tourism. In this sense, this study aims to connect the topics on welfare and tourism in rural areas to provide a theoretical starting point for future studies. The authors unravel essential elements in the pursuit of enhancing tourist's welfare on rural tourism by providing a visual conceptual model that provides guidelines for tourism stakeholders to

The chapter investigates the topic of happiness and quality of life in the slow-tourism context, with specific reference to the pilgrimage tourism phenomenon. The Camino de Santiago and the Via Francigena are analysed through explorative research with the aim of understanding the effects of pilgrimage on individuals' levels of well-being and its implications for peoples' lives. The results reveal several differences in motivations and benefits obtained according to the pilgrimage road undertaken, with a more existential connotation tied to personal happiness in the Camino de Santiago pilgrim group and a more experiential connotation tied to wellness in the Via Francigena pilgrim group. These findings provide new insights into contemporary pilgrimages in a comparative perspective.

Crisis, social discontent, and dissatisfaction among people lead to a greater interest in relaxing the soul, body, and mind as a refuge within tourism. Thus, there are changes and new trends in international health tourism's demand and supply conditions. Even before the deep crisis caused by the new coronavirus pandemic in 2020, people increasingly turned to health and wellness activities while traveling, especially using tourist activities to achieve health. Since consumption patterns are changing among travelers, such as the growing popularity of wellness tourism, new forms of tourism may stimulate future tourism activities, and the changes are expected to push businesses to reevaluate their service designs and distribution channels. When the recovery of the tourism sector is under discussion, it is worth questioning how wellness trends could help tourism and what the perspectives are for the future. Therefore, this chapter makes recommendations based on new tourists' demands and new processes and protocols for the operation of wellness tourism services.

such as tangible aspects, reliability, empathy, promptness, and security.

Foreword

The plurality of experiences presented in this book helps to understand the importance of the restructuration of the tourism industry in its correlation with happiness and quality of life. In a society that prioritizes the dimension of work, the opportunity to experience tourism when getting to know new cultures, landscapes, allowing the relationship with the local community, can perform the most important part of people's lives, denoting feelings deeply rooted in well-being and good health.

It seems apparent the recognition of tourism as part of the globalization process and involves markets around the world, affecting labour relations and cultural aspects held as part of a broad global movement. This has provided greater force for international organizations that control a substantial number of services in the tourism market, such as hotel chains, airlines, companies, and travel agencies.

Tourism is also an important basis for protecting natural, cultural and historic for its ability to mobilize economic and political. Activities which foresee both physical and social sustainability of tourism should be prior to the detriment of a meaningless economical exploit in the sector. Technology in tourism, for instance, is an element that influences the way people travel considering tourism has emerged as one of the most profitable economies in many countries once the access to travel is growing due to advances in technology. It is prospected for the future in the academic literature a kind of tourism that will exploit the possibilities of space tracks, resorts on the orbit and adventure activities performed on the moon.

There is a certain consensus in the academic literature that the term "happiness" symbolizes the state of well-being characterized by emotions ranging from contentment to intense joy. Happiness is directly connected with associated terms such as well-being and quality of life. The theme of well-being is gaining due importance and can be causally related to a series of global goals. Perhaps the most important of these at the present time is the Sustainable Development Goals (SDG), proposed by the United Nations (UN).

SDG is linked to the UN´2030 Agenda as a new driver for promoting sustainability among stakeholders in developed and developing countries being one of the most important action plans worldwide. It is a document with 17 goals, where the goal number #3 concerns on "Health and Well-Being", with the intention of ensuring a healthy life and promoting well-being for all, in all areas and ages.

Another important link to tourism, in its relation to happiness and quality of life, is "Leisure". Documents published by the World Tourism Organization (WTO) express that leisure trips suffers a huge influence of holidays, school vacations or climate, implying in an attachment to this market on the seasonality of the sector in which are offered distinct conditions for each season, even if it is high, mid, or low.

This Organization also recognized that the international tourism is strongly influenced by the segment "visiting friends and relatives", with travellers motivated by being with their friends and relatives and important as a precursor of trips that aims for the promotion of tourism in general conditions. In this scenario, the commercialization driven by consumption for both leisure and tourism sectors is sharpened, by the re-encounter with the values associated with the countryside or the beach, in the functional expectation of escape from large cities and the replenishment of energies.

More importantly, leisure is a universal component of human culture which takes different forms and meanings based on the characteristics of a given society. The Universal Declaration of Human Rights, promulgated in 1948 by the United Nations Organization, recognizes in its articles 24 and 27 the importance of occurrence of leisure for the right of citizenship and as a fundamental element for the personal and social development. The academic literature on leisure studies converges in recognizing the fact that culture plays a central role in the understanding of this social sphere, whatever the context in which it is found.

Understood as a social right, leisure must be articulated to a democratic and participative conjuncture beyond being excessively rooted in the logic of consumption. Leisure also promotes the opportunity to improve and expand the freedom of choices in the daily decisions. Thus, leisure is a powerful vehicle that has the potential to influence individuals in their journeys through life.

Both leisure and tourism cannot be understood without recognizing their reciprocal relations with the social, economic, cultural, environmental, and political fields. Knowing about the habits, attitudes and values that occurred in leisure and tourism are fundamental for planning, management and policies associated with the areas of economic development, cultural preservation, health and well-being, environmental management, among others.

Therefore, leisure stands out as an imperative field for human and social development related to happiness, well-being, and quality of life, in its necessary articulation with daily life and with the most diverse sectors of the broader society. When it's understood as a field of human promotion, besides the simple economical perspective, it's allowed to intent this environmental and social sustainability that we long for in a real possibility of value the tourism and understand it as an element of culture.

By bringing authors who contribute to the development of chapters related to Luxury Tourism, Medical Tourism, Well-being in Tourism, Public Policy in Tourism, Tourism and Hospitality, Entrepreneurship Activities in Tourism, New Technologies for the Tourism Industry, among others, this publication is committed to developing an important contribution to tourism studies in its interface with essential elements in today's society, such as well-being, quality of life, leisure and health.

I wish you all a pleasant reading, filled with moments of intense joy, relaxation, and happiness.

Ricardo Ricci Uvinha
University of Sao Paulo, Brazil

Ricardo Ricci Uvinha *is Full Professor at the School of Arts, Sciences and Humanities - University of Sao Paulo, Brazil. Coordinator of the World Leisure Center of Excellence WLCE/USP. Member of the International Academy for the Development of Tourism Research in Brazil – ABRATUR.*

Preface

Rebuilding and Restructuring the Tourism Industry: Infusion of Happiness and Quality of Life is an edited collection of papers that provides conceptual and empirical insights into well-being, happiness, quality of life and public welfare. This collection offers various epistemological perspectives, connecting scholars from different parts of the world. The book is suitable for students, researchers, and professionals from public and private sectors who wish to improve their understanding of tourism's role in the tourism actors' quality of life and happiness. The contributions of the papers selected cover a range of applications, from new approaches in academic research to public policy, destination management, and the development of new tourism services and products.

Travel is widely acknowledged as an activity that increases individuals' levels of well-being. Travel experiences are generally associated with positive emotions, such as joy and pleasure, generally contributing to tourists' well-being and happiness. Indeed, one of the most expected results of tourism is happiness. Nevertheless, the relationship between happiness and tourism is still a matter of investigation. Despite the overall beneficial effects of travel on happiness levels, tourism is not an activity with solely positive implications on peoples' lives. Therefore, it is crucial to understand how happiness manifests through tourism and how it can be expanded to different tourism actors, such as tourists, workers, and communities.

Therefore, in the first part of the book, six chapters examine happiness and quality of life, public welfare, subjective well-being and neurosciences in the context of tourism policies, destination management, human resources management, and academic research. In the second part of the book, opportunities for developing and improving tourism services and products are the chapter's central focus. We ensured that different approaches to well-being and the perspective of all tourism stakeholders - residents, entrepreneurs, workers and tourists - were taken into account.

SECTION 1: PUBLIC POLICIES, MANAGEMENT, AND ADVANCES ON RESEARCH AGENDA

In Chapter 1, Vicente de Paula Censi Borges analyzed happiness as the basis for the constitution of public policies in tourism, acknowledging the intimate relationship between tourism, happiness, and tourism development. The reflection undertaken on public policies in tourism is based on the capacities to generate happiness in society. In Chapter 2, Francisco Alexandre Araújo Barros, Jakson Renner Rodrigues Soares, Priscila Cembranel, and Laura Mary Marques presents a study about the impacts of tourism on the residents' quality of life of two tourist destinations in Ceará-Brazil, identifying positive and negative impacts. Valério Rodrigues de Souza Neto and Osiris Marques, in Chapter 3, discuss traveller's well-being and resident's welfare in a rural tourism context, providing a conceptual model with guidelines for rural tourism stakeholders. In Chapter 4, Şükran Sirkintioğlu Yildirim and Ülkühan Bike Esen, Çankırı examine the literature on Social entrepreneurship and discuss how the concept is handled in the field of tourism. Christian Ehrlich, Sashenka Milston and Robert van der Veen, in Chapter 5, describe evidence-based happiness techniques that are highly relevant for workers in the tourism and hospitality industries. The first section of the book ends with Chapter 6, where Verônica Feder Mayer, Carla C. L. Fraga and Leticia Cynara Santos Silva developed a bibliometric study to investigate the potential contributions of the neurosciences to the state of the art of tourism research, evaluating the emerging literature on subjective well-being in tourism.

SECTION 2: OPPORTUNITIES FOR TOURISM SERVICES AND PRODUCTS

The second section of the book starts with Chapter 7, where Mariana Brandão Cavalheiro, Verônica Feder Mayer and Aline Barbosa Tinoco Luz identify managerial, governmental and academic opportunities to develop nautical sports tourism in a post-pandemic scenario. In Chapter 8, Sabrina Latusi and Massimiliano Fissore investigate happiness and quality of life in the slow-tourism context, with specific reference to the pilgrimage tourism phenomenon in The Camino de Santiago and the Via Francigena. In Chapter 9, Leticia Cynara Santos Silva and Verônica Feder Mayer discuss wellness tourism trends, making recommendations based on new tourists' demands and new processes and protocols for the operation of wellness tourism services. Maria Francisca Casado-Claro and Marina Mattera, in Chapter 10, proposes a comprehensive approach to understand how Augmented Reality, Virtual Reality and Mixed Reality can contribute to enhancing tourism experiences, ensuring the safety of workers and visitors during travel restrictions and in preparation for a post-pandemic new normal. In the final Chapter 11, Beatriz Fernandes Duarte, Laurentina Cruz Vareiro, Bruno Barbosa Sousa and Victor Figueira present a study about Portugal's hostel sector, analyzing expectations and perceptions of the service quality provided.

We consider this book an excellent opportunity to get into the tourism field and to discover contemporary research topics, as it is suitable for academics (masters and doctorate students) and practitioners. Scholars from hospitality, tourism, destination management, economics, marketing, sociology, anthropology, and the health sciences will find valuable insights from this book. Moreover, professionals who want to improve their understanding of the relationship between tourism and happiness and well-being will find valuable visions to develop better and sustainable practices for the sector.

Section 1

Chapter 1
Happiness:
The Basis for Public Policy in Tourism

Vicente de Paula Censi Borges
Universidade Federal do Delta do Parnaíba, Brazil

ABSTRACT

This chapter presents happiness from the perspective of a public good, that is, of collective character understood as the basis for the constitution of public policies in tourism. Thus, the intimate relationship between tourism, happiness, and tourism development is recognized. The reflection undertaken on public policies in tourism is based on the analysis of the capacities to generate happiness in society and on the offer of a proposition that state interventions, whether of distributive, redistributive, regulatory, or constitutional nature, under pressure from collective needs, must attend to the tourism development process and sustained in a humanistic conception. Therefore, it is believed that tourism, in its various aspects, will only achieve the happiness of a tourist destination if it meets the premises of sustainable development to the extent that the citizen residing at the tourist destination feels part of the decision-making process and the formation of public policies.

INTRODUCTION

The world is dynamic and, whether from an economic or social point of view, transformations occur without people having any control over them. However, when we observe the collective context we realize that specific forces, namely formed by social and economic groups, such as financial institutions, multinational companies, wars, and also migratory movements, among others, have the ability to influence and sometimes dictate the course of events. In this context, conformity, predictability

DOI: 10.4018/978-1-7998-7239-9.ch001

Copyright © 2021, IGI Global. Copying or distributing in print or electronic forms without written permission of IGI Global is prohibited.

are characteristics of the past, because with the advent of technology, generating new communication and information processes, people and organizations find themselves in need of learning how to learn (Senge, 2018), that is, conceiving and knowing new methods and knowledge that allow them to achieve their individual and collective goals.

In the above perspective, the forces described affect people's lives, either at the individual or collective level, by the decisions of public agents in legislative assemblies and other public organizations. According to Frey & Stutzer (2017), these decisions set restrictions, define the type, quality, quantity, and form of access to public services, influencing individual well-being and, consequently, the perception of happiness.

The policy practiced by public agents is defined based on the interests intrinsic to the acts and plans that reflect the desires of social groups, companies, states and countries, depending on the area, focus and the interested group.

Nowadays, in several tourist destinations, based on the experiences and knowledge about large flows of tourists - overtourism - and on the travel limitations resulting from the coronavirus pandemic (Sars- Cov-2), could it be said that the decisions of public agents and organizations for tourism development reflect the needs of the indigenous populations? Perhaps in some countries the perception may be positive, but in most, the answer to this question would be negative (Müller & Friederich, 2020; Novy, cit in Zmyślony & Kowalczyk-Anioł, 2019). The fact that the needs of people and social groups are not met, in the desired way and at the desired time, leads us to reflect on how public policies are designed and implemented.

Based on the reasoning referenced above, this chapter aims to develop reflection on the constitution of public policies in tourism, analyzing them in their capacities to generate the perception of happiness in society and offer the proposition that the actions of State intervention, whether they are of distributive, redistributive, regulatory or constitutional character, under pressure of collective needs, with reflection at the core of society, must meet the tourism development process that is sustained in the humanistic conception of Max-Neef (2012), fundamentally from the perspective that human needs are the origin for public policies and, therefore, pertinently associated with the dimensions of the GNH – Gross National Happiness index (or also Gross Internal Happiness) proposed by the Bhutan Study Center (Ura & Galay, 2004), because it is understood that this is the way to satisfy society in its expectations, generate welfare and also value to tourist destinations, conquering, consequently, high levels of happiness.

Thus, it is understood that the concept of happiness has had special attention by society since classical antiquity. Aristotle (384 BC to 322 BC) understood happiness as man's greatest good, which is identified with living well and doing well (Marques, 2000).

Happiness

When looking at a tourist destination, seeing tourists circulating and consuming, the flow of vehicles and pedestrians, the urban transformations, the individual tends to imprint to his vision conceptions of aesthetics, organization, and also efficiency. Aesthetics is based on the individual's references, organization is based on the disposition and accessibility of what is of interest, and efficiency is based on the way spaces are implemented in favor of socialization and/or capital. In this way, when one observes the tourist territory and shares the perception, the construction of a collective perception is made possible.

Sharing the same vision about a certain territory, be it in an urban or rural environment, shows a social cohesion desired by many, but probably never fully achieved. Thus, this chapter brings a reflection about happiness as something possible to be shared in the social environment. Therefore, the central question is: is there any possibility of having public policies for tourism that will make a happy tourist destination?

The above question is pertinent for several reasons, one of them being the fact that in numerous writings by scientific thinkers of tourism and related areas one finds conflict and social tensions as primary characteristics of a territory with constant flows of people from different origins (Milano, Noveli & Cheer, 2019; Goodwin, 2017; Dodds & Butler, 2019; Postma & Schmuecker, 2017, Borges, 2014).

The societies in the world capitalist system tend to understand happiness from the point of view of having, that is, the one defended in the conservative utopia classified by Karl Mannheim (1976), associated to the idea of economic growth. For them, indexes and indicators such as the GDP – Gross Domestic Product – satisfy as an indication of success or failure, or, also, in a broader conception, that of economic development, which can be guided by the HDI - Human Development Index. The HDI is added to this limited vision of the attainment of happiness, since this index does not include issues related to the environment and sustainability, limiting itself to the variables of per capita income, health, and education. Along these lines, Stiglitz, Sen & Fitoussi (2009) warn that the GDP, like the HDI, has limitations for measuring the progress of societies, since it does not effectively encompass and measure the social welfare of a society. According to Lustosa & Melo (2010) the GDP measures only the economic transactions carried out, making no distinction between what represents costs and what represents benefits.

Thus, in tourism the measurement of happiness, as an indicator of development, in the view of many managers of tourist destinations, contemplates the context of the tourist-resident or tourist-destination relationship, seeking, erroneously, to perceive development through the optics of economic growth. The error, in this case, is the reductionism in the analysis of the observed reality, which fails to encompass issues inherent to the life of the resident in tourist destinations, such as education, governance, health, employment, among others. Similarly, the reduction of reality to utilitarian approaches presents itself in the misunderstanding of the meaning of economic growth and development, sometimes even encompassing the term progress to describe the status quo or the evolutionary process of the tourism sector (Heidmann & Salm, 2010).

It is worth mentioning that, in a mercantile vision, happiness becomes a value to be communicated to the tourist, explaining that the leisure time at the offered destination will provide the desired levels of satisfaction and well-being for moments of rest and/or extraordinary activities, and thus becomes part of the list of vacation motivations.

However, happiness in tourism is at the core of the interrelationships between tourists and residents, but more difficult to find as a focus of public policies and activities concerning public administration and private initiative. One example is the approach given in promotional materials for tourists, which always indicate that happiness is a place located in the destination featured, where positive feelings and experiences are offered.

In disagreement with the promotional approach to happiness in tourism (Csikszentmihalyi, 2013; Diener, 2000), it is stated that in the context of GNH (Gross National Happines), empirically, happiness is assessed as an attitude. Whereas in other approaches it is considered as an indicator of subjective well-being, referring to experiences.

The conception of happiness that will prevail, throughout the chapter, will be the one disseminated by Galinha and Ribeiro (2005), who proposed the notion of well-being relating the concepts of Satisfaction and Happiness from the Bottom/Up perspective - the immediate satisfaction of needs will produce happiness while the permanent unmet need will cause unhappiness - and Top/Down - where the degree of satisfaction needed to produce happiness will depend on the adaptation or level of desire, which is influenced, according to the authors, "by past experiences, comparisons with others, personal values, and other factors" (Galinha and Ribeiro, 2005, p.204).

Satisfaction in tourism here is understood in the conception of Hunt (1977 cited by Barros, 2008, p.44), who emphasizes that "satisfaction is not the pleasure taken from the consumption experience, it is the resulting evaluation that the experience was at least as good as it was supposed to be". Therefore, happiness will only be felt if expectations of the role played by each element intrinsic to the tourist experience are met or exceeded, increasing the perception of usefulness (Kahnemann, 1999). Thus, Katz and Kahn (1987 cited by Honório and Mattos, 2010) provide the understanding that expectations exist in the minds of those who experience a tourist activity and represent standards in whose terms one evaluates the performance of the other. These expectations are not only internalized, but communicated or transmitted, even in a non-continuous manner.

From this perspective, the experiences of tourists and the receiving community, in the generation of tourist happiness, are directed to the public arena, occurring in the interaction between the two, and also in the perception of the tourist and support structures and infrastructures. Thus, Ura (2009, p.3-4) subdivides the idea of happiness into two understandings: 1) the first one is based on the conception of happiness as a public good, but subjectively felt, not being possible to be left to private devices and efforts. Thus, one of the hypotheses, tested by Borges (2017), defines that the more unequal a society is, the higher the levels of unhappiness will be. In this case, happiness is perceived as a direct consequence of sensory stimuli.

Thus, understanding happiness as a public good, the chapter presents it as the basis for the constitution of public policies in tourism, considering the close relationship between tourism, happiness, and tourism development.

The connection between the themes mentioned is justified because it has been found that tourism, when perceived as an integral part of the life of the inhabitant of a tourist destination, is likely to meet local needs and generate more benefits than disturbances to daily life, provided it is properly planned and effectively implemented, leading to higher levels of satisfaction and well-being, that is, to a stage referred to in this chapter as a happy tourist destination.

HAPPINESS: MEASURABLE AND NECESSARY FOR SUSTAINABLE DEVELOPMENT

The society of having, which holds, more prominently, the characteristics of selfishness and hyperindividualism (Lipovetsky, 2011), does not present itself with the ability to fill the existential gaps of the human being. From this perspective, it should be noted that human needs are few, non-hierarchical and their forms of satisfaction subject to constant change (Max-Neef, 2012), making them specific to the social, cultural and economic context. Thus, based on Max-Neef (2012), the understanding of happiness is the result of another model of society, based on serving the human being and not the market. In the author's view, the economy, in other forms (domestic, community, popular, public, and variations), can come to better fill "voids created by the market institution" (Sampaio, 2019).

In the search for a new development model, new indicators and indexes arise to compose a substantive rationality, that is, one that supports a vision of serving the common good. Thus, in a counterpoint to economic indicators and, at the same time, considering it as an index to complement the existing ones, despite its amplitude and holism, the GNH - Gross National Happiness - arises as a systemic indicator of Happiness, created in 1972 by the 4th king of Bhutan, aiming to provide public managers with guidelines focused on the population's happiness (UN, 2002).

Unlike traditional indicators, which have economic growth as their measurement objective, the GNH is based on four pillars: the promotion of a sustainable and isonomic socioeconomic development; the preservation and promotion of cultural values; the conservation of the natural environment; and the establishment of good governance (Carvalho, 2010). Based on this foundation, the GNH was constituted with a distinct approach, emphasizing that to assess happiness it is necessary to consider nine domains/dimensions (Figure 1): economic standard of living, governance, education, health, community vitality, environmental resilience, access to culture, balanced time management, and psychological well- being, as these factors are considered to be universally common.

Figure 1. The nine GNH indicators
Source: Adapted from media.gnhcentrebhutan.org, 2016.

Therefore, according to Neto (2010) the concept of the Happiness Index is based on the principle that some development models are not easily measurable from an economic perspective. Thus, in a broader spectrum, the GNH measures and suggests that the development of societies occurs fully and equally when material and spiritual development happen at the same time, complementing and reinforcing each other.

The reign of Bhutan's 4th king guided Bhutan into an environment that was conducive to prepare the people for the introduction of decentralizing policies, generating laws and policies in line with GNH, which, at its core, comprises a set of values that promote collective happiness as the end-value of any development strategy, and is described as: (a) Holistic - recognizing all aspects of human needs, from those of spiritual, material, physical or social nature; (b) Balance - balancing progress towards GNH attributes; (c) Collectivity - happiness becomes a comprehensive collective phenomenon; (d) Sustainability - Achieving well-being for the current generation, managing to maintain it for future generations; (e) Equity achieving reasonable and isonomic distribution of the level of well-being.

Ura (2009) emphasizes that GNH in Bhutan is distinct from the Western understanding of happiness for two reasons. First by the fact that it is multidimensional, as it does not only contemplate subjective well- being, but includes other dimensions of human life. And second, because it internalizes responsibility and motivations explicitly.

According to Ura (2009), initially the GNH presents six propositions of use for the index: 1. To define an alternative development framework; 2. To provide indicators to guide sectors of society to development; 3. To allocate resources according to the GNH goals and tracking tools; 4. To measure people's happiness and well-being; 5. To measure progress over time; and 6. Compare progress across the country.

In tourism the use of GNH as a development indicator is not easily found, especially in scientific studies. The GNH methodology is not yet supported in tourism studies (Duro & Faré, 2015; Paula, 2013; Guedes, 2014; Elías, Fernandez &Castro, 2012), although issues such as income distribution and poverty are addressed, through issues such as positive and negative impacts of tourism on the economy and society.

However, looking separately at research on subjects related to Ura's (2009) propositions, one can see studies that approach tourism from the perspective of a transforming agent (Marcon, 2007) capable of generating a state of well-being, as well as other concerns more attentive to the economic impact on community life (Cunha & Cunha, 2005).

Lohmann and Bloom (2015, p.1) explain that "the search for the link between tourism and happiness seems to be a must for Bhutan," but the relevance for other tourist destinations in the world is not yet as evident, reflected in international and regional scientific studies of tourism, which generally focus on visitor experiences and conflicts arising from the interaction between tourists and communities.

The fields of tourist experience (see Figure 2), according to Pine II and Gilmore (1998), are divided into absorption and immersion. However, the visitors and the visited are expected to participate in the experience environment, whether actively or passively. From this understanding, it is possible to define the role of each one at the moment of interaction and development of tourism, allowing the anticipation and prediction of actions in the field of public policies to generate the best possible experience that meets the expectations generated.

Figure 2. Fields of experience
Source: adapted from Pine II and Gilmore, 1998.

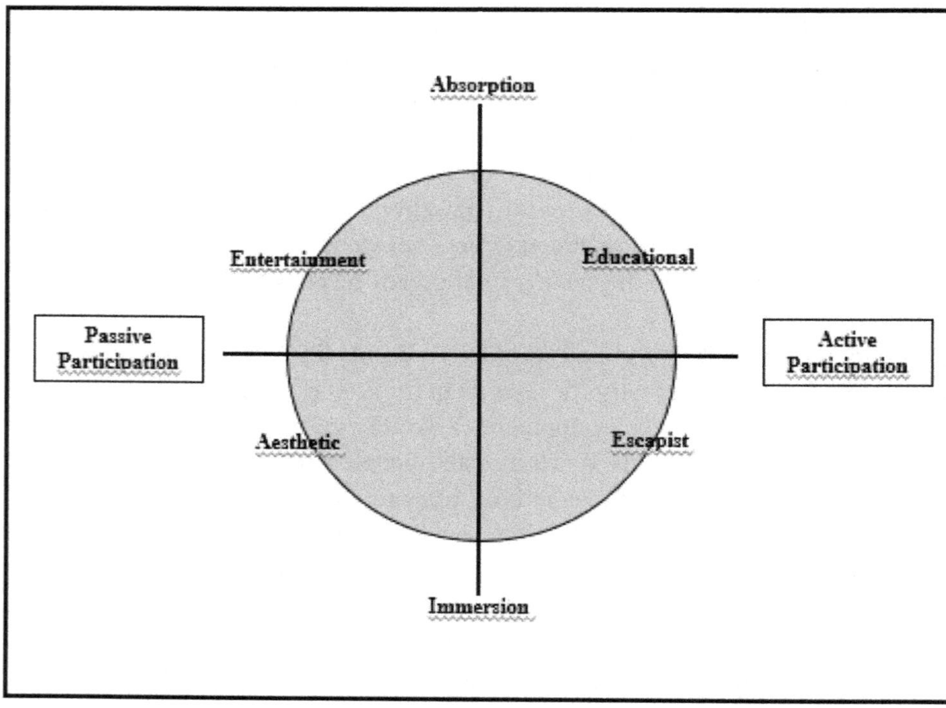

It can be seen that tourism fits into the Immersion field, since the understanding is that the tourist, being at the destination, undresses himself from the roll of ordinary activities and consumes throughout his stay the influences offered by stimulation; and in the Absorption field, which is characterized when active participation is required by passive activities, such as music shows, theater performances, among others, but in the aesthetic context active participation is not required, but immersion is fundamental.

With all the breadth of activities that fit into the tourism universe and that potentially generate happy experiences, even so it remains on the list of concerns of the development process of a tourist destination the generation of employment and income. Singer (2003) denotes the relevance of the subject, considering the individual and collective need to conquer the acquisitive capacity of goods and services that, consequently, allow, hypothetically, a virtuous cycle of income and economic growth. However, Desbiens and Ferreira de Lima (2004) consider that development should go beyond the economic expectation of wealth generation and aim at generating quality of life. However, Singer (2003) and Desbiens and Ferreira de Lima (2004) share the idea that for there to be a development in which the virtue, in this case the systemic virtue of isonomy, prevails, the process of development must necessarily guarantee social opportunities, transparency, and social security.

With this same understanding, the World Bank (2007) defines that to achieve solid and sustainable development, economic growth must come from a broad base of activities. To this end, and thinking about the improvement of living conditions and the generation of opportunities, labor becomes the main asset for people. In this way, making them more productive drives the process toward achieving poverty reduction. Thus, to sustain success, education and health are equally important, aiming at the collective increase of skills and capabilities, and well-being.

However, the World Bank's vision supported in this work is subject to criticism, in view of its utilitarian character and its lack of attention to the factors that generate inequality in the world capitalist system.

Therefore, it is fundamental to provide, from public development policies that encompass the social, economic, cultural, and environmental variables in tourism, greater amounts of moments of happiness, mainly related to tourism activities that may come to shape a destination more suitable for the reception and satisfaction of tourists' expectations.

Although attention has turned to alternative theories of development, public policies, forced by the hegemony of the neoliberal ideal, continue to prioritize the economic dimension and relegate secondary attention to the other dimensions of life in society. Sen (2000) questions the hegemonic development models in the West, describing them as a cruel development policy, as they tend to deplete natural resources and increase social inequalities. For the author, an effort must be made in the development process to combine the economic and sociocultural dimensions in a context of strengthening values and institutions, bearing in mind that economic development is the result of the sum total of technological innovation, technical progress associated with investments in R&D with, fundamentally, human resources and the accumulation of social capital.

Moreira (2013) warns that one often sees a recognition of the paradigm shift pertinent to the concept of development, noticing special attention to the character of sustainability for the territories. For the author, in tourism the increase in competitiveness of a tourist destination is, in addition to other variables, connected to a strategic vision based on four aspects:

(i) the integration of the principles of sustainable development in the processes (the philosophy of sustainability is a key factor in achieving long-term competitiveness); (ii) the emphasis that is placed on the future (the strategic vision requires this perspective of the future); (iii) the need for collaboration between the destination's stakeholders (in the planning and decision-making processes) and (iv) the flexibility that the process offers in changing environmental conditions (the strategic vision gives the destination the flexibility to adapt to environmental changes and future uncertainties that also characterize market conditions).

Tourism, according to Tarlombani da Silveira and Zem (2010, p.1), "imposes itself as an intense agent of transformation of the territory by exerting great influence on the socio-economic and environmental reality", and may be positive (generation of employment and income) or negative (environmental degradation, increase in social inequalities). Therefore, having a tourist destination with a low Gross National Happiness rate, considering the characteristics of the environment, can be the consequence of impacts that, in principle, do not present themselves as negative, but are not positive, that is, they do not come from negative impacts, but from the lack of positive impacts.

In another perspective, tourism can develop in a constant pressure to the population of the tourist destination, emerging with extraordinary economic results, but with strong impacts on the social environment through, for example, the feeling of exclusion and decreased accessibility to public spaces and/or significant for the community. Krippendorf (2001, p.19) explains that "the inhabitants of the regions visited are also beginning to feel a certain resentment towards the negative effects of the exodus of the tourist masses. These populations increasingly have the impression that they are invaded by this development and, at the same time, excluded from it". Such feeling is described by tourism researchers in different tourist destinations, either in Europe or South America, the most notorious cases being Barcelona (Catalonia), Lisbon (Portugal) and Florianopolis (Brazil).

Therefore, it is understood that there are other variables that, arising from social relations between visitors and the visited, influence the development of tourism and the perception of the indigenous population. When considering the amplitude and complexity of the dynamics of tourism activities, one realizes the need for the use of more holistic indices and indicators that contemplate the diversity and plurality, characteristic of tourism.

The use of GNH as an indicator for the generation of public policies has, in its domains and indicators, the ability to guide the public management of tourism towards a solidarity economy in contrast to the more conservative initiatives of development based on the economy of scale.

Moreover, by paying attention to the nine dimensions of the GNH, tourism managers may have an instrument that will not only provide the measurement of Gross National Happiness, but will also allow guiding public policies and development. Thus, the nine dimensions of the GNH for tourism are

1. **Psychological Well-Being**: To evaluate the degree of satisfaction and optimism that each individual (visitor and visited) has in relation to his or her own life. The indicators present data on the rate of positive and negative emotions, analyzing self-esteem, sense of competence, stress, and spiritual activities. Vinhas (2008, p.25) emphasizes that "tourist satisfaction is deeply dependent on the effort and the ability of cooperation between public and private organizations that contribute to the configuration of the products and destinations in question. Measuring tourist satisfaction objectively and accurately is of central importance in the process of strategic decision making".
2. **Health:** measure the efficiency of health policies, with criteria such as self-assessment of health, disability, risky behavior patterns, exercise, sleep, nutrition, among others. O'Donnell (1986 cited by Amendoeira, 2006) delimits the scope of health in five dimensions: emotional health, social health, intellectual health, spiritual health, and physical health. Appropriately dimensioning the health system will tend to allow for a high level of well-being in both tourists and the resident population.
3. **Time Use:** It is one of the most significant factors that influence quality of life, considering the time for leisure and socialization. Assessing balanced time management will make it easier to establish parameters that comprise the different times that make up the daily agenda of visitors and the community, such as: time in traffic, at work, in educational and recreational activities, and mealtimes.
4. **Community Vitality:** in tourism, especially tourism products that value the experience, depend on the relationships and interactions in communities and destinations. Therefore, examining the level of trust, the sense of belonging, the vitality of affective relationships, safety at home and in the community, the practice of giving and volunteering, becomes vital for building a sharing economy and, fundamentally, an environment of cooperation. (Bursztyn, Bartholo & Delamaro 2009) reveals that relationships centered on dialogue and interaction between visitors and visited, in which tourists do not submit to tourists and the visited do not become objects of consumerist instrumentalization, facilitating the exchange of experiences and egalitarian dialogue.
5. **Education:** To the generation of welfare is of great relevance to consider in tourism planning influencing factors such as: levels of participation in formal and informal education, skills and abilities, evolving in the educational process

values that permeate the respect for diversity, environmental education, among others that help to build a harmonious environment. Ansarah (2002, p.23) reiterates that "education in tourism should be directed towards multidisciplinary reflection and teamwork, contemplating multicultural contexts in which creativity combines traditional or local knowledge and applied knowledge of advanced science and technology".

6. **Culture:** Evaluate local traditions, festivals, core values, participation in cultural events, opportunities to develop artistic skills, and discrimination because of religion, race or gender, because tourism is a "social, human, economic and cultural achievement of the most interesting today, because it induces the proximity of relationships and stimulates them, so that, places, people, services and attractions are interdependent and create an intense network of sensations, situations and events" (Virginio, 2010, p.67).
7. **Environment:** To ensure the sustainable character of development, there is a need to measure the perception of citizens regarding the use of water, evaluating its quality and distribution, air quality, soil quality, and biodiversity. The indicators in the GNH include access to green areas, garbage collection systems, and other aspects related to the conservation and preservation of the environment. From the measurement and control tools, the "rational management of tourism can contribute greatly to the protection and development of the physical environment and cultural heritage, as well as to the improvement of the quality of life" (Lickorish, 2000, p.117-118).
8. **Governance:** Evaluating how the population perceives the government, the media, the judiciary, the electoral system and public safety, in terms of accountability, honesty and transparency, involving the measurement, in this perspective, of citizenship and citizen participation in decisions and political processes, tends to provide the organizational environment and in tourism development a greater transparency. Trentin (2014) explains that is related to the concept of governance, however there is no conceptual consensus, the ideas of power sharing, participation, multilevel integration, diversity and decentralization, deliberation, flexibility and reversibility, experimentation and knowledge, creation, autonomy and authority, networks and interdependence, cooperation and interaction.
9. **Standard of Living:** the GNH recognizes the relevance of economic indicators, but does not isolate them, but treats them in an integrative manner with the other indicators present in the methodology. Thus, the standard of living evaluates individual and family income, financial security, level of debt, quality of housing, and spending patterns.

Finally, the relevance of public policies based on data and information, even though they may still seem subjective to the dynamics of tourism, involving levels of complexity and diversification that challenge the creation of metrics to measure them, is located in the need to understand, at least as close as possible, the totality of the reality of tourism development in a given destination.

Thus, isolated indices and indicators provide few answers, in view of their own methodological limitations. Therefore, covering the largest number of dimensions of human life in the measurement process allows offering to public and private tourism managers, as well as the receiving communities, the reality of tourism and social, cultural and environmental life, enhancing the multiplying effects of tourism through more precise actions of the State and private initiative. It is in this perspective that the GNH and its set of indicators is perceived as the most appropriate index to represent the totality of interactions and dynamics of tourism in a given territory, expressing them in gross internal happiness.

PUBLIC POLICY IN TOURISM: THE POSSIBILITY OF A HAPPY TOURIST DESTINATION

The development of a society, whether that of a city, state, or country, occurs as a result of decisions thought out and implemented by governments, public managers, in conjunction with other social forces. Thus, among other understandings, public policies are the consequence of the actions or inactions of the ruler before the demands of society, the reactions and expectations of people, affected directly or indirectly by political decisions, which have an anticipatory effect for the political process of formulation and implementation, which makes the social environment in negotiating and conflicting (Borges, 2014).

In tourism, according to Beni (2003, p. 101), public policies are the "set of conditioning factors and basic guidelines that express the ways to achieve global objectives for tourism". This statement is part of a broad and unfinished discussion about the various meanings of the term policy.

For a better perception of the understandings about the term, Heidemann and Salm (2010, p. 28- 29) present some definitions, highlighting them in five categories or conceptions. In the first, politics encompasses everything related to the collective life of people in society and in their organizations. In the second conception, it is a set of "processes, methods, and expedients" that people and/or their social groups use to influence, conquer, and remain in power. The third one predicts that it is "also the art of governing and achieving the public good". In the fourth, a more operational definition is presented, in which it is understood as "actions, practices, political guidelines, founded in laws and undertaken as state functions by a government, to solve general and specific issues of society". And in the fifth and last one, politics is conceived as "the knowledge of phenomena related to the regulation and control of human life in society, as well as to the organization, ordering, and administration of political-administrative jurisdictions.

Therefore, and with more propriety, one realizes that the expression public policy, explained in the understanding of Beni (2003), presents itself, in the tourism environment, with a more operational and pragmatic aspect. In tourism, as in other areas, "public policy is everything that governments decide to do or not to do" (Dye 2005 cited by Heideman and Salm 2010, p. 30).

Based on Dye's (2005) understanding, it is understood that public policies have become over the years the main factor of compensation for popular desires, renewing themselves with each new claim, as a kind of bargaining chip. Thus, their formalization favors the understanding between government and citizens, placing them as a reference base. However, sometimes, from this perspective, they also have a character of legitimization of state power, that is, their own specific mechanisms created and used by the state to convince citizens of the need for acceptance and political domination (Rodrigues and Ferreira, 2006).

Lowi (1972 cited by Frey, 2000) presents the policy arena model, referring to the dynamics of conflict and consensus, distinguished by characteristics, be they distributive, redistributive, regulatory or constitutive (Table 1). Frey (2000, p. 223-224) lists and defines them as: 1. Distributive policies are characterized by a low degree of conflict in political processes, since policies of a distributive character only seem to distribute advantages and do not entail costs. 2. Redistributive policies, on the contrary, are conflict-oriented. The goal is the conscious detour and displacement of financial resources, rights, or other values between social strata and groups in society. Regulatory policies work with orders and prohibitions, decrees and ordinances. The effects concerning costs and benefits are not determinable in advance; they depend on the concrete configuration of policies. Constitutive policies determine the "rules of the game" and thereby the structure of political processes and conflicts, that is, the general conditions under which distributive, redistributive and regulatory policies are negotiated.

Public policies can go beyond the perspective of government policies, because a government, even with its organizational structure, is not the only one to intervene and serve the community, that is, to promote public policies. As the State does not meet society's expectations, society gets organized and tries to meet the existing need. These initiatives can be seen in several contexts, whether in the community or in business, through neighborhood associations, non-governmental organizations, single or cooperative companies, in which they perform a service and/or produce a public good, by virtue of a responsibility, even a business one, and their own interests.

Table 1. Classification and characteristics of public policies

CLASSIFICATION	PUBLIC POLICIES	IMPACT	FOMENT RESPONSIBILITIES
Distributive	Actions to implement programs, plans, laws and/ or decrees that aim to share benefits, such as: museum accessibility policy, emergency income policy in the COVID-19 pandemic, among others.	Improves the population's quality of life; Increases the perception of happiness in the social environment; Increases state investments.	Federal, state, and municipal governments; international, national, and regional development banks;
Redistributive	Actions that meet the demands of social groups, guaranteeing them benefits and rights that the social environment alone does not provide.	Decrease in inequality; Democratization of access to social structures; Social ascension; Poverty reduction.	Federal, state and municipal governments; International, national and regional development banks; Non-governmental organizations (NGOs)
Regulatory	Actions taken by the State (executive and legislative) and the judiciary that aim to regulate behavior in society and the organizations and individuals that participate in them.	Establishment of isonomic treatment; Greater insertion of the population in discussions and deliberations; Possibility of decentralizing decisions.	Federal, state, and local governments, legislative assemblies, and the judiciary.
Constitutive	Structuring actions that establish the social, technological, legal, economic, and cultural dynamics.	Meets the demands of society; establishes the mechanisms to satisfy the needs of organizations and individuals;	Federal, state and local governments and legislative assemblies

Source: Adapted from Borges, 2014.

Beni (2003) explains that tourism, in its various sectors, as a characteristically liberal environment, that is, with little regulation by governments, has as a predisposition when implementing a public policy the economic benefit, whether for private agents or for the social vigor of a city, state, or country, having to be felt in the results internally and externally, with the activation of tourism products and the increase in the inflow of foreign exchange.

In tourism, or in any other socio-economic sector, in a more restrictive context, the State generates public policies from the premise of the need to control the socio-economic order, in order to maintain legitimized its power and the use of force that is necessary, that is, to contain the growing claims within an acceptable limit for those involved. Therefore, there is only State action or inaction as public policies if there is enough social pressure, or lack of it, to demonstrate the primordiality of the intervention.

Thus, it can be seen that governments have motivations to intervene in tourism and other areas, due to economic relevance and significant external pressures from development (Elliot, 1997; Johnson; Thomas, 1992) or by political character, and may be both positive and negative. Similarly, Solha (2004, p. 9) states that "in tourism, concerns about establishing policies for the sector only appear when it acquires economic importance or when it begins to bring inconvenience" and impacts to populations and places.

Tourism is inserted in a diverse socio-economic context, because it is a multifaceted activity, dependent on the stakeholders involved in the provision of diversified services and activities, and it is the responsibility of the state to provide an adequate socio-political and legal environment in defense of the actions undertaken by the various stakeholders (Akama, 2002).

According to the WTO (2001), the main objective of the public administration is linked to the obligation to create and maintain adequate conditions to heat up the competitiveness of companies and tourist regions. In this sense, Silva (2000) indicates from WTO guidelines that in relation to tourism, the State has two basic functions: a) to execute society's preferences and b) to maximize the social benefits of tourism, facilitating access and prioritizing management. With this purpose, the World Code of Ethics for Tourism - Preparing for the Millennium (1999) lists the basic guidelines for the construction of public policies and private initiatives, which make explicit the need to develop equitable, responsible, and sustainable tourism, always for the shared benefit of all sectors of society. Thus, tourism is established as the channel that contributes to understanding and mutual respect between man and societies (art. 1).

Dias (2003) points out that the function of the State goes through some areas of involvement in the tourism sector, such as: coordination, planning, legislation and regulation, enterprises, incentive, social action and promotion of tourism. However, the UNWTO - United Nations World Tourism Organization, in the guide book Sustainable tourism for Development (2013), highlights the need to project the expected results of public policies, designing them to meet the current and future desires of society. In this document, the UNWTO (2013) explains that an essential requirement of tourism policies, when thinking of tourism as a development tool, is that these policies must be grounded in the principles of sustainable development.

The theme sustainable development, although widely debated, has not yet found echoes in governmental actions, nor has it become effective in clear public policies. In the specific literature associated with tourism one finds understandings of this statement (Egoshi, 2009; Gomes & Ferreira, 2018; Liu, 2003; Mundt, 2011; Leslie, 2012).

The barriers to the success of public policy initiatives for sustainable development reflect, as Sousa Santos & Meneses (2009, p. 7) state, social relations, which are "always cultural (intra-cultural or inter- cultural) and political (representing unequal distributions of power). In this way, Max-Neef (1993), conceives development on a human scale, in which all initiatives of collective impact and represented in public policies are derived from a constructive process that begins with the identification of needs, which, for the author, are fixed in four dimensions of social life: Being, Having, Doing, and Knowing.

Borges (2017) presents the result, by crossing the qualitative variables of the GNH with the spectrums of human needs of Max-Neef (1993), in which he obtained the perception that the happiness of a tourist destination - in the case of the author's study in the Brazilian municipalities of Ilha Grande, Cajueiro da Praia, Parnaíba, and Luís Correia, all in the northeastern state of Piauí - is based on the satisfaction of collective needs inherent to life in society (Table 2).

Table 2. Human needs matrix

FIB Dimensions	Be	Have	Be	Make
Psychological well- being	Balance/ solidarity	Low debt/	Family/ Community	Socialization/ participatio n
Health	Physical/mental health	Food/medical and hospital services/ sanitation	Nature/ home	Feed/rest
Time Use	Free	Leisure	Conviviality	Friends
Education	Sage/reader	Public Policies/ teachers	Schools/ universities	Study/ analyze
Diversity cultural and resilience	Bilingual	Artistic Skills	Square	Manual arts/ music
Governance	Reviewer	Voice/ participation/ rights	Association of residents/ municipality	Policy/ cooperating/ sharing
Communit y Vitality	Solidarity	Housing/ Security/ identity	Community /family	Contribute/ collective actions
Ecological Diversity	Conscious	Basic sanitation	Clean City	Environmental Education
Standard of living	Honest	Housing/income /work/leisure	Transportatio n	Physical exercises

Source: Borges, 2017.

Observing the table 2, which is the result of research in the four municipalities mentioned above, it can be seen that tourism, even in ambiguity in relation to the impacts caused, has the probable ability to align the needs for a balanced collective environment, because in terms of psychological well-being, tourism can enhance the needs of being (balance / solidarity) with the doing (socialization / participation), as well as the use of time. From the perspective of cultural diversity, it is evident, as a necessity, the use of public space for the development and exposure of the artistic skills of the local population. However, more specifically in relation to public services (transportation) and living conditions, the needs are direct and connected with more than one GNH dimension, taking into account that for a set of better quality public services, community vitality, governance, psychological well-being, ecological diversity, education, and health are fundamental, aiming at a better standard of living.

Thus, the process of developing a public policy for a happy tourist destination should be decentralized and focused on people (residents and tourists), paying attention to the set of collective needs and the cultural context (Figure 3).

Figure 3. Tourism public policy process (conception of collective happiness)
Source: Prepared by the author.

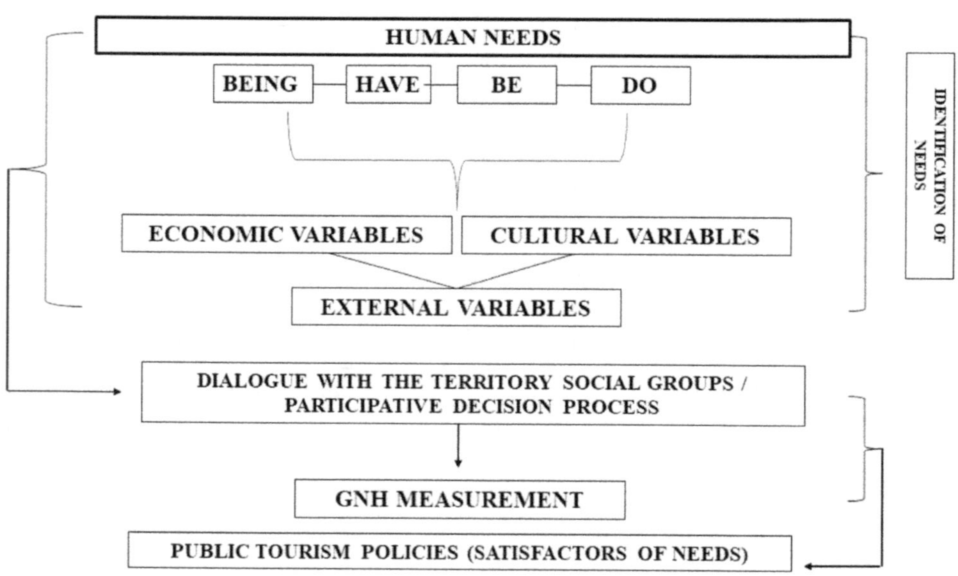

The process of developing public policies, understood here, is established on the premise of collective happiness, which is defined as a state of satisfaction of society expressed in contentment, territorial identity and perception of well-being with the territory. Thus, the steps that form the flow of actions for the development of public tourism policies are: 1. identifying human needs in the territory; 2. understanding the needs considering the internal and external variables, especially those of economic and cultural origin, which directly interfere in social dynamics; 3. establishing a channel of dialogue for the joint construction of knowledge of local tourism, allowing access and full participation of the population;

4. measuring the GNH and, finally, 5. developing public tourism policies, constituting them as tools to satisfy the needs identified.

The strategic management of a tourist destination is not only done in the search for alignments and adaptations required by the dynamics of the elements of the external environment, but also and primarily in the identification and fulfillment of the desires of the local population. Moreira (2013, p.119) clarifies that

"presupposes that through the management of available resources, a mode of action is defined which, by being pondered, allows for the achievement of certain objectives that at a given moment are of capital importance. In this sense, strategy is relativized, in time and geographic space, according to the contexts that give rise to it".

By observing the tourist destination from the perspective of happiness, as an area of knowledge, it allows the understanding of the dynamics of the destination in relation to the system in which it is inserted, as well as providing an opportunity for the public manager and tourist organizations to reduce the tensions and conflicts arising from the relationships between visitors and visitors from large tourist flows.

HAPPINESS PUBLIC POLICY: SUGGESTION AND CONCLUSION

The reflections undertaken here are based on the principle that happiness is a state of satisfaction of the individual and society, which is expressed in contentment, territorial identity and the perception of well-being with the territory. In this way, tourism loses when it focuses only on satisfying the desires of market agents, in view of the increase in tensions and territorial conflicts between the market, the tourist, and the local community.

It is believed, therefore, that tourism, in its various aspects, will only achieve the happiness of a tourist destination if it meets the premises of sustainable development to the extent that the citizens residing at the tourist destination feel part of the decision-making process and the formation of public policies.

After the reflections triggered in this chapter and the findings explained, it is understood that the subject of happiness still needs more attention from tourism researchers, as well as from other areas, given the lack of methodologies that allow the approximation between the theory and the political practices of intervention and management in tourist destinations.

REFERENCES

Akama, J. S. (2002). The Role of Government in the Development of Tourism in Kenya. *International Journal of Tourism Research*, *4*, 1-13.

Almond, J. (2006). *A shared biography of nursing. The second half of the twentieth century*. Coimbra: Formasau.

Ansarah, M. G. d R. (2002). Formação e capacitação do profissional em turismo e hotelaria: reflexões e cadastro das instituições educacionais no Brasil. São Paulo: Aleph.

Barros, J. L. G. (2008). *A satisfação com a qualidade de serviços na fidelização dos destinos turísticos* (Master's Dissertation). Faculdade de Economia da Universidade do Porto.

Beni, M. C. (2003). Structural analysis of tourism (9ª ed). São Paulo: Senac.

Borges, V. P. C. (2014). Políticas Públicas e o Desenvolvimento Segregador no Turismo: Uma discussão conceitual e reflexiva referente ao turismo de Parnaíba-Piauí. *Turismo: Estudos e Práticas*, *3*, 23–40.

Borges, V. P. C. (2017). *Política Pública e Planejamento Estratégico do Turismo no Território Polo Costa do Delta: Um Modelo de Desenvolvimento Turístico*. Tese de Doutoramento. Universidade de Coimbra.

Bursztyn, I., Bartholo, R., & Delamaro, M. (2009). Turismo pra quem? On development paths and alternatives for tourism in Brazil. In R. Bartholo (Ed.), *Turismo de base comunitaria: diversidade de olhares* (Vol. 144-162, pp. 76–107). Letra e Imagem.

Carvalho, M. B. (2010). *A Felicidade na agenda da administração e suas relações com conceitos organizacionais* (Dissertação de mestrado). Minas Gerais: FUMEC College.

Csikszentmihalyi, M. (2013). *Flow: The psychology of happiness*. Random House.

Cunha, S. K. & Cunha, J. C. (2005). Competitiveness and sustainability of a tourism cluster: a proposed systemic model for measuring the impact of tourism on local development. *Journal of Contemporary Administration*, *9*(2).

Desbiens, Y. & Ferreira de Lima, J. (2004). Cadrage du Develóppment Regional. *Revista Interfaces*, *4*, 179-192.

Diener, E. (2000). Subjective well-being: The science of happiness and a proposal for a national index. *The American Psychologist*, *55*(1), 34–43. doi:10.1037/0003-066X.55.1.34 PMID:11392863

Dodds, R., & Butler, R. (2019). The phenomena of overtourism: A review. *International Journal of Tourism Cities*, *5*(4), 519–528. doi:10.1108/IJTC-06-2019-0090

Duro, J. A., & Farré, F. X. (2015). Tourist Seasonality in Spanish Provinces: Measurement and Analysis. *Cuadernos de Turismo*, *36*, 157–174. doi:10.6018/turismo.36.230921

Dye, T. R. (2005). Understanding Public Policy (11th ed.). Pearson Education Inc.

Egoshi, K. (2009). The Fallacy of Sustainable Development. In A. J. H. Guevara (Ed.), *Consciousness and Sustainable Development in Organizations* (pp. 115–144). Elsevier.

Elías, S., Fernandez, M. R., & Castro, M. (2012). Proposal for a system of indicators to evaluate the impact of tourism on poverty in the town of Monte Hermoso. *Tourism and Society Yearbook*, *8*, 41–53.

Elliot, J. (1997). *Tourism: politics and public sector management*. Routledge.

Frey, B. & Stutzer, A. (2017). *Public Choice and Happiness*. CREMA Workpaper, nº2017-03. Center for Research Economics, Management and the Arts.

Frey, K. (2000). Public Policies; A Conceptual Debate and Reflections concerning the Practice of Public Policy Analysis in Brazil. *Planejamento e Políticas Públicas*, *21*.

Galinha, I., & Ribeiro, J. L. P. (2005). History and Evolution of the Concept of Subjective Well-Being. *Psychology. Health & Illness*, *6*(2), 203–214.

GNH Centre Bhutan. (n.d.). *The 9 domains of GNH*. Accessed on 21/01/2021. Available at http://www.gnhcentrebhutan.org/?page_id=833

Gomes, M., & Ferreira, L. J. (2018). Public policies and the goals of sustainable development. *Law And Development*, *9*(2), 155–178.

Goodwin, H. (2017). *The Challenge of Overtourism. Responsible Tourism Partnership*. Working Paper 4. Retrieved from http://www.millennium-destinations.com/uploads/4/1/9/7/ 41979675/rtpwp4overtourism012017.pdf

Guedes, A. S. (2014). *O Domínio Cultural no Turismo Organizado - O Caso Portugês: Da polarização territorial à valorização endógena e local* (Doctoral dissertation). Programa de Doctorado: New Resources and Sustainability in Tourism. Universidad de Salamanca.

Heidmann, F. & Salm, J. F. (2010). *Políticas públicas e desenvolvimento: bases epistemológicas e modelos de análise*. Brasília: UnB.

Honorio, J.B. & Mattos, P.L.C.L. (2010). Organizational roles: What the pragmatics of language leads us to think. *Alcance Journal - Electronic*, *17*(1), 22-33.

Johnson, P., & Thomas, B. (1992). Tourism research and policy. In P. Johnson & B. Thomas (Eds.), *Perspectives on tourism policy*. Mansell Publishing Limited.

Kahneman, D. (1999). Objective happiness. In D. Kahneman, E. Diener, & N. Schwarz (Eds.), *Well-being: The foundations of hedonic psychology* (pp. 3–25). Russell Sage Foundation.

Krippendorf, J. (2001). *Sociology of Tourism - for a new understanding of leisure and travel*. Aleph.

Leslie, D. (2012). *Tourism Enterprises and the Sustainability Agenda across Europe. Surray*. Ashgate Publishing.

Lickorish, L. J., & Jenkins, C. L. (2000). Introduction to Tourism. Rio de Janeiro: Campus.

Lipovetsky, G., & Serroy, J. (2011). *A cultura-mundo, respostas a uma sociedade desorientada*. Companhia das Letras.

Liu, Z. (2003). Sustainable Tourism Development: A Critique. *Journal of Sustainable Tourism*, *11*(6), 459–475. doi:10.1080/09669580308667216

Lohmann, M., & Bloom, J. (2015). *Happiness in a Tourism Context*. Paper presented to the International Conference on Gross National Happiness, Paro, Bhutan.

Lowi, T. (1972). Four Systems of Policy, Politics and Choice. *Public Administration Review*, *22*.

Lustosa, A. E., & Melo, L. F. (2010). Gross Internal Happiness (GNH) - Index of sustainable development. Brasília: Conjuntura econômica goiana.

Mannheim, K. (1976). *Ideologia e Utopia*. Zahar Editores.

Marcon, E. M. (2007). Tourism as an Agent of Social Development and the Guarani Community in the Jesuit Ruins of São Miguel das Missões. *Pasos Revista de Turismo y Patrimonio Cultural*, *5*, 343–352. doi:10.25145/j.pasos.2007.05.025

Marques, R. (2000). *O livro das virtudes de sempre*. Ed. Asa.

Max-Neef, M. (1993). *Desarollo a Escala Humana: Conceptos, Aplicaciones y Algunas Reflexiones*. ICARIA.

Max-Neef, M. (2012). *Development at the Human Scale: ConceptionApplication-Further Reflections*. EDFURB.

Milano, C., Novelli, M., & Cheer, J. M. (2019). Overtourism and degrowth: A social movements perspective. *Journal of Sustainable Tourism*, *27*(12), 1–19. doi:10.1080/09669582.2019.1650054

Moreira, C. C. O. (2013). *Turismo, Território e Desenvolvimento: Competitividade e Gestão Estratégica de Destinos* (Doctoral dissertation), Coimbra, Universidade de Coimbra.

Müller, C., & Friederich, D. B. (2020). A Dinâmica do Federalismo Brasileiro no Tema das Políticas Públicas, Controle Social e COVID-19. *Revista de Direitos Fundamentais & Democracia*, *25*(3), 49–77.

Mundt, L. (2011). *Tourism and Sustainable Development: Reconsidering a Concept of Vague Policies*. Berlin: ESV.

ONU - Organização das Nações Unidas. (2002). *Publicação Johannesbourg Summit. World Summit on Sustainable Development*. United Nations Department of Public Information.

Paula, A. T. (2013). *State role, poverty alleviation and tourism: theoretical relations with the National Tourism Plan 2007/2010* (Master's dissertation). Programa de Pós- Graduação em Turismo, Universidade de Brasília.

Pine, B. J. II, & Gilmore, J. H. (1998, July). Welcome to the Experience Economy. *Harvard Business Review*.

Plaice, K. T. (2004). *State Public Agencies and the Development ofTourism in Brazil* (Doctoral dissertation). Universidade de São Paulo. São Paulo.

Postman, A., & Schmuecker, D. (2017). Understanding and Overcoming Negative Impacts of Tourism in City Destinations: Conceptual Model and Strategic Framework. *Journal of Tourism Futures*, *3*(2), 144–156. doi:10.1108/JTF-04-2017-0022

Sampaio, C. A. C. (2019). Homage to a Barefoot Economist: Manfred Max-Neef. *Development and the Environment, 51*, 370–372.

Sen, A. (2000). Development as Freedom. *Revista Gaceta Ecológica*, 55.

Senge, P. (2018). *The Fifth Discipline: The Art and Practice of the Learning Organization*. Best Seller Publisher.

Silva, J. A. S. (2000). Pensando o planejamento face à intervenção do estado no turismo: A questão do sistema do informações. *Turismo: Visão e Ação*, *2*(5), 9–22.

Singer, P. (2003). *Globalização e Desemprego: Diagnóstico e Alternativas* (6th ed.). Contexto.

Sousa Santos, B., & Meneses, M. P. (2009). *Epistemologias do Sul*. Almedina.

Stiglitz, J.E., Sen, A., & Fitoussi, J.P. (2009). *The Measurement of Economic Performance and Social Progress Revisited*. Paris: OFCE - Centre de recherche en économie de Sciences Po.

Tarlombani da Silveira, M. A., & Zem, L. C. (2010). Impactos Ambientais do Turismo em Zonas Litorâneas: Um Estudo de Percepção Ambiental no Litoral do Paraná-Brasil. In *VI Latin American Seminar of Physical Geography*. Universidade de Coimbra.

Trentin, F. (2014). *Public Policies in Brazil: Decision Making and the Analysis of Governance Structures in the Inducer Destinations of Armação dos Búzios and Paraty, State of Rio de Janeiro, Brazil* (PhD Thesis). Department of Geography, Faculty of Arts, University of Coimbra.

UNWTO - World Tourism Organization. (2013). *Sustainable Tourism for Development*. European Commission, Gran Contract DCI-MULTI-2011/280-663.

Ura, K. (2009). *A proposal for GNH value education in schools*. Gross National Happiness Commission.

Ura, K., & Galay, K. (2004). *Gross National Happiness and Development*. The Centre for Bhutan Studies.

Vinhas, R. S. (2008). *Índice de Satisfação e Lealdade do Turista* (Dissertação de Mestrado). Instituto Superior de Estatística e Gestão da Informação da Universidade Nova de Lisboa, Dez.

Virginio, D. F. (2010). Tourism and Culture: A study on the museum qualification program for tourism. *Electronic Journal of Cultural Tourism*, *4*(1), 65–84.

World Bank. (2007). World Development Report: Development and the next generation. Washington, DC: Author.

WTO - World Tourism Organization. (1999). *World Code of Tourism Ethics: Preparation for the New Millennium*. General Assembly.

WTO - World Tourism Organization. (2001). *Introdução ao Turismo*. Rocca.

Zmyślony, P., & Kowalczyk-Anioł, J. (2019). Urban tourism hypertrophy: Who should deal with it? The case of Krakow (Poland). *International Journal of Tourism Cities*, *5*(2), 247–269. doi:10.1108/IJTC-07-2018-0051

KEY TERMS AND DEFINITIONS

Collective Happiness: It is a state of satisfaction of the individual and of society, which is expressed in contentment, territorial identity, and the perception of well-being with the territory.

Gross Domestic (National or Internal) Happiness: Index created in Bhutan in 1972 that measures happiness from nine dimensions of life.

Human Needs: These are individual or collective states of deprivation, which, in the social context, can be satisfied with the constitution of public policies.

Public Policies: Actions taken by the State and/or civil society organizations that mediate the relationship between State and Society.

Satisfaction: Perception that a need has been met, removing the individual or society from a state of deprivation.

Territorial Identity: This is a personal identity based on the specific characteristics of a given territory, be it a country, state, city, or a region.

Tourist Destination: A place whose characteristics attract people to stay temporarily as visitors.

Chapter 2
Tourism and Quality of Life:
Different Perceptions Within the Same Tourist Territory

Francisco Alexandre Araújo Barros
Universidade Estadual do Ceará, Brazil

Jakson Renner Rodrigues Soares
Universidade da Coruña, Spain & Universidade Estadual do Ceará, Brazil

Priscila Cembranel
Universidade Sociedade Educacional de Santa Catarina, Brazil

Laura Mary Marques Fernandes
Universidade Estadual do Ceará, Brazil

ABSTRACT

This chapter is about residents' perceptions of the social impacts caused by tourism. The objective of the study is to analyze how the impacts of tourism influence the quality of life of residents of tourist destinations. The proposed methodology is based on a quali-quantitative approach with field research through the application of a questionnaire with residents of three coastal areas of Fortaleza: Grande Pirambu, Beira Mar, and Praia do Futuro (Northeast of Ceará-Brazil). The results show that tourism impacts positively on the quality of life, especially for residents of Beira-Mar and Praia do Futuro; however, residents of Grande Pirambu pointed to job generation as the greatest of the positive impacts in the area. About the negative impacts, the highest percentage was identified in Beira Mar, corroborating the need to plan tourism to expand benefits and minimize costs of tourism. It concludes that the impacts of tourism on the quality of life of residents in the studied areas are diverse, and that the most affected dimensions were related to economic aspects and citizenship.

DOI: 10.4018/978-1-7998-7239-9.ch002

Copyright © 2021, IGI Global. Copying or distributing in print or electronic forms without written permission of IGI Global is prohibited.

INTRODUCTION

Tourism activity generates positive and negative impacts several in the tourism receiving centers. That is, benefits and costs that, as highlighted by Soares, Pazos and Gabriel (2021), which can be appreciated through economic opportunities, social integration, improvement of infrastructure and services that impact on the quality of life of the community receiving the tourist flow. For this reason, establishing preferences linked to the tourism economy and social development requires a careful analysis in terms of the quality of life of the residents of that territory (Queiroz, 2017).

For Lemos (2016), in the same way that tourism brings development, it also promotes several modifications in the lives of residents of tourist destinations. These consequences are called social impacts and are visualized, mainly, in areas that were not prepared for a large tourist flow due to environmental, structural and social vulnerabilities of the location. Thus, individual and collective changes of values and behavior patterns occur (Fernandez et al., 2012).

From all this, this work seeks to understand how the impacts of tourism affect the quality of life of residents of tourist destinations. It seeks to investigate the perception of residents of Grande Pirambu, Beira Mar and Praia do Futuro. The three areas are located in the city of Fortaleza in Northeast Brazil (Figure 1).

Figure 1. Fortaleza map and location of Grande Pirambu, Beira Mar, and Praia do Futuro.
Source: Google Maps, 2021.

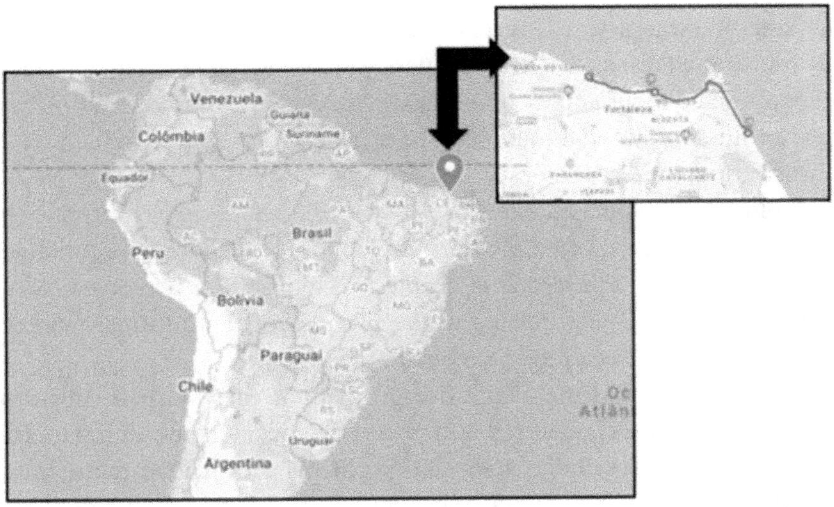

TOURISM IMPACTS AND QUALITY OF LIFE OF RESIDENTS

When talking about quality of life, it is understood that the development of a place should advocate the satisfaction of basic needs. Such as: education, health, decent housing, access to water, safety, and a and leisure, balanced environment (Stein et al., 2018). According Bâc (2012), the increase in tourism flow has the ability to generate through economic stimulation, improving the quality of life of residents.

According to Dias and Mendes (2002), the impacts related to the social dimension are regarding to relationships, economic balance, environment, citizenship, leisure, safety, and urban mobility. The quality of life is understood in its social perspective in respect to the local culture and the community preservation (Martins et al., 2002). In this sense, the social impacts caused by tourism promote great consequences in tourist places. These impacts refer to the relationship between residents and people who come from other places or countries, although changes in this sense are usually slow (Barreto, 2004). Finally, the development of the tourism sector will only be advantageous to the local population if it is planned (Quintero, 2004). Certainly, a planning that values the interests of residents and not only the interests of the market.

Quintero (2004) indicates that tourism can help improve the quality of life of residents of a given region thanks to its power to contribute to the provision of general infrastructure. In the same vein, Rosa and Kanikadam (2021) argue that the expansion of infrastructure caused by tourism has great importance in social impacts, because it stimulates, for example, building of large conglomerates of hotels, among other enterprises that are deployed to support the reception of tourist flows.

According to Lickorish (2002), the building of tourism enterprises causes social impacts. In the case of Fortaleza, disorderly use of tourist areas has occurred. Another situation that arises is the increase of prostitution and contamination by sexually transmitted diseases. The inefficiency of public policies and the worsening of problems related to the supply of urban services such as basic sanitation can also be observed. In addition, the unemployment of residents is highlighted when the tourist works or activities decrease.

For Coelho, Mota and Vasconcellos (2015), when a tourism enterprise begins, it is necessary to have qualified labor. And, due to the generation of employment, migratory movements to the receiving centers of tourism are perceived. A type of migration that occurs is also related to the increase of the informal market with street vendors, trading from food to souvenirs for visitors.

In order to understand the social transformations that tourism causes in the receiving areas and their residents, it is important to consider the cultural limitations or the lack of basic knowledge about the problems that plague these countries. These include environmental, economic, social, and political problems. There are also structural limitations, because when tourism is introduced in a place, there is a need for time, capital, and specific skills so that it can contribute to the development of the place. And, finally, there are operational limitations. These arise due to the centralization of administration and the imposition of restrictions on community actors (Stein et al., 2018). The aspects of the social dimension of quality of life and its impacts are presented in Table 1:

Table 1. Aspects of the social dimension of quality of life and its impacts

Social aspects of quality of life	Positive impacts	Negative impacts
Relationships and leisure	Cultural Exchange	Space Dispute
Economic balance	Foreign investment, investment in infrastructure, trade development	Real Estate speculation, dependence on Tourism
Environment	Environmental care plans	Excess, of tourists, pollution
Citizenship	Employment generation	Prostitution
Security and urban mobility	Increased policing in the streets	Crime

Source: elaborated by the authors, adapted from Dias and Mendes (2002).

In this way, there are several elements that make up the analysis of tourism impacts on the quality of life of the residents of the tourism receiving centers.

METHODOLOGY

This study was developed through a quali-qualitative approach. Besides the bibliographic survey, a field research was carried out using a structured form with 24 questions. The form was answered by 150 residents, 50 equally distributed per area: Grande Pirambu, Beira Mar and Praia do Futuro. The informants were randomly chosen in the study areas themselves. The researcher asked people to participate, presenting the research objectives and asking the initial filter question: How long have you lived in this area? Only those people who answered that they had lived in the area for more than 20 years were included in the target population of the study. The data obtained were analyzed using the *Statistical Package for Social Sciences* (SPSS) version 22.0 for Windows. The analyzes are presented in a descriptive way. To cite the open-ended responses obtained from the form, the research participants were named RES01, RES02, and so on.

DISCUSSION OF RESULTS

Fortaleza is the capital of the state of Ceará. The territory can be considered a metropolis that for years has been investing in the touristification of its coastline. For Silva (2015), the Secretary of Tourism of the State of Ceará (SETUR/CE) is fundamental in the development of tourism in the region. It is observed that tourism in the state of Ceará is considered, as a rule, as a public policy of the state, being verifiable through the various surveys that the secretariat does annually. It is noteworthy that Ceará has been defined as a tourist destination by the government since the 1980s. In this way, tourism is one of the inductive activities of the state's socioeconomic growth, because tourism policies are in the modernization process of Ceará that also invests in industry and agribusiness (Fernandes, 2014). The recent Municipal Secretariat of Tourism, created in 2005, reinforces the importance given to tourism.

Barreira (2012) states that Fortaleza has many tourist attractions, citing the Náutico Beach, which has a busy nightlife due to the sidewalks and the existing nightclubs and restaurants in this area. At Beira Mar of Fortaleza you can see the handcraft fair that sells typical products such as cashew nuts and cachaça, besides the handcraft that is also well known in other states of the country. Another point that should also be mentioned, according to Barreira (2012), is the cultural heritage of the city, especially those located in the streets of the historic center, where are the Passeio Público, the Ceará Museum, the Industry Museum, the Central Market and the Tourism Center, and also places like the Dragão do Mar Center of Art and Culture.

Praia do Futuro is one of the city's tourism highlights by offering tents by the sea with food service structure, musicals and comedy shows. Grande Pirambu is a popular area that receives tourists interested in the practice of sports related to the sea, such as surfing and kitesurfing (Medeiros, 2013 apud Barros, 2014). The public projects aimed at tourism produce the geographic space of Ceará changing, among other areas, the Metropolis Fortaleza. Ceará reinvented for the tourist activity was modernized and became socioeconomic highlight, but the activities related to tourism occur unevenly in the geographic space and tourist regions (Fernandes, 2014).

It is worth mentioning that Fortaleza has received investments and incentives for the development of the tourist activity for years. However, these are not evenly distributed. Thus, there is an imbalance of public investment in the Grande Pirambu when compared to the incentives in the areas of Beira Mar and Praia do Futuro area. The two most benefited areas concentrate tourist developments and residential neighborhoods geared towards the population with greater economic power. For this reason, it was decided to investigate the impacts of tourism in areas with distinct characteristics: Grande Pirambu, Beira Mar, and Praia do Futuro. In view of this, a paradoxical question arises: does tourism help develop the region or does the development of the region attract tourism?

Regarding the participants of the research, the residents of the localities are distributed as follows: Grande Pirambu with 23 women and 27 men, Beira-Mar with 32, women and 18 men and Praia do Futuro with 28 women and 22 men. Regarding age, in Grande Pirambu the majority of the respondents are between 40-49 years old (26 people), between 50-59 years old (19 people), and between 30-39 years old (5 people). In Beira Mar the ages were: between 50-59 years old (18 people), 40-49 years old (13 people), 60-69 years old (11 people) and 30-39 years old (8 people). Finally, in Praia do Futuro, the ages of the respondents corresponded to: 50-59 years (17 people), 40-49 years (16 people), 30-39 years old (12 people) and over 79 years old (1 person). With regard to education, in Grande Pirambu, more than half of the respondents have at least completed high school. In Beira mar, practically all have a high school education. It is noteworthy that half of the respondents have a post-graduate degree. Finally, in Praia do Futuro, almost two thirds of the respondents have a graduate degree in progress or completed.

Initially, the residents were asked about the impacts of tourism on the quality of life in the community where they live. In Grande Pirambu, it was noticed that for 42% the respondents there is an impact, for 38% there is not, and 20% are unclear about the impacts. In Beira mar, 86% said there is, 10% said there is not, and 4% did not know the impacts. And, in Praia do Futuro, 80% believe there are impacts, 14% see no difference in their quality of life, and 6% did not know. In view of the data, it was noticed that the residents of the Grande Pirambu area pointed out the lowest percentages of impact. This number of services and tourist flow is small, may be associated with the fact that tourism is incipient in Grande Pirambu. The remaining residents mostly agree that there are changes that affect their quality of life

From this, people were asked about the benefits of tourism to their lives. In Grande Pirambu area, 44% said that their lives had improved due to job creation resulting from tourism activities. Also mentioned were: cultural exchange (with 18% of the answers), increased quality of life in general (with 16% of the answers), commerce development (with 14% of the answers), and investment in infrastructure (with 8% of the answers).

At Beira Mar, 28% mentioned the increase in quality in general, without highlighting a specific point. Also highlighted were the impacts related to foreign investment (18% of the answers), investment in infrastructure (16% of the answers), development of commerce and job generation (both with 14% of the answers each), and cultural interchange (with 10% of the answers). In Praia do Futuro, the increase in the quality of life in general (with 24% of the answers), the generation of jobs and foreign investment (with 22% of the answers each), investments in infrastructure (with 12% of the answers), commerce development and cultural exchange (with 10% of the answers each).

From the analysis of the percentage of the three areas, it can be seen that the research participants identify benefits due to tourism activity, since 69% answer positively to the questioning. This shows that the benefits generated by this activity in the areas are experienced by most of those surveyed.

The highest percentage of residents stated that the generation of jobs is the most noticeable positive impact, a result that corroborates with studies by Maracajá and Pinheiro (2020), in the state of Rio Grande do Norte, which had as one of the objectives to understand the perception of the actors linked to tourism. In this sense, although conducted in different contexts, the researches pointed out this benefit of tourism.

In the case of the residents of Grande Pirambu, job creation was the most recalled benefit, 44%. Among this percentage, RES04 stated that his son always worked at the beach restaurants located in the area, because he could not give him a good education and, consequently, he had difficulty getting a "better" job. This means that although the job opportunity is evaluated as positive, RES04 understands that it is the possible job, but not necessarily the desirable one. In this sense, at the same time that the tourist activity can generate a direct benefit for the receiving community as the creation of jobs and income, one cannot forget that tourism is characterized by seasonality and, many times, the jobs generated are precarious and are not able to improve the quality of life of its workers. In this same line of labor seasonality, RES05 also responded as being positive about job generation. However, he stated that the worst part is during the rainy season, when the tourism flow in the restaurants decreases and, consequently, the salary at the end of the month decreases. This demonstrates the precariousness of the jobs generated in this region.

In the Beira Mar area, the highest percentage was projected in relation to the increase in quality of life, 28%. According to the report by RES06, it was due to the events held in the surroundings that there was greater investment in areas that he affirmed to be of great importance, such as, for example, the offer of public transportation and a greater quantity of security professionals. In the Praia do Futuro area, the highest percentage of respondents also stated that tourism provided a better quality of life, 24%. For RES07, the transformation can be verified by the several modifications that have existed in the avenues with several duplications, a greater sense of security was also felt since both the public and private power invest in this area in a stronger way, because, according to the resident, a very violent city scares away tourists. That is, improvement in infrastructure, generation of employment and income, improvement in safety and supply of complementary services derived from public actions are seen as benefits of tourism that have a positive impact on their quality of life.

The second group of questions concerns the costs of tourism activity on residents' quality of life. Respondents were asked to explain which negative impacts affect their quality of life. In Grande Pirambu, 30% perceived an increase in prostitution, increase in the cost of living (16%), increase in crime (14%), dependence on tourism (12%), pollution of public spaces and real estate speculation (with 8% of the answers each), spaces that are too crowded and too many tourists (with 6% each). In Beira mar the following stood out: the increase in criminality (26%), prostitution (18%), dependence on tourism (16%), higher cost of living (12%), pollution (10%), disputed spaces (8%), real estate speculation and excess of tourism (with 6% and 4% respectively). In Praia do Futuro, the highlights were the increase in crime (30%), prostitution (20%), real estate speculation and dependence on tourism (with 14% each), disputed spaces and pollution (with 8% each) and increase in the cost of living (6%).

Analyzing the negative impacts most present in the perception of residents, the themes linked to the increase in crime, 23.3%, and prostitution, 22.7%, were the two most recalled. These costs of tourism coincide with the results of Machado (2012). He states that the lack of security brings fear and terrifies residents. And, depending on the location, tourists are not well regarded by the community because the more visitation to an area increases, the greater the possibility of events linked to theft and robbery occurring. There is here an analysis of the questions related to the costs caused by the tourist activity. Residents were asked if they perceive that tourism provides some kind of impact. In all three areas surveyed, the answer "yes" was the highest percentage, with the highest numbers in Beira Mar, 86%, and Praia do Futuro, 80%.

The negative impacts, most noticeable, by residents varied. A large part of the residents of Grande Pirambu, 30%, perceived an increase in prostitution. It is interesting to note that prostitution is common in Beira Mar and secondly in Praia do Futuro, but these two areas had lower percentages compared to Grande Pirambu in relation to prostitution. According to some residents of Grande Pirambu, it is not prostitution itself that occurs in the area, but the consequences of this practice which, according to the residents, are experienced by many families when, for example, the daughter comes home in the morning and starts fighting with her parents or grandparents. It appears that the territory of prostitution of Grande Pirambu residents is outside the area of residence.

It was noticed during the field research that some residents seemed to dislike tourism mainly because of the negative impacts. The increase in crime had a very low percentage in Grande Pirambu, probably due to the fact that the area has high crime rates independent of tourism. Or it is simply due to the fact that the tourist activity is still in a very incipient moment in this area.

In Beira Mar, crime was considered by most residents, 26%, to be the impact most experienced and suffered by them. RES08, who suffers from insomnia, said that she always used to go for a walk during the early morning hours, because there weren't many people around, but with the increase in criminality, she stopped exercising. In her speech, she admits that there is safety on the avenue, however, at night there is a decrease of police officers increasing the feeling of insecurity. This is unrelated to tourism, the main object of this chapter. The insecurity derived from tourism is not noticed in this area, what is noticed is that the residents of the city perceive a little sense of security in Fortaleza, something that can harm the tourist activity.

In the Praia do Futuro area, the percentage of people who noticed that tourism changes their quality of life was 30%. The increase in crime is more perceived by the residents than what occurs in Beira Mar. For some of the residents, crime is greater at night when most of the tents are closed and consequently the flow of visitors decreases, causing thefts and especially robberies along the beach promenade, even with the good lighting. RES09 affirmed that the low number of police officers also favored the increase in crime. Thus, for living in a tourism-inducing territory, the residents understand that security would be a primary aspect to develop such area. However, assuming that we may not have asked the question in the best way, generating a limitation of our research, the resident indicates that there is no security in the area, however, we do not see that this lack of security is a negative impact of tourism.

However, regarding the influence provided by tourism coming, 58.7% of the responses admit that this influence occurred in a positive way. In the survey conducted by Gomes, Pinto, and Almeida (2017), residents felt their lives were influenced in a positive way in the vast majority. They identified, according to their results, that tourists coming from various locations both nationally and internationally influenced aspects linked to culture or environment preservation, for example. In the Grande Pirambu area the percentage that affirmed that it was a negative influence on their quality of life was 46%, while in the other two areas a high percentage of residents affirmed that it was a positive influence. Beira Mar (68%) and Praia do Futuro the percentage was higher than Beira Mar with 74%.

For those who stated that tourism had positively influenced their lives, the increase in job offers, regardless of whether formal or informal, in the areas under study was one of the most talked about points among the surveyed residents. RES14 from Grande Pirambu responded that tourism offered her son his "first job". She reported that because he did not finish high school, many doors were closed to her son and it was through the tourist activity, selling coconut candy (typical sweet in the region of Ceará), on the beaches of Grande Pirambu that he was able to finish his studies and help at home. The only area studied that had a higher negative percentage than the positive one was in Grande Pirambu. From the answers, it was found that many tourists when they are visiting the beaches in their locality do not want to stay in simple restaurants.

Final Considerations

With regard to the analysis of the influence of tourism impacts on the quality of life of the residents of tourist destinations, there are different perceptions of these impacts. In general, the perceptions of positive impacts were lower than the perception of negative impacts. This reinforces the understanding of the unequal distribution of public policies in the areas studied. The positive impacts are related to the increase in the quality of life due to the generation of jobs from tourism activities, to cultural interchange, to commerce development, and to the investment in infrastructure and attraction of foreign investments. The negative impacts were: increase in prostitution, cost of living, dependence on tourist activities, pollution, spaces disputed by the excess of tourists, and real estate speculation.

It was observed that the most affected dimensions were related to economic balance and citizenship issues. For, although there is the generation of employment caused by the improved economic balance in the territory, there is also an increase in prostitution and crime related to security aspects. In addition, the residents' lives are affected in terms of relationships and leisure. In a positive way by having cultural exchange, and in a negative way by having to deal with too many tourists. The collapse of the environmental system can also be observed, as more solid waste, pollution, and higher living costs for residents are generated.

An important theoretical contribution of this research is related to the study of the perception of tourism impacts on the quality of life of residents in different areas within the same tourist territory. Thus, understanding how residents relate to tourism, from different perspectives (neighborhoods with higher or lower tourist load), offers the academic community insights into the different perceptions that exist about such an important economic activity. In this sense, we conclude that the receiving community perceives the benefits and costs of tourism according to the presence of this activity in its area. That is, the perception of these impacts of tourism depends to a large extent on where the local community resides in the tourism territory. Finally, people who live in tourist districts will have different perceptions from people who live in non-tourist districts, even seeing the benefits and costs of this activity more keenly. While people who do not reside in areas with large tourism flows tend to idealize the benefits and sometimes even minimize the costs of the activity.

The study is limited because it is only a sample of the reality researched and is based on a non-probabilistic methodology. As future lines of investigation, it is understood that one could deepen the study of tourism impacts in order to understand the differences between the different neighborhoods of the city, and not only stick to the areas of influence of tourism.

REFERENCES

Bâc, D. P. (2012). The Impacts Of Tourism On Society. *The Annals of the University of Oradea, 21*(1), 500-506. Retrieved from: https://url2.cl/jpZNE

Barreira, I. (2012). *Cidades narradas: memória, representações e práticas de turismo*. Pontes Editores.

Barreto, M. (2004). Manual de iniciação ao estudo do Turismo (13th ed.). Campinas/SP. Publisher: Papirus.

Coelho, S. D., Mota, K. C. N., & Vasconcelos, F. P. (2015). The vision of the community in the implementation of urban development projects for tourism and quality of life: Public policies in the Grande Pirambu neighborhood, Fortaleza-CE. Revista Turismo - Visão e Ação. *Balneário Camboriú, 17*(1), 210–240.

Dias, E. C., & Mendes, R. (2002). *Contribution to the expansion and operationalization of the concepts of occupational health promotion and healthy workplaces for informal sector workers*. Mimeographed.

Fernandes, L. M. M. (2014). O Ceará turístico: Regionalization policy and governance in inducing destinations. *Journal GeoUECE, 3*(5), 213–214.

Fernandez, M. J., Gullette, G. S., Luque, A. C., Mendoza-Otiveros, M., & Monterrubio, J. C. (2012). Social impacts of tourism as perceived by state-planned tourism destination residents: The case of Huatulco, Mexico. *International Journal of Tourism Anthropology, 2*(1), 34–52. doi:10.1504/IJTA.2012.046094

Gomes, R.S.E., Pinto, H.E.R.S.C., & Almeida, C.M.B.R. (2017). Residential tourism in the Algarve: the perception of public sector managers. *Revista Bras. Pesq. Tur., 11*(2), 197-217.

Lemos, C. C. (2013). Tourism planning at the federal level: an analysis of the instruments used and investments in the sector. *Revista de Administração Pública, 47*(6), 1.401-1.427.

Lickorish, L. J., & Jenkins, C. L. (2002). *Introduction to tourism*. Campus.

Machado, M. B. T. (2012). Social fear and tourism in Rio de Janeiro. *Tourism & Management Studies, Faro*, (8), 48–54.

Maracajá, K. F. B.; & Pinheiro, I. F. S. (2020). Environmental perception in tourism planning: proposals for the development of Seridó, RN. *Revista Observatório de Inovação do Turismo, 14*(1), 41-59.

Martins, S. R. O. (2002, September). Local development: Conceptual and methodological issues. *International Journal of Local Development, Campo Grande*, *3*(5), 51–59.

Queiroz, A. (2017). Planning and metropolization of maritime leisure in Fortaleza-Ceará, Northeast Brazil. Eure. *Santiago de Chile*, *43*(128), 153–173.

Quintero Santos, J. L. (2004). Los impactos económicos, socioculturales y medioambientales del turismo y sus vínculos con el turismo sostenible. *Anales del Museo de América*, (12), 263–274.

Rosa, M. E. T., & Kanikadan, A. Y. S. (2021). Tourism in Cape Verde: development for whom? Turismo: Estudos & Práticas. *Mossoró*, *10*(1), 1–19.

Silva, M. E. M., & Soriane-Sierra, E. J. (2015). Analysis of visitor perception on the revitalization of the seafront: The Calçadão Linear Park model, Itapema - Santa Catarina, Brazil. *Brazilian Journal of Tourism Research.*, *9*(1), 76–96.

Soares, J. R. R., Pazos, A. P., & Gabriel, L. P. M. C. (2021). Involucrando la comunidad anfitriona en la gestión y desarrollo del turismo. In *Turismo y desarrollo: Contextos Diversos*. Thomson Reuters Aranzadi.

Stein, A. C., Hack Neto, E., & Dechechi, E. C. (2018). Reflections on the socioeconomic impacts of tourism in the municipality of Foz do Iguaçu. Turismo: Estudos & Práticas. *Mossoró*, *7*(2), 26–55.

Chapter 3
Rural Tourism Fostering Welfare Through Sustainable Development:
A Conceptual Approach

Valério Rodrigues de Souza Neto
Federal Fluminense University, Brazil

Osiris Marques
Federal Fluminense University, Brazil

ABSTRACT

In the past decade, we could see a myriad of efforts in fostering rural development. Two critical elements in travel, tourism, and leisure activities are the traveler's wellbeing and resident's welfare, which are inherent factors of rural tourism. Despite the importance of rural areas to tourism development, the focus of the studies on wellbeing usually is resident or employee oriented. There is a lack of inquiries trying to understand the relationship between rural tourism and welfare on tourism. In this sense, this study aims to connect the topics on welfare and tourism in rural areas to provide a theoretical starting point for future studies. The authors unravel essential elements in the pursuit of enhancing tourist's welfare on rural tourism by providing a visual conceptual model that provides guidelines for tourism stakeholders to enhance tourists' welfare while mitigating negative crisis impacts. They expose the central elements of the graphical model and the actions required to enhance welfare on rural tourism as well as their interrelationships.

DOI: 10.4018/978-1-7998-7239-9.ch003

Copyright © 2021, IGI Global. Copying or distributing in print or electronic forms without written permission of IGI Global is prohibited.

INTRODUCTION

In the past decade, we could see a myriad of efforts in fostering rural development (United Nations [UN], 2016). These guidelines are mirrored in tourism activities since 2020 was labelled as the International Year of Tourism and rural development (United Nations World Tourism Organization [UNWTO], 2020a) and with initiatives such as the 'rural tourism competition' that seeks to source new startups and entrepreneurs that advance the contribution of tourism to rural development and support recovery (UNWTO, 2020b).

Two critical elements in travel, tourism, and leisure activities (Filimonau et al., 2017) are the traveller's wellbeing (Mayer et al., 2019) and resident's welfare (An & Alarcón, 2020; Bramwell, 2009), which are inherent factors of rural tourism (Sharpley & Jepsen, 2011). These two particularities are more intense and necessary in current times since the global mobility restrictions (Göslling, Scott, & Hall, 2021) caused by the coronavirus disease 2019 (COVID-19) (Sigala, 2020). This pandemic generated several negative social (e.g., xenophobia [Kock et al., 2020]), economic (e.g., an increase of extreme poverty [World Bank, 2020]) and personal impacts on society (e.g., despair; tension, panic, stress, crying, nervousness, agony, to name a few (See Mayer & Coelho [2021] for a detailed study), and it is changing tourist's and communities' psyches perhaps towards a "new equilibrium" (Kock et al., 2020, p. 1).

Despite this tourist's psychosomatic swift change and the importance of rural areas to tourism development, the focus of the studies on wellbeing applied to rural tourism usually is resident's (Boukas, 2019; Wang et al., 2021) or employees oriented (Xia et al., 2020). There is a lack of inquiries trying to understand the relationship between rural tourism and welfare on tourism, and this deficiency is even more prominent when it comes to search this theme under the sustainability lens, which appears to be one of the dominant paradigms in tourism research currently (Moyle et al., 2020; Ruhanen et al., 2015). In this sense, this study aims to connect the topics on welfare and tourism in rural areas to provide a theoretical starting point for future studies. We unravel essential elements in the pursuit of enhancing tourist's welfare on rural tourism by providing a visual conceptual model that provides guidelines for tourism stakeholders to enhance tourists' welfare while mitigating negative crisis impacts.

WELFARE TOWARDS SUSTAINABILITY AGENDA

The term welfare is used in numerous disciplines with several connotations (Bergson, 1954) and constantly applied as a synonym for 'utility', 'satisfaction', 'happiness' or 'preference' (Brouwer et al., 2008). Partly of this polysemy is that the term's usage always has a "normative political dimension that makes it difficult to derive a commonly accepted meaning" (Hudson, 2013, p. 4). Hudson complements his idea by saying that welfare "will always remain, in part at least, political ones, reflecting different values, ideologies and world views" (ibid. p. 4). The term is often used "to refers to both people's wellbeing, and systems which are designed to provide for people" (Spicker, 1995, p. 66).

Moreover, Van Praag (1989, p. 1) argues that researchers generate fragmented concepts (that explains partially or not at all the observed phenomena) due to the impossibility of understanding the whole phenomena. To express the multidimensionality of the term that might apprehend part of the complexity inherent from welfare, we present the twelve welfare values categorized into three dimensions (Sumner, 1999):

1. **Having**: Income; Housing; Employment; Health[1]; Education;
2. **Loving**: Community attachment; family attachment; friendship patterns;
3. **Being**: Personal prestige; Insubstitutability; Political resources; Doing interesting things.

In a simplistic manner, Van Praag (1989, p. 2) presents welfare as "an evaluation by the individual of his situation" (p. 2). From an ethical and etymological view, Sumner (1996, p. 1), says that welfare is a "condition of faring or doing well. It is in this sense that welfare attaches pre-eminently to the lives of individuals, and a person's welfare is more or less the same as her wellbeing or interest or [...] her good" (p. 1). In line with our argument, it is possible to identify, from studies with economic perspective, that welfare is not explicitly defined (Kaneno & Nakamura, 1979).

As a matter of fact, welfare is more than a jargon used across fields. It is necessary to understand the cultural and historical context in which the term is embedded to comprehend its 'true meaning' (Greve, 2008). The term is regularly fused to depict subtly distinct variations in specific contexts (Hudson, 2013). Under this perspective, it becomes imperative to present the theoretical lens under which we understand people's welfare. Our perspective comes from a recent global sustainability trend derived from the 2030 agenda (UN, 2015).

The term 'sustainability' was used for the first time in 1972 (Goldstein et al., 1972; Fragoso, Kidd, 1992). However, the concept, as we know, dates back to 1987 with the publication of the report 'Our common future', worldly known as the "Brundtland report". Its "release sought multilateral support for enhancing awareness, cooperation, and a more balanced approach to development" (Ruhanen et al., 2015, p. 517). In this report, the World Commission on Environment and Development (WCED) define sustainable development as "development that meets the needs of the present without compromising the ability of future generations to meet their own needs"

(WCED, 1987, n.p.). The document states that there are two key concepts related to this term: "the concept of 'needs' [...] [and] the idea of limitations imposed by the state of technology and social organization on the environment's ability to meet present and future needs" (ibid). Furthermore, the report (WCED, 1987) states that:

the goals of economic and social development must be defined in terms of sustainability in all countries - developed or developing, market-oriented or centrally planned. Interpretations will vary, but must share certain general features and must flow from a consensus on the basic concept of sustainable development and on a broad strategic framework for achieving it. [...] development involves a progressive transformation of economy and society. A development path that is sustainable in a physical sense could theoretically be pursued even in a rigid social and political setting. But physical sustainability cannot be secured unless development policies pay attention to such considerations as changes in access to resources and in the distribution of costs and benefits. Even the narrow notion of physical sustainability implies a concern for social equity between generations, a concern that must logically be extended to equity within each generation (n.p)

As WCED expose, sustainability embraces social, cultural, political, environmental domains. It is presented as an alternative to social and environmental entropy. After the 27 principles defined on the Rio Declaration (UN, 1992), it expanded to a "development that integrates economic growth and environmental protection" (Basiago, 1995, p. 110). Basiago (1995) sees sustainability as a protocol to maintain the earth's resource and a methodology to secure the maximization of the social and environmental systems' vitality. He also states, as a methodology, that we need to apply four criteria that will "act as constraints on untoward forms of development" (p. 119): they are:

- futurity, i.e., a concern for the welfare of future generations;
- equity, i.e., the fair sharing of economic benefits and burdens within and between generations;
- global environmentalism, i.e., a recognition of the global dimension of ecological problems associated with use or depletion of natural capital by one or some at the cost of others; and
- biodiversity, i.e., the maintenance of the integrity of ecological processes and system.

In this sense, for rural destinations "to keep pace and adapt to the vigorous globalized economy" (Okech, Haghiri, & George, 2012, p. 37) they need to develop a management culture aiming sustainability (Blancas et al., 2011) fostering welfare.

This could constitute a practical problem since sustainability studies on tourism lack theoretical and methodological progress (Moyle et al., 2020).

RURAL TOURISM AND SUSTAINABLE DEVELOPMENT

The importance of tourism is recognized in all three domains of sustainability. From the economic perspective, Brida, Gómez and Segarra (2020) identified an empirical relationship between 80 countries economic development and countries' tourism increasing. Regarding the social aspect, Figueiredo and Franco (2018) demonstrate the importance of wines cooperatives towards achieving social missions of common prosperity. From the environmental viewpoint, Souza et al. (2018) concluded that for each dollar that the Brazilian government invest in environmental safeguard of protected areas, they could receive seven dollars in return.

In rural areas, the tourism's bonuses are even higher since the activity can stop rural depopulation (Polukhina et al. 2021), preservation of culture and traditional ways of living (Nagaraju & Chandrashekara, 2014), along with many others. The UNWTO (2021) defines tourism in rural areas, i.e., rural tourism as "activity in which the visitor's experience is related to a wide range of products generally linked to nature-based activities, agriculture, rural lifestyle/culture, angling and sightseeing." For them, the areas which rural tourism takes place are with: i. low population density; ii. the use of the land and its scape dominated by forestry and/or agriculture; and iii. traditional's lifestyle and social structure.

Greffe (1994) presents the evolution of the term under two-way viewpoints: i. from the supply; and ii. the demand side. Nagaraju and Chandrashekara (2014) expose the multifaceted characteristic of rural tourism that entail ecotourism, cultural tourism and so forth. Ayazlar and Ayazlar (2015) present the complexity of rural tourism's background by synthesizing several academic and governmental definitions of the topic. They address six intrinsic characteristics, named: 1. It needs to be located in rural areas; 2. It needs to be based on rural tourism's features; 3. It needs to have buildings and settlements on a small scale; 4. It needs to connect tourists with local people and their families; 5. It needs to be based on villages and small towns; 6. It needs to represent complex economic, environmental and historical patterns.

Although there is no consensus on the definition of rural tourism due to its heterogeneity (Polukhina et al., 2021) and all of those views could not apprehend the whole rural tourism's complexity, the authors express the socio-environmental nature of this type of tourism in which rural tourism can meet people's (tourists and autoctone population) by attending the needs, enhancing life, happiness and inner peace (Wang et al. 2021). Therefore it can be considered a key tool for rural development (Naghiu, Vásquez, & Georgiev, 2005).

Rural tourism's becomes even more important in a pandemic scenario. In fact, the World Tourism Organization enacted 2020 as the year of Tourism and Rural development. Moreover, UNWTO foresaw a growth in domestic tourism that puts rural communities with protected livelihoods on the recovery's front (UNWTO, 2020a). UN is spending several efforts to diminish disparities between rural and urban areas (UN, 2015). They also create a startup competition that seeks to source new enterprises to advance the contribution of tourism to rural development and support recovery (UNWTO, 2020b). Additionally, "several national governments have instituted sector planning following the definition of [rural] tourism model" (Blancas et al., 2011, p. 28). For instance, Brazil developed an applied research project in the area of family farming and tourist circuits (Experiences from Rural Brazil / Experiências do Brasil Rural[2]) that seeks to foster and support the development of tourism in rural areas through work focused on tourism-related products that are within tourist routes and belong to the agri-food chains of cheese, wine, beer and Amazon fruits (Universidade Federal Fluminense [UFF], 2021; Brazilian Ministry of Tourism, 2021). We can also visualize initiatives to digitalize rural tourism destinations in Latin American countries such as Ecuador, Costa Rica and the Dominican Republic (Food and Agriculture Organization of the United Nations [FAO], 2021)

More recently, UNWTO launched a "guide to making tourism an effective tool for rural development" (UNWTO, 2020c), with the technical support of important international entities such as the Organization for Economic Cooperation and Development (OECD), World Bank, amongst others. This document (UNWTO, 2020c) sets five dimensions (table 1) and thirty-seven recommendations "to drive development for rural communities as one of the socio-economic activities engaging across and throughout communities and leaving no one behind" (ibid, p. 5).

Table 1. Five dimensions for rural development.

DIMENSION
i. Placing tourism as a strategic pillar in policies for rural development
ii. Tourism for the benefit and wellbeing of rural communities: Building a fair, inclusive, resilient and sustainable sector
iii. New opportunities for tourism and rural development: Innovation, technology and digitalization
iv. Product development and value chain integration for a sustainable and enhanced travel experience
v. Fostering sustainable policies and practices in rural destinations

Source: Adapted from UNWTO (2020c)

These five dimensions stress the need for fortification of the public sector to put tourism as a strategic pillar in the territories' development, enhance public-private partnerships, to build a strong and integrated rural community while enhancing technological development, strengthening tourism products and services to generate memorable tourism experiences and maintaining social, economic and environmental sustainability in destinations.

Even though stimulate tourism in rural areas seems like the logical approach, Ortiz (2020) calls attention to the need to think of rural tourism as multiple spaces with infinite possibilities to recreate a life that needs to renegotiate the relationships between rural and urban worlds derived from the late capitalism. Ortiz also argues that we should rethink what are the rural area needs, and if it is ready and/or wants to be part of this reactivation. In spite of this view, rural tourism presents as a way out from extreme poverty and that will reignite tourism, because this segment benefits (in comparison with others) due to restrictions imposed in the context of COVID-19 (Korstanje, 2020).

A CONCEPTUAL MODEL ON SUSTAINABILITY WELFARE IN RURAL TOURISM

Past studies on sustainability struggled with theoretical and methodological progress (Moyle et al., 2020). Moreover, welfare is an assumed topic that researchers lack critical discussion. In this sense, the current research intends to provide guidelines for tourism resumption grounded on these three theoretical bases (welfare, sustainability and rural tourism). To achieve this, we elaborate a conceptual model that aims to enhance welfare through several stakeholders' efforts. The model also minimizes external factors derived from the crisis. Those externalities, such as the COVID-19 pandemic, can drastically impact the economy (Chinazzi et al., 2020; UNWTO, N.D.a), people's response and perception towards tourism (Kock et al., 2020).

Our conceptual model (Figure 2) consists of four primary elements, in which three are possible to extend to a post-COVID-19 scenario. The elements are the:

1. top-down government inputs: in which local, state and federal generate efforts to support the reignition of tourism;
2. bottom-up society inputs: in which local and regional stakeholders such as the private sector and tourist communities generate efforts to support the reignition of tourism;
3. rural tourism environment: where the welfare enhancement (or reduction) happens; and

4. Crisis collateral effects: External factors that affect tourism activity that can not be detained but can be mitigated.

The model (Figure 1) represents a contribution to knowledge in both rural tourism literature and welfare on tourism studies. The examination of literature, government documents and grey literature (such as reports on tourism resumption in COVID-19 context) revealed the frailty that rural tourism might have and the opportunities for its own social-spatial nature, e.g., inherent characteristics of tourism in rural areas.

Figure 1.

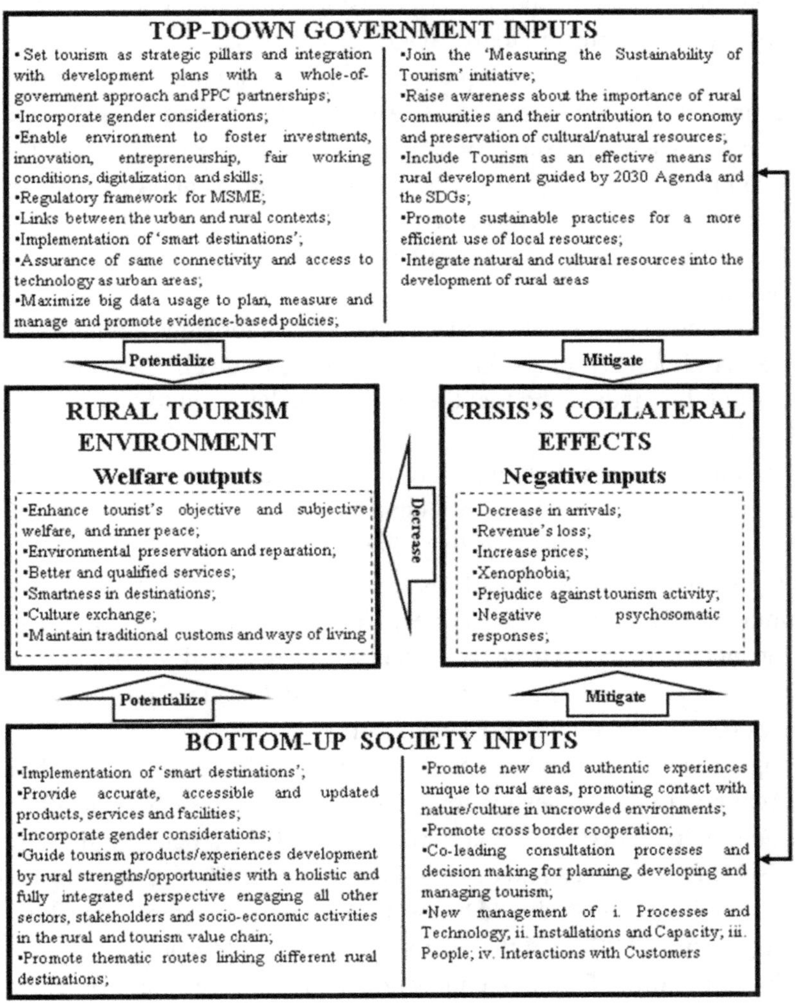

The model's utility for tourism stakeholders is twofold: reading it from bottom to top indicates private and communities efforts in more sustainable tourism development; and when reading from top to bottom, it indicates the public sector efforts enhance welfare as well. Both of them are connected because they mutually stimulate the development of each other.

Top-Down Government Inputs

In the first element of our model, we present thirteen actions that will elevate rural tourism's organization, in line with UNWTO instructions (2020c). All of the activities traverse the social, economic, and environmental dimensions from sustainability, emphasizing the need for a good interrelationship. All of the actions are in line with the Sustainable Development Goals (SDG) (UN, 2015).

As the figure shows, the first topic states that tourism must be a strategic pillar for the destinations, and managers must perform joint activities with diverse public entities (whole-of-government approach). According to Kubickova & Campbell (2020), economies that adopt the whole-of-government approach are the most successful ones in tourism performance.

We stress the need for gender equality opportunities and generate an open environment that fosters innovation and entrepreneurship initiatives. Those initiatives need to be grounded on a solid regulatory framework for Micro, Small & Medium Enterprises. Additionally, decision-makers should not dwell under the dichotomy of 'urban-rural, but understand that they are facets that need to be treated together (UNWTO, 2020c).

In a similar manner, urban areas are usually economical, techno and human deficient that hinder the adoption of technologies (Oliveira & Zouain, 2011). For rural destinations to be competitive, they should provide the same access that tourist has in the urban context. For this to happen, tourism destinations need to focus on becoming smarter destinations (Boes, Buhalis, & Inversini, 2015; Buhalis & Amaranggana, 2013), in this way, guaranteeing sustainable development, enriching communities' quality of life and enhancing tourists' welfare by providing richer experiences (Gretzel et al., 2015; Blanco, 2015).

Although, becoming smart also means that the decisions are evidence-based (e.g. resource reallocation), i.e., destinations will apply the use of big data correctly to support their decisions (Cacho et al., 2016) and to have a better understanding of their tourists and their needs and desires (Law, Chen, & Chan, 2019). In a nutshell, private and public managers should follow the rural tourist's trend about being more connected (Ballina, 2020), and the visitors should have this need met.

To achieve the title of welfare enhancer destination, managers should measure the sustainability in tourism sectors. The methodology from UNWTO is a proper "statistical framework for [measurement of] the Sustainability of Tourism [that integrates] statistics on economic, environmental and social dimensions" (UNWTO, N.D.b). By doing so, they can raise awareness about the economic impact of rural communities and the possibility to preserve cultural and natural resources through tourism. These initiatives, of course, are related to the 2030 agenda (UNWTO, N.D.a), therefore using efficiently local resources. Lastly, it is essential to integrate the natural and cultural resources that still lack connection with the main destination's attractiveness.

Bottom-Up Society Inputs

Regarding the 'bottom-up society inputs', our approach also presents societies' actions needed to develop rural tourism and tourist's welfare. We highlight that the first three actions are in both top-down and bottom-up inputs. Under this perspective, it is inexorable for the tourism sector to adapt to the new technologies (Buhalis, 2000). Information and Communication Technology (ICT) plays a key "role in boosting the competitiveness of the organizations (Návio-Marco, Ruiz-Gomez, & Sevilla, 2018). Therefore, "service management inevitably has been influenced by recent technological revolutions and smartness"(Buhalis et al., 2019, p. 485). Communities and private sectors should make their environments smarter and provide accurate information and preferably with real-time updates (Buhalis & Synarta, 2019)

The last topic identified in both approaches is the necessity to target gender inequality. The gender differences are more evident in the rural context (Tonini & Dolci, 2020). The development of new solutions is necessary to counteract the loss attributable to COVID-19 (Malisch et al., 2020) to produce better results aligned with SDG 5's target goals (Achieve gender equality and empower all women and girls). These reinforced lines of actions produce an enabling environment to enhance welfare.

Furthermore, service providers must engage stakeholders and collectively produce products and experiences with theme routes that will contribute to the destination's singularities and provides memorable tourism experiences (Coelho, Gosling, & Almeida, 2018). The uncrowdedness's and openness characteristics of rural areas favors the non-urban destinations in comparison with urban attractions. In the COVID-19 context, tourists are caring more about their safety (Kim et al., 2021; Pan et al., 2021), so managers should harness these opportunities to foster tourist demand by adopting all existing sanitary measures and. Otherwise, these initiatives might backfire.

Cross-border cooperation between entities and stakeholders from municipalities in the same territory is crucial to businesses' survival (OECD, 2020). For this to happen, regional "leadership should be participatory and democratic, giving equal voice to the various actors involved in regional tourism" (Valente, Dredge, & Lohmann, 2015, p. 129).

Concluding, the sanitary crisis evidentiates the need to redesign the process, the use of technologies, infrastructure and collaborators activities and their interaction with tourists (Coelho & Mayer, 2020). Until a welfare culture grounded on a sustainable tripod be adopted, rural tourism destinations might struggle across time.

Crisis's Collateral Effects

Regarding the negative impacts of externalities, such as COVID-19 pandemics, extant studies and technical reports show that the externalities produce a decrease in arrivals (Gössling, Scott, & Hall, 2021). Hence, the destination's revenue will force businesses to shut their activities (UN News, 2021), raise prices (European Central Bank, 2021), lower price and lose quality due to less staff working to maintain the enterprises alive. A wide-range of top-down and bottom-up activities are in our model that might mitigate these collaterals. Moretheless, the pandemic also brings xenophobia towards tourists and people from destinations whose countries were afflicted with higher rates of covid or the new virus's variants (BBC, 2020; He et al., 2021; Kock et al., 2020).

These negative feelings generate a prejudice against tourists, not only from individuals but also from public governments (Roberto, Johnson, & Rauhaus, 2020). Communities' responses like these increase tension, despair, and other frightening feelings regarding travelling (Mayer & Coelho, 2021), thus producing immobilities. As our model presents and restates, these collaterals significantly impact tourism activity. However, they might be mitigated with proper governance and collective efforts. In an ideal scenario, they need to be targeted from both top-down and bottom-up inputs. The current goal is to produce the same amount of confidence and safety in travelling, that the tourist might transcend the reality and 'travel back' to the old normal, even if it is for an ephemeral moment.

Rural Tourism Environment

The desired 'rural tourism environment' is where the tourist's welfare rises (instead of decreasing) when proper strategies are put into practice. Not only the subjective welfare need to be managed, but also the objective measures of welfare through sustainable measurement need to be checked. Proper measures following our conceptual model will generate confidence and safety for tourists, and as a result, it will be possible to produce information for advertising and marketing. Global initiatives such as 'Safe travels' stamps are pro-health labels in which destinations must align protocols before receiving them (World Travel & Tourism Council

[WTTC], N.D.). A survey study shows that tourist perception regarding certified crisis prepared travel destination is a significant predictor of travel's likelihood (Pennington-Gray et al., 2014).

All of the strategies above are indicated for achieving higher levels of tourists' welfare and improve tourism performance, and we provide guidelines for any kind of tourism governance (Hall, 2013). The importance for tourism destinations and stakeholders to follow our model enhance during crisis context because they are possible solutions to mitigate the negative impacts. However, collective efforts and higher levels of cooperation are crucial to best harness this theoretical tool.

CONCLUSION

The current study presents a discussion about how rural tourism might foster tourists' welfare. It starts with the problematization related to welfare's concept and the adoption of a sustainable view regard welfare. This is followed by a discussion on rural tourism and sustainable development, which unveil the inner relationship between rural tourism and tourists' welfare. Concluding with a graphical conceptual model on sustainable rural tourism development to foster tourists' welfare. We expose the central elements of the graphical model and the actions required to enhance welfare on rural tourism as well as their interrelationships. Our model counteracts the multidimensionality and complexity inherent in rural tourism while understanding the weaknesses and threats to rural destinations.

The major limitation of our model is that it lacks empirical validation. Future studies should empirically test it. Moreover, our study does not intend to present the whole set of components, strategies, and actions necessary to fully address our attention to tourists' welfare enhancer key components but rather present a response from the sustainable lens.

Our model holistically expands the studies on welfare by applying a sustainability approach to foster welfare in tourism. In this attempt, we connect the welfare framework and sustainable development to foster rural tourism development. Regarding the managerial contributions, this research presents both top-down and bottom-up strategies that will provide starting points to develop tourism in rural areas, hence, improving welfare while reducing side-effects derived from the crisis. It is important to reinforce that "only through collective action and international cooperation [managers] will we be able to transform tourism, advance its contribution to the 2030 Agenda [...] [and], embraces local values and communities and creates decent job opportunities for all" (UNWTO, N.D.a).

Funding

This work was supported by the project Experiências do Brasil Rural, funded by Brazilian Minsitry of Tourism, Brazilian Ministry of Agriculture, Livestock and Supply, and Fluminense Federal University (Ministério do Turismo do Brasil/ Ministério da Agricultura, Pecuária e Abastecimento e Universidade Federal Fluminense).

REFERENCES

An, W., & Alarcón, S. (2020). How Can Rural Tourism Be Sustainable? A Systematic Review. *Sustainability*, *12*(18), 7758. doi:10.3390u12187758

Ayazlar, G., & Ayazlar, R. A. (2015). Rural tourism: A conceptual approach. In Tourism, Environment and Sustainability. St. Kliment Ohridski University Press.

Ballina, F. J. (2020). Is there rural smart tourism? A Spanish experience. *Management Theory and Studies for Rural Business and Infrastructure Development*, *42*(3), 369–380. doi:10.15544/mts.2020.37

BBC. (2020). *Londoners 'experiencing racism over coronavirus'*. BBC News. Retrieved from: https://www.bbc.com/news/av/uk-england-london-51361930

Bergson, A. (1954). On the concept of social welfare. *The Quarterly Journal of Economics*, *68*(2), 233–252. doi:10.2307/1884448

Blancas, F. J., Lozano-Oyola, M., González, M., Guerrero, F. M., & Caballero, R. (2011). How to use sustainability indicators for tourism planning: The case of rural tourism in Andalusia (Spain). *The Science of the Total Environment*, *412*, 28–45. doi:10.1016/j.scitotenv.2011.09.066 PMID:22033358

Blanco, J. (2015). *Libro blanco de los destinos turísticos inteligentes*. Estrategias y soluciones para fomentarla innovación en el turismo digital. Madrid, Spain. LID Editorial Empresarial.

Boes, K., Buhalis, D., & Inversini, A. (2015). Conceptualizing smart tourism destination dimensions. In *Information and communication technologies in tourism 2015* (pp. 391–403). Springer. doi:10.1007/978-3-319-14343-9_29

Boukas, N. (2019). Rural tourism and residents' wellbeing in Cyprus: Towards a conceptualized framework of the appreciation of rural tourism for islands' sustainable development and competitiveness. *International Journal of Tourism Anthropology*, *7*(1), 60–86. doi:10.1504/IJTA.2019.098105

Bramwell, B. (1994). Rural tourism and sustainable rural tourism. *Journal of Sustainable Tourism, 2*(1-2), 1–6. doi:10.1080/09669589409510679

Brazililian Ministry of Tourism. (2021). *Ministério do Turismo seleciona roteiros para o Projeto Experiências do Brasil Rural.* Retrieved from: https://www.gov.br/turismo/pt-br/assuntos/noticias/ministerio-do-turismo-seleciona-roteiros-para-o-projeto-experiencias-do-brasil-rural

Brouwer, W. B., Culyer, A. J., van Exel, N. J. A., & Rutten, F. F. (2008). Welfarism vs. extra-welfarism. *Journal of Health Economics, 27*(2), 325–338. doi:10.1016/j.jhealeco.2007.07.003 PMID:18179835

Buhalis, D. (2000). Tourism and information technologies: Past, present and future. *Tourism Recreation Research, 25*(1), 41–58. doi:10.1080/02508281.2000.11014899

Buhalis, D., & Amaranggana, A. (2013). Smart tourism destinations. In *Information and communication technologies in tourism 2014.* Springer. doi:10.1007/978-3-319-03973-2_40

Buhalis, D., Harwood, T., Bogicevic, V., Viglia, G., Beldona, S., & Hofacker, C. (2019). Technological disruptions in services: Lessons from tourism and hospitality. *Journal of Service Management, 30*(4), 484–506. doi:10.1108/JOSM-12-2018-0398

Buhalis, D., & Sinarta, Y. (2019). Real-time co-creation and nowness service: Lessons from tourism and hospitality. *Journal of Travel & Tourism Marketing, 36*(5), 563–582. doi:10.1080/10548408.2019.1592059

Cacho, A., Mendes-Filho, L., Estaregue, D., Moura, B., Cacho, N., Lopes, F., & Alves, C. (2016). Mobile tourist guide supporting a smart city initiative: a Brazilian case study. *International Journal of Tourism Cities.* doi:10.1108/IJTC-12-2015-0030

Chinazzi, M., Davis, J. T., Ajelli, M., Gioannini, C., Litvinova, M., Merler, S., Pastore y Piontti, A., Mu, K., Rossi, L., Sun, K., Viboud, C., Xiong, X., Yu, H., Halloran, M. E., Longini, I. M. Jr, & Vespignani, A. (2020). The effect of travel restrictions on the spread of the 2019 novel coronavirus (COVID-19) outbreak. *Science, 368*(6489), 395–400. doi:10.1126cience.aba9757 PMID:32144116

Coelho, M. F., de Sevilha Gosling, M., & de Almeida, A. S. A. (2018, December). Tourism experiences: Core processes of memorable trips. *Journal of Hospitality and Tourism Management, 37*, 11–22. doi:10.1016/j.jhtm.2018.08.004

European Central Bank. (2021). *Prices for travel during the COVID-19 pandemic: is there commonality across countries and items?* Retrieved from: https://www.ecb.europa.eu/pub/economic-bulletin/focus/2021/html/ecb.ebbox202101_06~bcb28cb255.en.html

FAO, Food and Agriculture Organization of Nations. (2021). *Launch of the Digital Villages initiative in Latin America and the Caribbean*. Retrieved from: http://www.fao.org/americas/eventos/ver/en/c/1395198/

Figueiredo, V., & Franco, M. (2018). Wine cooperatives as a form of social entrepreneurship: Empirical evidence about their impact on society. *Land Use Policy*, *79*, 812–821. doi:10.1016/j.landusepol.2018.09.022

Filimonau, V., Lemmer, C., Marshall, D., & Bejjani, G. (2017). 'Nudging'as an architect of more responsible consumer choice in food service provision: The role of restaurant menu design. *Journal of Cleaner Production*, *144*, 161–170. doi:10.1016/j.jclepro.2017.01.010

Fragoso, R. M. D. S. (2015). Habitat Features and Strategies for the Sustainable Development in the Alentejo Region. *Revista de Economia e Sociologia Rural*, *53*(suppl 1), 103–114. doi:10.1590/1234-56781806-94790053s01008

Goldstein, E., Allen, R., Allaby, M., Davoll, J., & Lawrence, S. (Eds.). (1972). *A blueprint for survival. The Ecologist, 2(1)* .

Gössling, S., Scott, D., & Hall, C. M. (2021). Pandemics, tourism and global change: A rapid assessment of COVID-19. *Journal of Sustainable Tourism*, *29*(1), 1–20. doi:10.1080/09669582.2020.1758708

Gowdy, J. (2005). Toward a new welfare economics for sustainability. *Ecological Economics*, *53*(2), 211–222. doi:10.1016/j.ecolecon.2004.08.007

Greffe, X. (1994). Is rural tourism a lever for economic and social development? *Journal of Sustainable Tourism*, *2*(1-2), 22–40. doi:10.1080/09669589409510681

Gretzel, U., Sigala, M., Xiang, Z., & Koo, C. (2015). Smart tourism: Foundations and developments. *Electronic Markets*, *25*(3), 179–188. doi:10.100712525-015-0196-8

He, L., Zhou, W., He, M., Nie, X., & He, J. (2021). Openness and COVID-19 induced xenophobia: The roles of trade and migration in sustainable development. *PLoS One*, *16*(4), e0249579. doi:10.1371/journal.pone.0249579 PMID:33831012

Hudson, J. (2013). Welfare. In An introduction to social policy. Sage Publications. doi:10.4135/9781526435835.n1

Kaneko, M., & Nakamura, K. (1979). The Nash social welfare function. *Econometrica*, *47*(2), 423–435. doi:10.2307/1914191

Kidd, C. V. (1992). The evolution of sustainability. *Journal of Agricultural & Environmental Ethics*, *5*(1), 1–26. doi:10.1007/BF01965413

Kim, J., Lee, J., Jhang, J., Park, J., & Lee, J. C. (2021). The impact of the COVID-19 threat on the preference for high versus low quality/price options. *Journal of Hospitality Marketing & Management*, 1–17. doi:10.1080/19368623.2021.1884163

Kock, F., Nørfelt, A., Josiassen, A., Assaf, A. G., & Tsionas, M. G. (2020). Understanding the COVID-19 tourist psyche: The evolutionary tourism paradigm. *Annals of Tourism Research*, *85*, 103053. doi:10.1016/j.annals.2020.103053 PMID:32921847

Korstanje, M. E. (2020). El COVID-19 y El turIsmo rural: una perspectiva antropológica. *Dimensiones Turísticas, 4*, 179-196. doi:10.47557/CKDK5549

Kubickova, M., & Campbell, J. M. (2020). The role of government in agro-tourism development: A top-down bottom-up approach. *Current Issues in Tourism*, *23*(5), 587–604. doi:10.1080/13683500.2018.1551338

Law, R., Sun, S., & Chan, I. C. C. (2019). Hotel technology: A perspective article. *Tourism Review*, *75*(1), 286–289. Advance online publication. doi:10.1108/TR-05-2019-0150

Malisch, J. L., Harris, B. N., Sherrer, S. M., Lewis, K. A., Shepherd, S. L., McCarthy, P. C., Spott, J. L., Karam, E. P., Moustaid-Moussa, N., Calarco, J. M. C., Ramalingam, L., Talley, A. E., Cañas-Carrell, J. E., Ardon-Dryer, K., Weiser, D. A., Bernal, X. E., & Deitloff, J. (2020). Opinion: In the wake of COVID-19, academia needs new solutions to ensure gender equity. *Proceedings of the National Academy of Sciences of the United States of America*, *117*(27), 15378–15381. doi:10.1073/pnas.2010636117 PMID:32554503

Mayer, V. F., & Coelho, M. D. F. (2021). Interrupted dreams: Memories and emotions of travel experiences during Covid-19 breakthrough. *Revista Brasileira de Pesquisa em Turismo*, *15*(1). Advance online publication. doi:10.7784/rbtur.v15i1.2192

Mayer, V. F., Machado, J. D. S., Marques, O., & Nunes, J. M. G. (2020). Mixed feelings?: Fluctuations in wellbeing during tourist travels. *Service Industries Journal*, *40*(1-2), 158–180. doi:10.1080/02642069.2019.1600671

Moyle, B., Moyle, C. L., Ruhanen, L., Weaver, D., & Hadinejad, A. (2020). Are we really progressing sustainable tourism research? A bibliometric analysis. *Journal of Sustainable Tourism*, *29*(1), 106–122. doi:10.1080/09669582.2020.1817048

Nagaraju, L. G., & Chandrashekara, B. (2014). Rural tourism and rural development in India. *International Journal of Interdisciplinary and Multidisciplinary Studies*, *1*(6), 42–48.

Naghiu, A., Vázquez, J. L., & Georgiev, I. (2005). Rural development strategies through rural tourism activities in romania: Chance for an internal demand? *International Review on Public and Nonprofit Marketing*, *2*(1), 85–95. doi:10.1007/BF02893253

Nash, J. F. Jr. (1950). The Bargaining Problem. *Econometrica*, *18*(2), 155–162. doi:10.2307/1907266

Navío-Marco, J., Ruiz-Gómez, L. M., & Sevilla-Sevilla, C. (2018, December). Progress in information technology and tourism management: 30 years on and 20 years after the internet-Revisiting Buhalis & Law's landmark study about eTourism. *Tourism Management*, *69*, 460–470. doi:10.1016/j.tourman.2018.06.002

OECD - Organization for Economic Co-operation and Development. (2020). *The territorial impact of COVID-19: managing the crisis across levels of government.* Retrieved from: https://www.oecd.org/coronavirus/policy-responses/theterritorial-impact-of-covid-19-managing-the-crisis-across-levels-ofgovernment-d3e314e1/

Okech, R., Haghiri, M., & George, B. (2012). Rural tourism as a sustainable development alternative: an analysis with special reference to Luanda, Kenya. *CULTUR, 6*(3).

Oliveira, C. T. F., Zouain, D. M. (2011). Turismo rural e agricultura familiar: desafios e perspectivas para o campo. *Revista Acadêmica Observatório de Inovação do Turismo, 6*(2).

Pan, T., Shu, F., Kitterlin-Lynch, M., & Beckman, E. (2021). Perceptions of cruise travel during the COVID-19 pandemic: Market recovery strategies for cruise businesses in North America. *Tourism Management*, *85*, 104275. doi:10.1016/j.tourman.2020.104275

Pennington-Gray, L., Schroeder, A., Wu, B., Donohoe, H., & Cahyanto, I. (2014). Travelers' perceptions of crisis preparedness certification in the United States. *Journal of Travel Research*, *53*(3), 353–365. doi:10.1177/0047287513496470

Polukhina, A., Sheresheva, M., Efremova, M., Suranova, O., Agalakova, O., & Antonov-Ovseenko, A. (2021). The Concept of Sustainable Rural Tourism Development in the Face of COVID-19 Crisis: Evidence from Russia. *Journal of Risk and Financial Management*, *14*(1), 38. doi:10.3390/jrfm14010038

Roberto, K. J., Johnson, A. F., & Rauhaus, B. M. (2020). Stigmatization and prejudice during the COVID-19 pandemic. *Administrative Theory & Praxis*, *42*(3), 364–378. doi:10.1080/10841806.2020.1782128

Ruhanen, L., Weiler, B., Moyle, B. D., & McLennan, C. L. J. (2015). Trends and patterns in sustainable tourism research: A 25-year bibliometric analysis. *Journal of Sustainable Tourism*, *23*(4), 517–535. doi:10.1080/09669582.2014.978790

Sharpley, R., & Jepson, D. (2011). Rural tourism: A spiritual experience? *Annals of Tourism Research*, *38*(1), 52–71. doi:10.1016/j.annals.2010.05.002

Sigala, M. (2020). Tourism and COVID-19: Impacts and implications for advancing and resetting industry and research. *Journal of Business Research*, *117*, 312–321. doi:10.1016/j.jbusres.2020.06.015 PMID:32546875

Spicker, P. (2000). *The welfare state: a general theory*. Sage Publications.

Tonini, H., & Dolci, T. S. (2020). Turismo Rural e Novos Mercados para Produtos Alimentares Agroecológicos: Estudo de Caso da Rota Via Orgânica/Rural Tourism and New Markets for Agro-Ecological Food Products: The Case Study of Via Organica Route. *ROSA DOS VENTOS-Turismo e Hospitalidade*, *12*(3), 537–554. Advance online publication. doi:10.18226/21789061.v12i3p537

Töppig, S. E., & Moital, M. (2020). An exploration of the techniques, outputs and outcomes of circulation management at exhibitions. *International Journal of Event and Festival Management*, *11*(3), 311–335. Advance online publication. doi:10.1108/IJEFM-10-2019-0050

UFF, Universidade Federal Fluminense. (2021). *Apresentação Experiências Do Brasil Rural*. Retrieved from: http://pesquisaturaf.uff.br/apresentacao-do-projeto/

UN News - United Nations News. (2015). *One in three travel destinations still fully closed to international tourists: UN agency*. Retrieved from: https://news.un.org/en/story/2021/03/1086712

UN - United Nations. (2015). *Resolution adopted by the General Assembly on 1 September 2015*. 69/315. Draft outcome document of the United Nations summit for the adoption of the post-2015 development agenda. Retrieved from: https://www.unescwa.org/sites/www.unescwa.org/files/un_resolutions/a_res_69_315_e.pdf

UN - United Nations. (2016). *Resolution adopted by the General Assembly on 19 September 2016*. A/RES/71/1, 3 October 2016 (The New York Declaration). Retrieved from: https://www.unescwa.org/sites/www.unescwa.org/files/un_resolutions/a_res_69_315_e.pdf

UNWTO - United Nations World Tourism Organization. (2020a). 2020 year of 'tourism and rural development'. Background note.

UNWTO - United Nations World Tourism Organization. (2020b). *Rural tourism competition*. Retrieved from: https://www.unwto.org/unwto-rural-tourism-competition

UNWTO - United Nations World Tourism Organization. (2020c). *UNWTO recommendations on tourism and rural development: A Guide to Making Tourism an Effective Tool for Rural Development*. Retrieved from: https://www.e-unwto.org/doi/epdf/10.18111/9789284422173

UNWTO - United Nations World Tourism Organization. (2021). *Rural Tourism*. Retrieved from: https://www.unwto.org/unwto-rural-tourism

UNWTO - United Nations World Tourism Organization (n.d.a). *Secretary-General's Policy Brief on Tourism and COVID-19: tourism and covid-19 – unprecedented economic impacts*. Retrieved from https://www.unwto.org/tourism-and-covid-19-unprecedented-economic-impacts

UNWTO - United Nations World Tourism Organization (n.d.b). *Measuring the Sustainability of Tourism: on measuring the sustainability of tourism: MST*. Retrieved from: https://www.unwto.org/standards/measuring-sustainability-tourism

Valente, F., Dredge, D., & Lohmann, G. (2015). Leadership and governance in regional tourism. *Journal of Destination Marketing & Management*, *4*(2), 127–136. doi:10.1016/j.jdmm.2015.03.005

Van Praag, B. (1989). *The relativity of the welfare concept*. WIDER Working Paper Series. World Institute for Development Economic Research (UNU-WIDER).

Wang, R., Dai, M., Ou, Y., & Ma, X. (2021). Residents' happiness of life in rural tourism development. *Journal of Destination Marketing & Management*, *20*, 100612. doi:10.1016/j.jdmm.2021.100612

WCED. (1987). *Report of the World Commission on Environment and Development: Our common future*. Retrieved from: http://www.un-documents.net/our-common-future.pdf

World Bank. (2020). *Poverty and Shared Prosperity 2020: Reversals of Fortune*. The World Bank. doi:10.1596/978-1-4648-1602-4

WTTC - World Travel & Tourism Council. (n.d.). *'Safe Travels': Global Protocols & Stamp for the New Normal*. Retrieved from: https://wttc.org/COVID-19/Safe-Travels-Global-Protocols-Stamp

Xia, Y. U., Zheng, Y., Xiaofeng, S. U., & Zheng, Q. (2020). A Study on the Correlation among Social Interaction, Career Self-Efficacy and Well-Being of Employees in Rural Tourism. *Revista de Cercetare si Interventie Sociala*, *70*, 1.

ENDNOTE

1 Experiences from Rural Brazil website: http://pesquisaturaf.uff.br/

Chapter 4
Social Entrepreneurship Activities in the Tourism Sector:
Bibliometric Analysis and Social Network Analysis

Şükran Sirkintioğlu Yildirim
Kastamonu University, Turkey

Ülkühan Bike Esen
https://orcid.org/0000-0002-0967-2140
Çankırı Karatekin University, Turkey

ABSTRACT

Social entrepreneurship is a frequently encountered and increasingly popular concept. Although there are some differences and difficulties in defining the concept, the main reason why it has become such a frequently heard concept today is the social purpose inherent in it. The ability of social entrepreneurship activities to find a place in the tourism sector will make it easier for both social entrepreneurship activities and the tourism sector to achieve their goals. Therefore, in this study, it is aimed to reveal the required fields of study by determining how the concept of social entrepreneurship is handled in the field of tourism with bibliometric analysis. In line with the purpose, 43 studies including the determined keywords were reached. When the methodologies of the articles were examined, it was concluded that empirical studies were in the majority and most of these empirical studies were prepared by using the qualitative research method.

DOI: 10.4018/978-1-7998-7239-9.ch004

Copyright © 2021, IGI Global. Copying or distributing in print or electronic forms without written permission of IGI Global is prohibited.

INTRODUCTION

The entrepreneur is one of the production factors and is defined by Schumpeter as "the person who sets up a company to produce a new product or to produce an existing product in a new method". Marx, on the other hand, emphasizes the aim of the entrepreneur to make profit, and for this purpose, he states that the entrepreneur acts with the motive of constantly following innovations, improving and developing the existing, which is the basis of social change. Although they define the entrepreneur from different angles, both Schumpeter and Marx emphasized the importance of entrepreneurship in order to create innovations and therefore they argued that entrepreneurs should be protected, promoted and encouraged (cited in Tiryaki, 2012: 14-16).

There are three main dimensions of entrepreneurial activity that the entrepreneur carries out with the aim of making a profit or bringing an innovation: Innovation, risk taking and proactivity (Morris, Avila & Allen, 1993). Combinations of these dimensions in different proportions have led to the emergence of many different types of entrepreneurship such as opportunity entrepreneurship, innovative entrepreneurship, follower entrepreneurship (Tekin, 2009: 3-4), internal entrepreneurship (Pinchot, 1985: 4), green entrepreneurship (Berchicci, 2009: 1), and virtual entrepreneurship (Chandra & Leenders, 2012). Social entrepreneurship, which deals with entrepreneurship socially, is one of these new entrepreneurship types. Austin, Stevenson and Wei-Skillern (2006) define social entrepreneurship as "an innovative, social value creation activity that can take place in non-profit organizations, the private sector or the public sector". It is stated that the concept of corporate social responsibility was first used in Bowen's book Social Responsibilities of the Businessman published in 1953 (Aktan & Börü, 2007: 2). The concept of social entrepreneurship has emerged with the merger of corporate social responsibility and entrepreneurship. Studies on social entrepreneurship, whose main purpose is to create social value and for which entrepreneurs generally do not act with the aim of making a profit (Abu Saifan, 2012: 23), are increasing day by day. Studies in which social enterprise and social entrepreneurship are tried to be defined in different ways (Dees, 1998; Mort, Weerawardena & Carnegie, 2002; Roper & Cheney, 2005; Martin and Osberg, 2007), Catford's (1998) study emphasizing the necessity of a suitable environment for the development of social entrepreneurs, article of Barki et al. (2015) in which they examine social entrepreneurship retrospectively and prospectively, articles and reports examining social entrepreneurship activities in different countries and regions (Kerlin, 2006; Bulsara, Gandhi & Chandwani, 2015; Wulleman and Hudon, 2015; İnci, 2016) are a few examples of studies conducted on social entrepreneurship. Studies dealing with social entrepreneurship efforts in different sectors are also included in the literature (Souitaris and Zerbinati, 2005;

Ziegler et al., 2014; Loosemore, 2015). Tourism is one of the sectors subject to social entrepreneurship studies. The intersection point of these two concepts is that social entrepreneurship aims to create social value and tourism has a positive effect on social welfare. This intersection set, called tourism social entrepreneurship, is described by Sheldon, Pollock, and Daniele (2017: 7) as: "A process that uses tourism to create innovative solutions to immediate social, environmental and economic problems in destinations by mobilizing the ideas, capacities, resources and social agreements, from within or outside the destination, required for its sustainable social transformation".

Tourism is a sector that has an impact on the income generation and employment creation levels of the countries. So much so that, between 2009 and 2019, the real growth in international tourism revenues exceeded the growth in world GDP by 10% (UNWTO, 2020). According to OECD data (2020), tourism directly contributes 4.4% of GDP, 6.9% of employment and 21.5% of service related exports to OECD countries. On the other hand, social entrepreneurship is a creative, innovative and social process in which opportunities are followed to create social value and meet social needs (Besler, 2010: 3). Both tourism and social entrepreneurship contribute greatly to the economies of countries with the social and economic developments they provide. Therefore, social entrepreneurship activities in tourism are an issue that should be emphasized by countries that want to increase their income and development levels. The aim of this study is to determine how much social entrepreneurship activities in tourism are studied in which country and which issues are lacking through the literature. For this purpose, first the concepts were defined and then the studies in the literature were examined with bibliometric analysis and social network analysis.

LITERATURE REVIEW

Thompson, Alvy and Lees (2000: 328), stating that there is a need for social enterprise as much as a business enterprise, define a social entrepreneur as a person who realizes unmet demands and need opportunities, brings together the necessary resources to meet these wants and needs and makes a difference by using these resources. Swanson and Zhang (2012) asked if the participants saw themselves as social entrepreneurs in their study, in which they interviewed 202 people, most of whom were senior managers (92%) and the others were business owners. 83% of the participants stated that they see themselves as social entrepreneurs. While 33 people classify themselves as social entrepreneurs because they meet social needs, 25 people defined themselves as social entrepreneurs because they are associated with the society, 15 as they contribute to the development of the social structure, and 14 because they improve their quality of life. Other reasons, at decreasing rate, are listed as providing services to those who are ignored by the society, giving importance to social needs, considering young people and financial reasons. Therefore, the primary goal of social entrepreneurship is to meet the needs of society. It accomplishes this by creating new initiatives or managing existing organizations in an innovative way, through activities aimed at discovering, identifying and using opportunities to improve social welfare (Zahra, Gedajlovic, Neubaum and Shulman, 2009: 519).

Although social entrepreneurship, which forms the basis of charity, is not a new concept, it has started to be used more frequently today (Thompson et al., 2000: 328). Despite its frequent use, there is no single consensus and generally accepted definition of social entrepreneurship. While some authors define social entrepreneurship in terms of non-profit enterprises, some define it as an enterprise that a for-profit business takes as a result of social responsibility, and some define it as a joint venture that removes the boundaries between for-profit and non-profit sectors (Dees, 1998: 1; Sagawa & Segal, 2000). Therefore, it can be stated that social entrepreneurship exists in overlapping areas of public, private and voluntary. The reasons for the emergence of social entrepreneurship are also an indicator of this. Social entrepreneurship occurs for these three reasons: (1) Increased interest in solving social problems, (2) the existence of service areas that are not subject to the public sector and that do not attract the attention of the private sector, and (3) the fact that initiatives to increase social welfare are appreciated (Abdul Kadir and Sarif, 2016). Regardless of the reason it may arise, the aim of social entrepreneurship is to identify the injustices suffered by those who lack financial assets or political influence in the society, to make efforts to eliminate these injustices, and to provide a better future for the targeted population- also for the society in general (Martin and Osberg, 2007). Tan, Williams and Tan (2005) define social entrepreneurship as a risky and innovative process that aims to make profit for the benefit of a segment of the society. According to Santos (2012), social entrepreneurship is "the pursuit of sustainable solutions to problems of neglected positive externalities.". The role of social entrepreneurs in providing economic opportunities includes not only providing access to financial resources, but also using social innovation to solve social problems, empowering people and creating business areas (Sijabat, 2015). As a result of successful social entrepreneurship, the life quality of the society increases, productivity increases, social sustainability and / or economic growth occurs. This change can be large-scale or small-scale, the important thing is to make a difference or to see that you can (Popoviciu and Popoviciu, 2011: 47). In the interviews conducted by Sloan et al. (2014) with seven social entrepreneurs from six different countries, social entrepreneurs stated that there was varying degrees of improvement in the income, quality of life, education and employment levels of local employees due to their work. As a result of their studies, the authors stated that social entrepreneurship can "provide the inspiration that companies of all sizes seeking to be more responsible to society".

Tourism is a sector where various entrepreneurial activities can be carried out. Social entrepreneurship activities in the field of tourism are also needed for sustainable social development (Aquino, Lück and Schanzel, 2018). Sub-sectors of tourism such as accommodation, food and beverage services, tour organizations etc. provide opportunities for creative and risk-taking individuals to use these abilities. These opportunities also attract social entrepreneurs who want to make the world a better place. These entrepreneurs, called tourism social entrepreneurs, are defined as people who strive to solve social problems for the transformation and development of tourism areas (Sheldon, Pollock and Daniele, 2017: 3-7). The concept of tourism, which has types such as eco-tourism, geo-tourism, community-based tourism, is a tool that social entrepreneurs can use both in protecting the environment and combating poverty (Sloan, Legrand and Simons-Kaufmann, 2014: 53). Therefore, studies about tourism and social entrepreneurship diversify due to changing tourism types. For example, Situmorang and Mirzantia (2012) studied the relationship between ecotourism and social entrepreneurship; Sloan, Legrand and Simons-Kaufmann (2014), Mulindwa (2015), Erbaş (2019), Aji (2020) worked on community-based tourism; Peredo and Wurzelmann (2015) investigated indigenous tourism; Peng and Lin (2015), Simatupang (2018), Naderi, Vosta, Ebrahimi, Jalilvand (2019) discussed rural tourism; Celebi, Pirnar and Eris (2020) analyzed gastronomy tourism with social entrepreneurship. The effects of social entrepreneurship on tourism, no matter what type of tourism is applied, have been gathered under five headings (De Lange and Dodss, 2017):

- Social entrepreneurship helps the tourism sector to grow with the tourism brand it will create and the employment it will provide.
- Social entrepreneurship emerging innovations increase the pressure on existing firms and force them to offer similar innovations.
- Social entrepreneurship activities enable other profit-making activities to emerge.
- Social entrepreneurship facilitates implementation of environmental and social regulations.
- Social entrepreneurship enables the development of the local economy through the tourists it attracts to the region and at the same time increases the recognition of the region.

METHODOLOGY

The aim of the study is to systematically analyze the pulishings (articles) in which tourism and social entrepreneurship variables are discussed together, to determine the gap in the field and to make suggestions for future studies. Among the quantitative research methods, bibliometric analysis and social network analysis were used in the study. Bibliometric analysis offers the opportunity to determine the focus of studies in a particular field, to reveal the level of development and to examine it in detail (Heberger et al., 2010). With the social network analysis, the network structure between the keywords used in the articles was visualized. The study aims to answer the following questions.

- What is the distribution of studies over the years?
- What are the journals in which the studies are published?
- What is the number of authors of the studies?
- What is the type of studies?
- What research approaches have been used?
- What are the keywords used?

In the study, only full text articles were considered within the scope of "publication". Book chapters, abstract / full text papers presented at congresses, and writings presented as short notes are excluded. "Web of Science", "Emerald", "EBSCHO", "Elsevier-Scopus", "Science Direct" and "Google Scholar" are databases used to search journals and access publications. The binary variables "tourism" and "social entrepreneurship", "tourism" and "social enterprise", "tourism" and "social entrepreneur", "tourism" and "social business" were determined to be scanned in databases. These variables were scanned together in the title of the article (TI: title) in order to comply with the purpose of the study. In the databases, keywords related to entrepreneurship was added to the first line in the search field, and the keyword "tourism" was added to the second line by using the function "and". Thus, listing the studies that were not included in the study was prevented. 43 articles meeting the criteria determined in accordance with the purpose of the study were reached (bibliometric analysis was conducted in Jaunary 2021). All of the articles reached were published in full text in refereed journals. The information about the articles discussed within the scope of the study (year, author (s), country (ies) of author (s), title of study, keywords, type of study, research method, journal name) was transferred to the Excel table (Appendix A) and the tables and graphics used in the study were created by using the Excel program.

FINDINGS

43 articles obtained as a result of scanning the keywords of tourism and social entrepreneurship (social entrepreneur, social enterprise, social business) were examined in terms of content.

Distribution of Studies by Years

The study, which deals with the concepts of tourism and social entrepreneurship together within the scope of the research, was first published in 2008. As seen in Figure 1; there is no study conducted in 2009, 2010, 2011 and 2013. 2016 is the year with the most studies with nine articles. In the second place, 2020 and 2018 were the years with the most studies with eight and seven articles each.

Figure 1. Number of publications by year

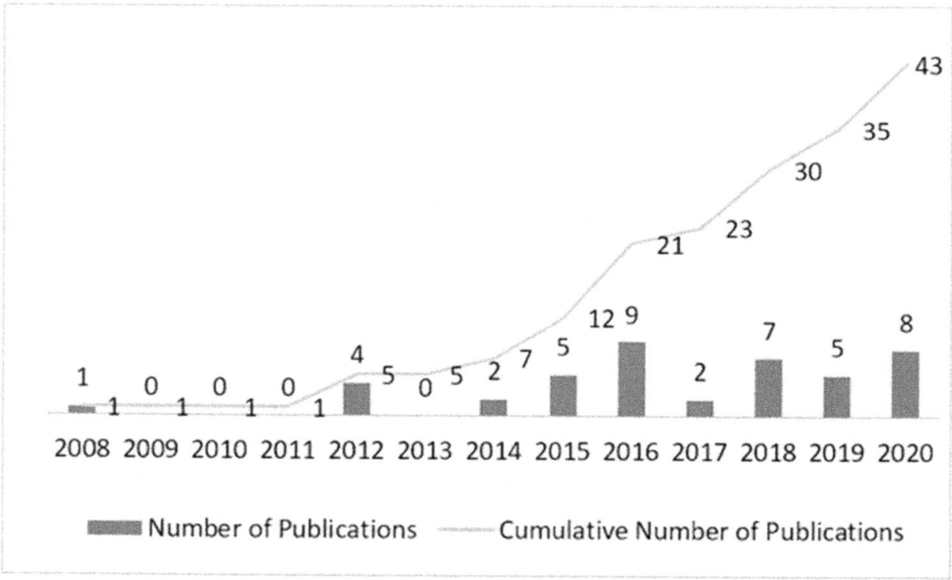

Journals Where the Studies Are Published

According to the list of journals in Table 1; 16% of the studies (seven articles) were published in the “International Journal of Contemporary Hospitality Management”. The other 35 articles were published in various journals in the field of tourism. The study, which handles these two concepts together, was published in "Education, Knowledge & Economy" journal in 2008.

Co-Authorship Status

When the studies are examined in terms of co-authorship status (Figure 2), the number of articles with a single author is 13, number of articles with two authors is 12, number of articles with three authors is 11, and the number of articles with four or more authors is also seven.

Research Method Used in Studies

In Figure 3, it is seen that the qualitative research method was used in 28 of the articles and the quantitative research method was used in five of the articles. Both qualitative research method and quantitative research method were used in all three articles. The remaining seven studies were prepared as a literature review.

Type of Study

Most of the articles reviewed consist of empirical studies. Only eight articles consist of theoretical studies. All of the theoretical studies are literature reviews. 29 of the 35 empirical studies consist of qualitative research method (most of them case study), three of them are mixed, and two of them are quantitative research method. Since one of the variables is the concept of tourism, most of the studies using the qualitative research method consist of case studies addressing a region (e.g. Chilean, Rural Island, Bolivian Amazon, Lisinia Doğa, Siem Reap, Cambodia, Mangalajodi, Bejiharjo, County Donegal).

Authors' Countries

When the articles are examined in terms of the countries where the universities, in which the authors work, are located; Indenosia (7) and United Kingdom (6) share the first place, followed by Australia (4), Turkey (4) and the USA (3).

SOCIAL NETWORK ANALYSIS

Social network analysis is a quantitative type of analysis. Social network analysis examines the relationships between people, groups and organizations. In social network analysis, data are collected at the individual level, but the analysis is carried

Table 1. List of journals

Journal Name	Number of Publications
International Journal of Contemporary Hospitality Management	7
Journal of Sustainable Tourism	2
Procedia Economics and Finance	2
ARA Journal of Tourism Research	1
CAUTHE	1
Chinese Journal of Population Resources and Environment	1
Education, Knowledge & Economy	1
E-Journal of International and Comparative Labour Studies	1
Entrepreneurship & Regional Development: An International Journal	1
Euroasia Journal of Social Science and Humanities	1
European Journal of Tourism, Hospitality and Recreation	1
Earth and Environmental Science	1
GeoJournal of Tourism and Geosites	1
Innovative Issues and Approaches in Social Sciences	1
International Journal of Innovation, Creativity and Change	1
International Journal of Leisure and Tourism Marketing	1
International Journal of Recent Technology and Engineering	1
International Journal of Scientific Management and Development	1
International Journal of Sociology and Social Policy	1
International Journal of Entrepreneurial Behavior & Research	1
International Journal of Humanities Social Sciences and Education	1
Journal of China Tourism Research	1
Journal of Hospitality and Tourism Management	1
Journal of Social Sciences and Humanities Researches	1
Journal of Tourism and Gastronomy Studies	1
Journal of Corporate Governance, Insurance, and Risk Management	1
MIMBAR (Jurnal Sosial dan Pembangunan)	1
PEOPLE: International Journal of Social Sciences	1
The International Indigenous Policy Journal	1
The TQM Journal	1
Think India Journal	1
Tourism Geographies	1
Tourism: An International Interdisciplinary Journal	1
Tourism, Leisure and Global Change	1
Worldwide Hospitality and Tourism Themes	1

Figure 2. Number of authors of the studies

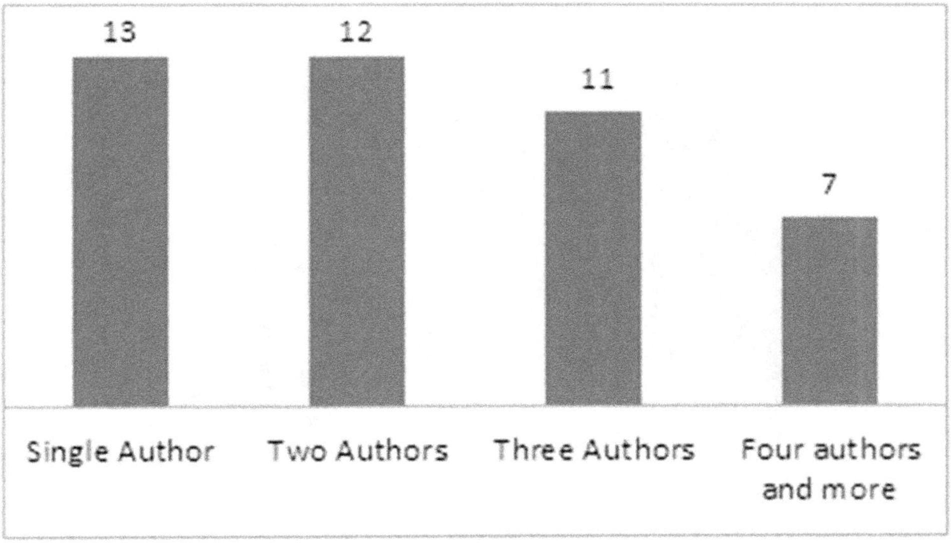

out at the structural level. The data collected for analysis are entered into the Excel file to form a square matrix. If there is a relationship between the data, it is indicated

Figure 3. Type of research method

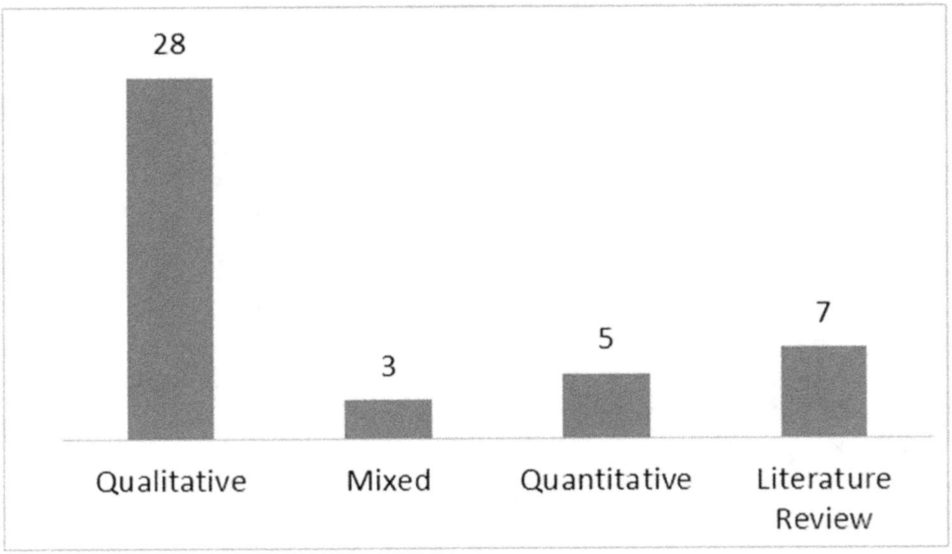

Figure 4. Type of study

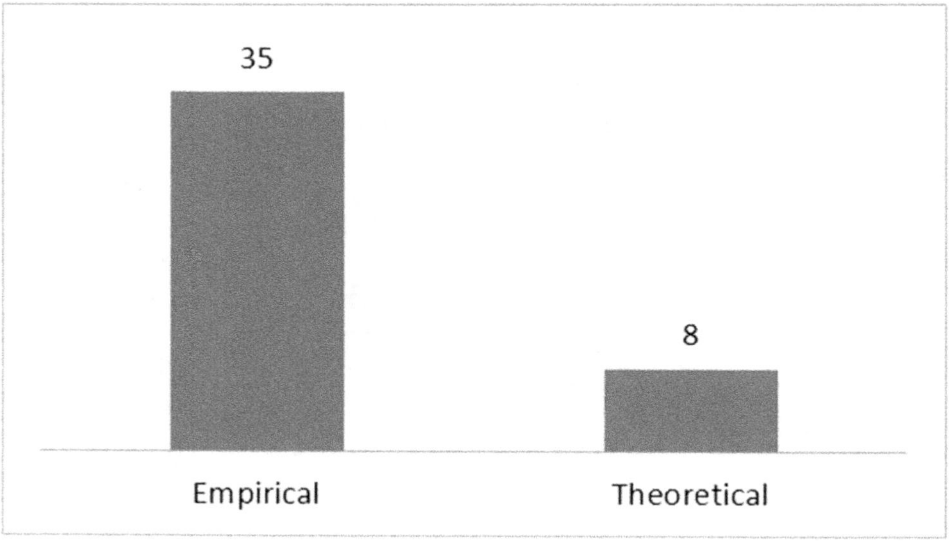

Figure 5. Authors' countries

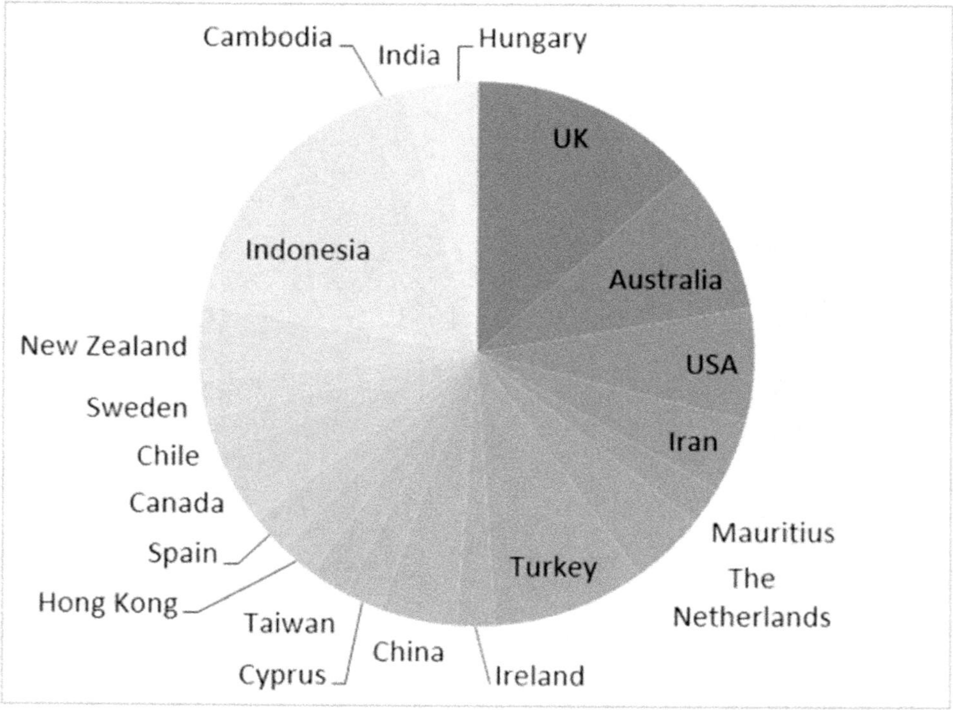

with “1”, if not, with “0”. Then, analysis is performed with the file transferred to the UCINET program (Hawe, Webster and Shiell, 2004).

When the keywords used in the articles were examined, a total of 176 keywords were used in 34 articles. When duplicate keywords were eliminated, 111 keywords included in the social network analysis were found (Appendix B). These keywords are primarily written in the Excel file to form a 111 * 111 matrix. Then, according to the use of keywords together, “1” for keywords used together and “0” for keywords that are not used together were coded. The created Excel table was transferred to UCINET 6 program and the network visual in Figure 6 was obtained. The most used keyword is “social entrepreneurship”. Keywords Social Enterprise, Social Innovation, Social Entrepreneur, Rural Tourism, Sustainability, Tourism, Responsible Tourism, Island Tourism, Entrepreneurship are also frequently used keywords. These concepts point to the most emphasized topics in the articles. As can be understood from the network visual, the concept of “social entrepreneurship” is at the center of the network and provides connections with other keywords. “Supply Chain” is the least used keyword. In order to save the network visual from the complex structure, keywords with centrality below five are not included.

Figure 6. Network visual

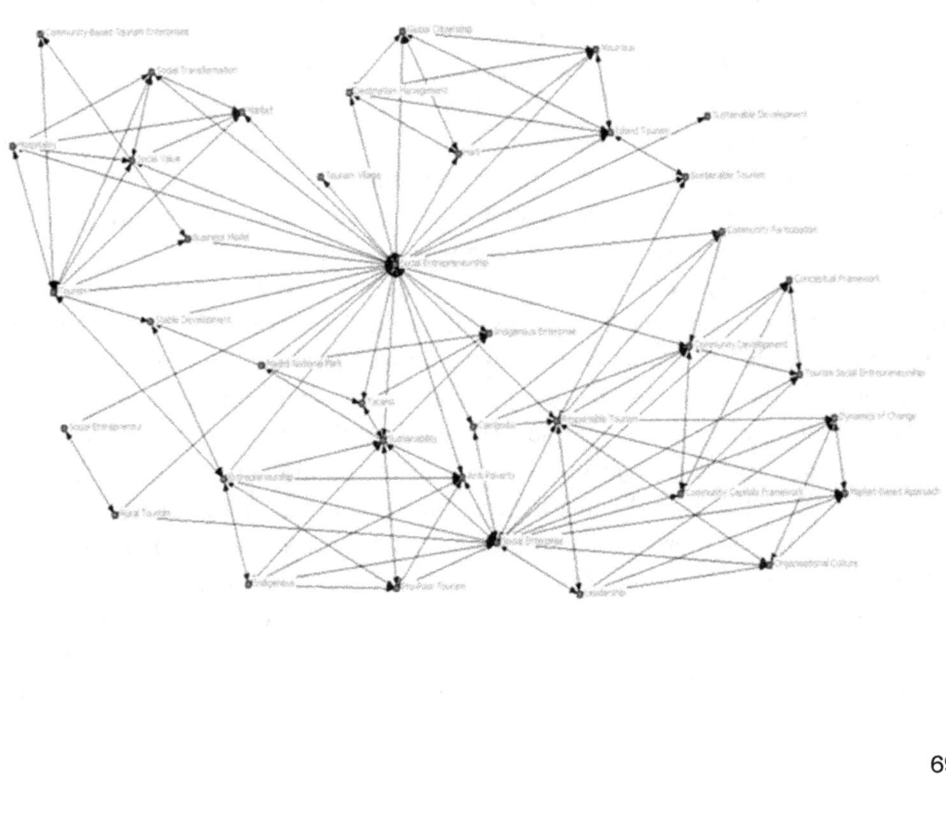

FUTURE RESEARCH DIRECTIONS

The limitation of the study is that it is not searched outside the databases determined as research field and only articles are included in the research. Conference papers, book chapters etc. are excluded from the scope of the research. In future studies, the scanned databases can be expanded and more different results can be achieved by including the written sources other than the articles in the research.

CONCLUSION

Social entrepreneurship, which incorporates society with "social", and innovation with "entrepreneurship", aims to create social value. It mediates economic development while achieving this goal and this means that it is on the same ground with the tourism sector. The actions taken in tourism sector in order to increase the welfare of society, employment creation and income generation will reach the desired level only if they are combined with the creative and innovative processes of social entrepreneurship. This reveals why social entrepreneurship activities in tourism are necessary and important. The aim of this study is to shed a light on how social entrepreneurship activities in tourism sector were mentioned in the literature and which of those have a great impact on the social and economic development of countries. For this purpose, bibliometric analysis and social network analysis have been made.

Bibliometric studies offer researchers the opportunity to have in-depth knowledge of the field and to see the shortcomings of the field. With this research method, the studies are examined in depth and the current situation is revealed. With this study, it is aimed to reveal studies that deal with the concepts of social entrepreneurship and tourism together. With bibliometric analysis, important issues and missing points in the field were determined. With the analysis made, the subjects that researchers study more frequently within the framework of these concepts and the subjects that are less mentioned were revealed. As a result of the literature review, it is seen that there are few studies that address these two concepts together. This demonstrates to academics that the subject is an open area for new research and can be developed.

According to the findings obtained as a result of the social network analysis, it has been revealed that the subjects of social innovation, rural tourism, sustainability, responsible tourism, island tourism are mostly studied in the field of social entrepreneurship and tourism. The results of the study are supported. The following typologies are found in the articles; Structuration Theory, Business Model, Market - Based Approach, Social Business Models, Sustaniable Livelihoods Approach and Women's Empowerment Models. As a result of the bibliometric analysis, it was concluded that 70% of the reviewed articles were designed using the case study research method. It is thought that the reason for this is that most of the studies in the field of tourism deal with a certain region. In future studies, different fields of study can be explored in different geographies by using different research methods other than quantative studies and case studies. The increase in the studies in the field of tourism and social entrepreneurship since 2016 shows that this area is open to development and can be enriched by using different methods.

REFERENCES

Abdul Kadir, M. A. B., & Sarif, S. M. (2016). Social entrepreneurship, social entrepreneur and social enterprise: A review of concepts, definitions and development in Malaysia. *Journal of Emerging Economies and Islamic Research, 4*(2), 1–16. doi:10.24191/jeeir.v4i2.9086

Abu-Saifan, S. (2012). Social entrepreneurship: Definition and boundaries. *Technology Innovation Management Review, 2*(2), 22–27. doi:10.22215/timreview/523

Aji, R. R. (2020). Tourism social entrepreneurship in community-based tourism: A case study of Pentingsari tourism village. *Earth and Environmental Science, 447*, 1–9. doi:10.1088/1755-1315/447/1/012009

Aktan, C. C., & Börü, D. (2007). Kurumsal sosyal sorumluluk. In C. C. Aktan (Ed.), *Kurumsal Sosyal Sorumluluk*. İktisadi Girişim ve İş Ahlakı Derneği Yayını.

Alegre, I., & Berbegal-Mirabent, J. (2016). Social innovation success factors: Hospitality and tourism social enterprises. *International Journal of Contemporary Hospitality Management, 28*(6), 1155–1176. doi:10.1108/IJCHM-05-2014-0231

Aquino, R. S., Lück, M., & Schänzel, H. A. (2018). A conceptual framework of tourism social entrepreneurship for sustainable community development. *Journal of Hospitality and Tourism Management, 37*, 23–32. doi:10.1016/j.jhtm.2018.09.001

Austin, J., Wei-Skillern, J., & Stevenson, H. (2006). Social and commercial entrepreneurship: Same, different, or both? *Entrepreneurship Theory and Practice, 30*(1), 1–22. doi:10.1111/j.1540-6520.2006.00107.x

Barki, E., Comini, G., Cunliffe, A., Hart, S., & Rai, S. (2015). Social entrepreneurship and social business: Retrospective and prospective research. *Revista de Administração de Empresas, 55*(4), 380–384. doi:10.1590/S0034-759020150402

Berchicci, L. (2009). *Innovating for sustainability: green entrepreneurship in personal mobility*. Routledge.

Besler, S. (2010). Sosyal girişimcilik. In S. Besler (Ed.), *Sosyal Girişimcilik* (pp. 3–30). Beta Yayıncılık.

Biddulph, R. (2017). Social enterprise and inclusive tourism. Five cases in Siem Reap, Cambodia. *Tourism Geographies*, 1–20.

Boukas, N., & Chourides, P. (2016). Niche tourism in Cyprus: Conceptualising the importance of social entrepreneurship for the sustainable development of islands. *International Journal of Leisure and Tourism Marketing*, *5*(1), 26–43. doi:10.1504/IJLTM.2016.077128

Budisusila, A., Sarungu, J. J., Wiyono, V. H., & Soesilo A.M. (2019). Supply chain analysis of tourism social enterprises in Bejiharjo tourism village. *International Journal of Recent Technology and Engineering, 8*(2S9), 305-313.

Bulsara, H. P., Gandhi, J., & Chandwani, S. (2015). Social entrepreneurship in India: An exploratory study. *International Journal of Innovation*, *3*(1), 7–16. doi:10.5585/iji.v3i1.20

Castellani, P., Bonfanti, A., Canestrino, R., & Magliocca, P. (2020). Dimensions and triggers of memorable tourism experiences: Evidence from Italian social enterprises. *The TQM Journal*, *32*(6), 1115–1138. doi:10.1108/TQM-11-2019-0259

Catford, J. (1998). Social entrepreneurs are vital for health promotion but they need supportive environments too. *Health Promotion International*, *13*(2), 95–97. doi:10.1093/heapro/13.2.95

Çelebi, D., Pirnar, I., & Eriş, E. D. (2020). Bibliometric analysis of social entrepreneurship in gastronomy tourism. *Tourism: An International Interdisciplinary Journal*, *68*(1), 58–67. doi:10.37741/t.68.1.5

Chandra, Y., & Leenders, M. A. A. M. (2012). User innovation and entrepreneurship, in the virtual world: A study of second life residents. *Technovation*, *32*(7-8), 464–476. doi:10.1016/j.technovation.2012.02.002

Christoph, G., Laeis, M., & Lemke, S. (2016). Social entrepreneurship in tourism: Applying sustainable livelihoods approaches. *International Journal of Contemporary Hospitality Management*, *28*(6), 1–32.

Cooke, G. B., Burns, J. K., Mann, S. L., Vardy, K. W. J., & Cass, B. L. (2016). The case of Ireland's County Donegal: Stimulating rural labour markets via training, tourism and nurturing social enterprise and entrepreneurship. *E-Journal of International and Comparative Labour Studies*, *5*(1), 1–37.

Dahles, H., Khieng, S., Verver, M., & Manders, I. (2019). Social entrepreneurship and tourism in Cambodia: Advancing community engagement. *Journal of Sustainable Tourism*, 1–18.

De Lange, D., & Dodds, R. (2017). Increasing sustainable tourism through social entrepreneurship. *International Journal of Contemporary Hospitality Management*, *29*(7), 1–57. doi:10.1108/IJCHM-02-2016-0096

Dees, J. G. (2001). *The meaning of "social entrepreneurship"*. Draft Paper. http://www.redalmarza.cl/ing/pdf/ TheMeaningofsocialEntrepreneurship.pdf

Erbaş, E. (2019). Social business model and sharing economy for community based tourism development: A case study of Lisinia Doğa. *Journal of Tourism and Gastronomy Studies*, *7*(4), 2399–2417. doi:10.21325/jotags.2019.478

Ergül, M., Johnson, C., & Kleinrichert, D. (2012). Social entrepreneurship: A hospitality and tourism perspective. *European Journal of Tourism. Hospitality and Recreation*, *3*(3), 115–126.

Gowreesunkar, G. B. V., Van der Sterren, J., & Séraphin, H. (2015). Social entrepreneurship as a tool for promoting global citizenship in island tourism destination management. Academy for Touris, 7-23.

Günlü, E. (2015). The conceptual analysis of social entrepreneurship and social entrepreneurship in tourism. *Journal of Social Sciences and Humanities Researches*, *35*, 23–42.

Hawe, P., Webster, C., & Shiell, A. (2004). A glossary of terms for navigating the field of social network analysis. *Journal of Epidemiology and Community Health*, *58*(12), 971–975. doi:10.1136/jech.2003.014530 PMID:15547054

Heberger, E. A., Christie, A. C., & Alkin, C. M. (2010). A bibliometric analysis of the academic influences of and on evaluation theorists' published works. *The American Journal of Evaluation*, *31*(1), 24–44. doi:10.1177/1098214009354120

Iorgulescua, M., & Ravara, A. S. (2015). The contribution of social enterprises to the development of tourism. The case of Romania. *Procedia Economics and Finance*, *32*, 672–679. doi:10.1016/S2212-5671(15)01448-3

Kerlin, J. (2006). Social enterprise in the United States and Europe: Understanding and learning from the differences. *Voluntas*, *17*(3), 247–263. doi:10.100711266-006-9016-2

Khangarot, G. (2019). An overview of social entrepreneurship in tourism. *Think India Journal,* 110-117.

Kummitha, H. R. (2020). Stakeholders involvement towards sustaining ecotourism destinations: The case of social entrepreneurship at Mangalajodi ecotourism trust in India. *Geo Journal of Tourism and Geosites*, *29*(2), 636–648. doi:10.30892/gtg.29220-495

Lange, D. L., & Dodds, R. (2017). Increasing sustainable tourism through social entrepreneurship. *International Journal of Contemporary Hospitality Management*, *29*(7), 1977–2002. doi:10.1108/IJCHM-02-2016-0096

Loosemore, M. (2015). Barriers to social enterprise in the UK construction industry. In A. B. Raidén, & E. Aboagye-Nimo (Eds.), *Procs 31st Annual ARCOM Conference*. Association of Researchers in Construction Management.

Martin, R. L., & Osberg, S. (2007). Social entrepreneurship: The case for definition. *Stanford Social Innovation Review*, 1–17.

McCarthy, B. (2012). From fishing and factories to cultural tourism: The role of social entrepreneurs in the construction of a new institutional field. *Entrepreneurship & Regional Development: An International Journal*, *24*(3-4), 259–282. doi:10.1080/08985626.2012.670916

Mdee, A., & Emmott, R. (2008). Social enterprise with international impact: The case for fair trade certification of volunteer tourism. *Education, Knowledge & Economy*, *2*(3), 191–201. doi:10.1080/17496890802426188

Moein, R. (2014). Investigation and prioritization of social entrepreneurship components based on tourism industry development approach using TOPSIS technique. *International Journal of Scientific Management and Development*, *2*(11), 602–613.

Morris, M. H., Avila, R. A., & Allen, J. (1993). Individualism and the modern corporation: Implications for innovation and entrepreneurship. *Journal of Management*, *19*(3), 595–612. doi:10.1177/014920639301900305

Mort, G., Weerawardena, J., & Carnegie, K. (2002). Social entrepreneurship: Towards conceptualisation. *International Journal of Nonprofit and Voluntary Sector Marketing*, *8*(1), 76–88. doi:10.1002/nvsm.202

Mottiar, Z. (2016). Exploring the motivations of tourism social entrepreneurs: The role of a national tourism policy as a motivator for social entrepreneurial activity in Ireland. *International Journal of Contemporary Hospitality Management*, *28*(6), 1137–1154. doi:10.1108/IJCHM-06-2014-0278

Muarifuddin, & Kriswanto, H. D. (2020). Social entrepreneurship learning based on community empowerment of tourism village. *PEOPLE: International Journal of Social Sciences, 6*(1), 81-92.

Naderi, A., Vosta, N. L., Ebrahimi, A., & Jalilvand, M. R. (2019). The contributions of social entrepreneurship and transformational leadership to performance insights from rural tourism in Iran. *The International Journal of Sociology and Social Policy, 39*(9/10), 719–737. doi:10.1108/IJSSP-06-2019-0124

OECD. (2020). *OECD Tourism Trends and Policies 2020.* OECD Publishing., doi:10.1787/6b47b985-

Peng, K., & Lin, P. M. C. (2016). Social entrepreneurs: Innovating rural tourism through the activism of service science. *International Journal of Contemporary Hospitality Management, 28*(6), 1225–1244. doi:10.1108/IJCHM-12-2014-0611

Peredo, B., & Wurzelmann, S. (2015). Indigenous tourism and social entrepreneurship in the Bolivian Amazon: Lessons from San Miguel del Bala. *International Indigenous Policy Journal, 6*(4), 1–26. doi:10.18584/iipj.2015.6.4.5

Pinchot, G. (1985). *Intrapreneuring*. Harper and Row.

Popoviciu, I., & Popoviciu, S. (2011). Social entrepreneurship, social enterprise and the principles of a community of practice. *Revista de Cercetare si Interventie Sociala, 33*, 44–55.

Radić, M. M., Jukić, I., & Roje, A. (2020). Increasing tourism through social entrepreneurship – the case of Croatia. *Journal of Corporate Governance, Insurance, and Risk Management, 7*(1), 25–38. doi:10.51410/jcgirm.7.1.3

Roper, J., & Cheney, G. (2005). The meanings of social entrepreneurship today. *Corporate Governance International Journal of Business in Society, 5*(3), 95–104. doi:10.1108/14720700510604733

Sagawa, S., & Segal, E. (2000). Common interest, common good: Creating value through business and social sector partnership. *California Management Review, 42*(2), 105–122. doi:10.1177/000812560004200201

Sam, I., John, O. O., & Bernadett, K. (2016). Social Entrepreneurship in Hospitality Industry. The Case of Operators at the Coastal Tourism Circuit in Kenya. *International Journal of Humanities Social Sciences and Education, 3*(2), 115–121.

Santos, F. M. (2012). A positive theory of social entrepreneurship. *Journal of Business Ethics, 111*(3), 335–351. doi:10.100710551-012-1413-4

Sarıbaş, Ö., Kömürcü, S., & Akbaba, A. (2020). The concept of social entrepreneurship in tourism: A bibliometric analysis for tourism literature. *Euroasia Journal of Social Science and Humanities*, *7*(4), 51–58.

Sheldon, P., Pollock, A., & Daniele, R. (2017). Social entrepreneurship and tourism: Setting the stage. In P. J. Sheldon & R. Daniele (Eds.), *Social entrepreneurship and tourism: Philosophy and practice* (pp. 1–18). Springer International Publishing. doi:10.1007/978-3-319-46518-0_1

Shen, H., Liu, X., Li, M., & Ji, M. (2018). Development of social enterprises in rural island tourism in China. *Journal of China Tourism Research*, 1–21.

Sigala, M. (2016). Learning with the market: A market approach and framework for developing social entrepreneurship in tourism and hospitality. *International Journal of Contemporary Hospitality Management*, *28*(6), 1245–1286. doi:10.1108/IJCHM-06-2014-0285

Sijabat, R. (2015). The role of social entrepreneurship in enabling economic opportunities for the poor: A synthesis of the literature and empirical works. *International Journal of Business and Social Science*, *6*(11), 35–41.

Simatupang, J. (2018). How is social entrepreneurship as an asset to implement rural tourism at Muarajambi herritage? *Innovative Issues and Approaches in Social Sciences*, *11*(2), 106–115. doi:10.12959/issn.1855-0541.IIASS-2018-no2-art6

Situmorang, D. B. M., & Mirzantia, I. R. (2012). Social entrepreneurship to develop ecotourism. *Procedia Economics and Finance*, *4*, 398–405. doi:10.1016/S2212-5671(12)00354-1

Sloan, P., Legrand, W., & Simons-Kaufmann, C. (2014). A survey of social entrepreneurial community-based hospitality and tourism initiatives in developing economies. *Worldwide Hospitality and Tourism Themes*, *6*(1), 51–61. doi:10.1108/WHATT-11-2013-0045

Souitaris, V., & Zerbinati, S. (2005). Entrepreneurship in the public sector: A framework of analysis in European Local Governments. *Entrepreneurship and Regional Development*, *17*(1), 43–64. doi:10.1080/0898562042000310723

Suminar, T., Budiartati, E., & Anggraeni, D. (2019). The effectiveness of a women's empowerment model through social entrepreneurship training to strengthen a tourism village program. *International Journal of Innovation. Creativity and Change.*, *5*(5), 324–338.

Swanson, L. A., & Zhang, D. D. (2012). Social Entrepreneurship. In T. Burger-Helmchen (Ed.), *Entrepreneurship - Gender, Geographies and Social Context*. InTech.

Tan, W. L., Williams, J., & Tan, T. M. (2005). Defining the 'social' in 'social entrepreneurship': Altruism and entrepreneurship. *The International Entrepreneurship and Management Journal, 1*(3), 353–365. doi:10.100711365-005-2600-x

Tekin, M. (2009). *Girişimcilik*. Günay Ofset Matbaacılık.

Tiryaki, A. (2012). İktisat teorisinde girişimcilik. In Z. Erdoğan (Ed.), *Girişimcilik ve KOBİ'ler*. Ekin Basım Yayın Dağıtım.

Tresiana, N., & Duadji, N. (2018). Social entrepreneur model for tourism development in Kiluan Bay. *MIMBAR, 34*(1), 83–92. doi:10.29313/mimbar.v34i1.3141

UNWTO. (2020). *Tourism Highlights*. Retrieved from https://www.e-unwto.org/doi/epdf/10.18111/9789284422456

Wang, C., Duan, Z., & Yu, L. (2016). From nonprofit organization to social enterprise. *International Journal of Contemporary Hospitality Management, 28*(6), 1287–1306. doi:10.1108/IJCHM-05-2014-0230

Weppen, J., & Cochrane, J. (2012). Social enterprises in tourism: An exploratory study of operational models and success factor. *Journal of Sustainable Tourism, 20*(3), 497–511. doi:10.1080/09669582.2012.663377

Wulleman, M., & Hudon, M. (2015). *Models of social entrepreneurship: empirical evidence from Mexico*. CEB Working Paper, 15/024, 1-41.

Zahra, S. A., Gedajlovic, E., Neubaum, D. O., & Shulman, J. M. (2009). A typology of social entrepreneurs: Motives, search processes and ethical challenges. *Journal of Business Venturing, 24*(5), 519–532. doi:10.1016/j.jbusvent.2008.04.007

Zebryte, I., & Jorquera, H. (2017). Chilean tourism sector "B corporations": Evidence of social entrepreneurship and innovation. *International Journal of Entrepreneurial Behaviour & Research, 23*(6), 866–879. doi:10.1108/IJEBR-07-2017-0218

Zeng, B. (2018). How can social enterprises contribute to sustainable pro-poor tourism development? *Zhongguo Renkou Ziyuan Yu Huanjing, 16*(2), 1–12. doi:10.1080/10042857.2018.1466955

ADDITIONAL READING

Bornstein, D., & Davis, S. (2010). *Social entrepreneurship what everyone needs to know*. Oxford University Press. doi:10.1093/wentk/9780195396348.001.0001

Certo, S. T., & Miller, T. L. (2008). Social entrepreneurship: Key issues and concepts. *Business Horizons*, *51*(4), 267–271. doi:10.1016/j.bushor.2008.02.009

Corner, P. D., & Ho, M. (2010). How opportunities develop in social entrepreneurship. *Entrepreneurship Theory and Practice*, *34*(4), 635–659. doi:10.1111/j.1540-6520.2010.00382.x

Dacin, P. A., Dacin, M. T., & Matear, M. (2010). Social entrepreneurship: Why we don't need a new theory and how we move forward from here. *The Academy of Management Perspectives*, 36–56.

Grassl, W. (2012). Business models of social enterprise: A design approach to hybridity. *ACRN Journal of Entrepreneurship Perspectives*, *1*(1), 37–60.

Karnani, A. (2011). Social entrepreneurship: Beyond the hype. *Innovations*, *6*(2), 99–116. doi:10.1162/INOV_a_00072

Luke, B., & Chu, V. (2013). Social enterprise versus social entrepreneurship: An examination of the 'why' and 'how' in pursuing social change. *International Small Business Journal*, *31*(7), 764–784. doi:10.1177/0266242612462598

Thompson, J. (2002). The world of the social entrepreneur. *International Journal of Public Sector Management*, *15*(5), 412–431. doi:10.1108/09513550210435746

Weerawardena, J., & Mort, G. (2006). Investigating social entrepreneurship: A multi-dimensional model. *Journal of World Business*, *41*(1), 21–35. doi:10.1016/j.jwb.2005.09.001

KEY TERMS AND DEFINITIONS

Entrepreneur: A person who brings together the factors of production to run a business and undertakes the risk.

Social: Being related to society.

Social Enterprise: A business established for social purposes.

Social Entrepreneur: An entrepreneur who carries out entrepreneurial activity for social purposes.

Social Entrepreneurship: Entrepreneurial activities to achieve social goals.

Social Network Analysis: A type of data analysis conducted on a social network, which aims to reveal the hidden relationship between entities and to calculate the parameters and indicators using nodes, edges, and communities.

Tourism: It is an important economic sector that refers to the short-term movements of people to different locations.

APPENDIX 1

Table 2. Data on articles

Year	Author(s)	Country/ies of Author/s	Title of Study	Type of Study	Research Method	Study Site
2008	*Anna Mdee, *Richard Emmott	UK	Social enterprise with international impact: The case for Fair Trade certification of volunteer tourism	Empirical	Qualitative	United Kingdom
2012	*Breda McCarthy	Australia	From fishing and factories to cultural tourism: The role of social entrepreneurs in the construction of a new institutional field	Empirical	Qualitative	Ireland
2012	*Dohar Bob M. Situmorang, *Isti Raafaldini Mirzantia	Indonesia	Social entrepreneurship to develop ecotourism	Empirical	Qualitative	Indonesia
2012	*Janina von der Weppen, *Janet Cochrane	UK	Social enterprises in tourism: an exploratory study of operational models and success factors	Empirical	Qualitative	Multiple
2012	*Mehmet Ergül, *Colin Johnson, *Denise Kleinrichert	USA	Social entrepreneurship: A hospitality and tourism perspective	Empirical	Qualitative	N/A
2014	*Rahim Moein	Iran	Investigation and prioritization of social entrepreneurship components based on development tourism ındustry approach using TOPSIS technique	Empirical	Quantitative	Iran
2014	*Philip Sloan, *Willy Legrand, *Claudia Simons-Kaufmann	Germany	A survey of social entrepreneurial community-based hospitality and tourism initiatives in developing economies	Empirical	Quantitative	N/A
2015	*Bernardo Peredo, *Samuel Wurzelmann	UK, USA	Indigenous tourism and social entrepreneurship in the Bolivian Amazon: Lessons from San Miguel del Bala	Empirical	Mixed Method	Bolovia
2015	*Vanessa GB Gowreesunkar, *Jos Van der Sterren, *Hugues Séraphin	Mauritius, The Netherlands, UK	Social entrepreneurship as a tool for promoting global citizenship in island tourism destination management	Empirical	Qualitative	Mauritius, Haiti
2015	*Ebru Günlü	Turkey	Sosyal girişimcilik olgusunun kavramsal analizi ve turizmde sosyal girişimcilik (The conceptual analysis of social entrepreneurship and social entrepreneurship ın tourism)	Theoretical	Literature Review	N/A
2015	*Dirisa Mulindwa	England	Social Entrepreneurship through Community Based Tourism in a Small Village in Uganda: The Case of KAFRED in Bigodi, Western Uganda	Empirical	Qualitative	Uganda
2015	*Maria-Cristina Iorgulescua, *Anamaria Sidonia Ravara	Romania	The contribution of social enterprises to the development of tourism. The case of Romania	Empirical	Quantitative	Romania
2016	*Ziene Mottiar	Ireland	Exploring the motivations of tourism social entrepreneurs The role of a national tourism policy as a motivator for social entrepreneurial activity in Ireland	Empirical	Mixed Method	Ireland
2016	*Chunlei Wang, *Zhaowen Duan, *Larry Yu	China, USA	From nonprofit organization to social enterprise: The paths and future of a Chinese social enterprise in the tourism field	Empirical	Qualitative	China
2016	*Marianna Sigala	Australia	Learning with the market a market approach and framework for developing social entrepreneurship in tourism and hospitality	Theoretical	Literature Review	N/A
2016	*Nikolaos Boukas *Pieris Chourides	Cyprus	Niche tourism in Cyprus: Conceptualising the importance of social entrepreneurship for the sustainable development of islands	Theoretical	Literature Review	Cyprus

continued on following page

Table 2. Continued

Year	Author(s)	Country/ies of Author/s	Title of Study	Type of Study	Research Method	Study Site
2016	*Kang-Lin Peng, *Pearl M.C. Lin	Taiwan, Hong Kong	Social entrepreneurs: Innovating rural tourism through the activism of service science	Empirical	Qualitative	Taiwan
2016	*Gabriel Christoph Maria Laeis Stefanie Lemke	N/A	Social entrepreneurship in tourism: Applying sustainable livelihoods approaches	Empirical	Qualitative	South Africa
2016	*Ines Alegre, *Jasmina Berbegal-Mirabent	Spain	Social innovation success factors: Hospitality and tourism social enterprises	Empirical	Qualitative	Spain
2016	*Gordon B. Cooke, *Jennifer K. Burns, *Sara L. Mann, *Kyle W.J. Vardy, *Bronwyn L. Cass	Canada	The case of Ireland's County Donegal: Stimulating Rural Labour Markets via Training, Tourism and Nurturing Social Enterprise and Entrepreneurship	Empirical	Qualitative	Ireland
2016	*Ikwaye Sam, *Ogembo O. John, *Kiarie Bernadett	Kenya	Social Entrepreneurship in Hospitality Industry. The Case of Operators at the Coastal Tourism Circuit in Kenya	Empirical	Qualitative	Kenya
2017	*Ieva Zebryte *Hector Jorquera	Chile	Chilean tourism sector "B Corporations": Evidence of social entrepreneurship and innovation	Empirical	Qualitative	Chile
2017	*Deborah de Lange, *Rachel Dodds	Canada	Increasing sustainable tourism through social entrepreneurship	Empirical	Qualitative	Canada
2018	*Robin Biddulph	Sweden	Social enterprise and inclusive tourism. Five cases in Siem Reap, Cambodia	Empirical	Qualitative	Cambodia
2018	*Richard S. Aquino, *Michael Lück, *Heike A. Schänzel	New Zealand	A conceptual framework of tourism social entrepreneurship for sustainable community development	Theoretical	Literature Review	N/A
2018	*Han Shen, *Xiao Liu, *Mimi Li *Mingjie Ji	China, UK	Development of social enterprises in rural island tourism in China	Empirical	Qualitative	China
2018	*Benxiang Zeng	Australia	How can social enterprises contribute to sustainable pro-poor tourism development?	Empirical	Qualitative	Taiwan, China
2018	*Johannes Simatupang	Indonesia	How is social entrepreneurship as an asset to implement rural tourism at Muarajambi heritage	Empirical	Qualitative	Indonesia
2018	*Novita Tresiana, *Noverman Duadji	Indonesia	Social entrepreneur model for tourism development in Kiluan Bay	Empirical	Qualitative	Indonesia
2018	*Richard S. Aquino, *Michael Lück, *Heike A. Schänzel	New Zeland	Tourism Social Entrepreneurship for Sustainable Community Development: Review and Conceptual Framework	Theoretical	Literature Review	N/A
2019	*Emre Erbaş	Turkey	Social business model and sharing economy for community-based tourism development: A case study of Lisinia Doğa	Empirical	Qualitative	Turkey
2019	*Heidi Dahles, *Sothy Khieng, *Michiel Verver, *Ireen Manders	Australia, Cambodia, The Netherlands	Social entrepreneurship and tourism in Cambodia: Advancing community engagement	Empirical	Qualitative	Cambodia
2019	*Antonius Budisusila, *JJ Sarungu, *Vincent Hadi Wiyono, *A.M. Soesilo	Indonesia	Supply chain analysis of tourism social enterprises in Bejiharjo tourism village	Empirical	Qualitative	Indonesia
2019	*Ahmad Naderi, *Leila Nasrolahi Vosta, *Abolghasem Ebrahimi, *Mohammad Reza Jalilvand	Iran	The contributions of social entrepreneurship and transformational leadership to performance insights from rural tourism in Iran	Empirical	Quantitative	Iran
2019	*Tri Suminar, *Emmy Budiartati, *Dewi Anggraeni	Indonesia	The effectiveness of a women's empowerment model through social entrepreneurship training to strengthen a tourism village program	Empirical	Mixed Method	Indonesia

continued on following page

Table 2. Continued

Year	Author(s)	Country/ies of Author/s	Title of Study	Type of Study	Research Method	Study Site
2020	*Garima Khangarot	India	An overview of social entrepreneurship in tourism	Theoretical	Literature Review	N/A
2020	*Duygu Celebi, *Ige Pirnar, *Engin Deniz Eris	Turkey	Bibliometric analysis of social entrepreneurship in gastronomy tourism	Empirical	Qualitative	N/A
2020	*Muarifuddin, *Hendra Dedi Kriswanto	Indonesia	Social entrepreneurship learning based on community empowerment of tourism village	Empirical	Qualitative	Indonesia
2020	*Harshavardhan Reddy Kummitha	Hungary	Stakeholders involvement towards sustaining ecotourism destinations: The case of social entrepreneurship at Mangalajodi ecotourism trust in India	Empirical	Qualitative	India
2020	*Paola Castellani, *Angelo Bonfanti, *Rossella Canestrino, *Pierpaolo Magliocca	Italy	Dimensions and triggers of memorable tourism experiences: Evidence from Italian social enterprises	Theoretical	Qualitative	Italy
2020	*Mijana Matošević Radić, *Ivona Jukić, *Antonija Roje	Croatia	Increasing tourism through social entrepreneurship – the case of Croatia	Empirical	Qualitative	Croatia
2020	*Özgür SARIBAŞ, *Simge KÖMÜRCÜ, *Atilla AKBABA	Turkey	The Concept of Social Entrepreneurship in Tourism: A Bibliometric Analysis for Tourism Literature	Theoretical	Literature Review	N/A
2020	*R R Aji	Indonesia	Tourism social entrepreneurship in community-based tourism: A case study of Pentingsari tourism village	Empirical	Qualitative	Indonesia

Table 3. Keywords used in articles

Community Capitals Framework	Pro-Poor Tourism	Cambodia
Anti-Poverty	Business Model	Samin Community
Nonprofit Organization	Sharing Economy	Women's Empowerment
1kg.org	Sustainable Tourism	Social Capital
Structuration Theory	Niche Tourism	Creativity
China	Island Tourism	Transformational Leadership
Strategic Change	Cyprus	Supply Chain
Neo-Institutionalism	Social Transformation	Ecotourism
Cultural Tourism	Hospitality	Stakeholders Involvement
Motivations	Social Value	Social Business Models
Entrepreneurial	Market	Sustainability Entrepreneurship
Entrepreneurs	Market Segment	Corporate Social Responsibility
Ireland	Components of Social Entrepreneurship	Service Experience Engineering
Tourism in China	Stable Development	Service Science
Influencing Factors	Tourism	Community Participation
Island Destination	Sustainability	Hospitality and Tourism Sector
Rural Tourism	Madidi National Park	Tourism Village
B Corporation	Tacana	Community Empowerment
Community Development	Indigenous Enterprise	Learning
Entrepreneurship	Sustainable Hospitality and Tourism	Haiti
Indigenous	Sustainable Development	Mauritius
Conceptual Framework	Heritage Branding	Destination Management
Social Enterprise	Canada	Global Citizenship
Social Innovation	Conservation Activities	Community Tourism
Tourism Social Entrepreneurship	Heritage Value	Fair Trade
Social Entrepreneurship	Community Marketer	Service Innovation
Social Entrepreneur	Community	Local Community
Gastronomy Tourism	Community-Based Tourism Enterprises	Dynamics of Change
Bibliometric Analysis	Inclusive Development	Leadership
Latin America	Inclusive Tourism	Organisational Culture
Market-Based Approach	Responsible Tourism	Volunteer Travel
Voluntourism	Collaborative	Tourism Research
Case of Croatia	Competitive Advantage	Social Function
Authenticity	Unconventionality	Tourist Experience
Tourism and Hospitality Management	Development Regions	Social Economy
Exclusion	Empowerment	Community Based Tourism
Developing Countries	Sustainable Hospitality	Local Communities

Chapter 5
Selected Happiness-Increasing Interventions for Tourism and Hospitality Workers

Christian Ehrlich
Oxford Brookes University, UK

Sashenka Milston
https://orcid.org/0000-0003-2226-6363
Oxford Brookes University, UK

Robert van der Veen
Oxford Brookes University, UK

ABSTRACT

This chapter describes evidence-based happiness techniques that are highly relevant for workers in the tourism and hospitality industries. Although happiness creates success for many stakeholders, there is limited evidence on how to increase the happiness of workers in these industries as the focus has predominantly been on the happiness of the customers. The authors fill this gap in the literature by presenting three proven interventions that are particularly relevant to these sectors: job crafting, acts of kindness, and gratitude exercises. The chapter explains what these concepts are and how they work. It also provides specific examples of how they can be implemented into tourism and hospitality organisations.

DOI: 10.4018/978-1-7998-7239-9.ch005

Copyright © 2021, IGI Global. Copying or distributing in print or electronic forms without written permission of IGI Global is prohibited.

INTRODUCTION

Research has conclusively demonstrated that a happy workforce is associated with a plethora of relevant organisational outcomes, including organisational profit (Lyubomirsky, King, & Diener, 2005). For example, positive psychological well-being of employees has been shown to be associated with reduced turnover intentions and increased customer loyalty – both of which impact strongly on an organisation's bottom line (e.g., Amin & Akbar, 2013; Gordon et al., 2019; Lyubomirsky et al., 2005; O'Neill & Davis, 2011). In fact, not only do the benefits of employee well-being positively impact individuals (employees, customers, and colleagues) and the business, but they have even been shown to extend out to the wider society (see Zhang et al., 2014). However, Chia and Chu (2016) note that, despite Lyubomirsky, King, and Diener's seminal findings which indicate that happiness creates success for many stakeholders, there is limited evidence on how to increase the happiness of tourism and hospitality workers. For example, happiness of tourism and hospitality workers does not seem a common factor in the study of turnover intentions – the conventional antecedents tend to be job satisfaction, organisational commitment, emotional exhaustion, or perceived supervisor or organisational support (Park & Min, 2020). Indeed, for industries that focus on making guests feel welcome and happy, there is a surprising lack of research into accomplishing the same for its workers (Melief et al., 2020).

This conceptual chapter aims to bridge that gap by introducing three selected and well-established concepts that are particularly suitable for the context of tourism and hospitality workers. With so many organisations being negatively impacted by the pandemic, this seems even more pertinent now. The three concepts are job crafting, acts of kindness, and gratitude exercises. The focus of this chapter on research with practical application has lent itself to easy and accessible interventions that may hopefully enable a quicker recovery of these industries. The interventions selected all focus on the individual effort each employee can do, and therefore are closely aligned with what is commonly labelled as "intentional activity" – engaging in specific actions that would otherwise not occur and require a degree of effort. Intentional activities are one very important factor in extending people's long-term happiness (Lyubomirsky, Sheldon, & Schkade, 2005). Along with explanations of these interventions, practical examples of how they can be applied in the tourism and hospitality sectors are provided.

JOB CRAFTING

Doing Meaningful Work-Tasks We Enjoy

One factor that contributes to people's happiness is the engagement in meaningful activities which are experienced as pleasurable or enjoyable (Ehrlich, 2018; Ehrlich, 2019; Sheldon & Elliot, 1999). At first glance, this seems obvious and most people would probably agree with that. However, there is more to the notion of doing something fun that is worthwhile mentioning. According to the broaden-and-build theory (Fredrickson, 2004), positive emotions not only make people feel good whilst performing the enjoyable activity, but also have positive spillover effects extending beyond finishing the activity. Fredrickson explains that these positive emotions can create an upward spiral as, once we are in a positive state of mind, we feel more resourceful, more optimistic, and more creative. This mindset then positively impacts the next thing we do. It is like falling in love where people see the whole world in a positive way. Thus, helping employees to modify their work-tasks in such a way as to make them more enjoyable and meaningful seems vital. However, the question that arises is: How can this be done? This chapter suggests that there is one concept that firmly satisfies this brief: Job crafting.

Definition of Job Crafting

Job crafting implies that employees are not merely passive performers of their assigned roles or tasks at work. Employees, in varying degrees, also proactively mould their tasks at work to fit with their own preferences. This is typically seen as not being managed top-down, but rather as a bottom-up process initiated by the employees themselves. According to Slemp and Vella-Brodrick (2013, p. 126), "job crafting is a promising yet relatively unexplored approach that, potentially, employees can use to heighten their job satisfaction and wellbeing." In essence, job crafting is defined as "the physical and cognitive changes individuals make in the task or relational boundaries of their work" (Wrzesniewski & Dutton, 2001, p. 179). One of the most famous studies in this context was done by Wrzesniewski et al. (2003) whereby they studied the work-related behaviour of cleaning staff who started to interact and care for patients and their family members in a way beyond their cleaning duties to create more meaningful and more enjoyable jobs. These behaviours included, for example, chatting with lonely patients, walking family members back to their car, or rearranging the pictures on the wall in the hope that creating a more stimulating environment for comatose patients might have some positive effect.

According to Wrzesniewski et al. (2013), job crafting can be executed on three levels. Firstly, it can happen on a task level. This can be accomplished through adding new tasks, dropping some tasks, or adjusting the time or effort spent on each task. Job crafting can also happen on a relationship level whereby employees redefine how and with whom they want to interact. This typically includes spending more time with preferred individuals and avoiding or reducing the time spent with others. The third method of job crafting is on a cognitive level whereby employees reframe the meaning or purpose of their tasks and relationships in a way that makes their job more significant. Wrzesniewski et al. (2013) provide the example of a janitor who reframes their job as enabling clean, distraction-free classrooms to enable students to learn better. Regardless of how it is done, this personal shaping of one's job, or perception of one's job, gives employees a greater sense of control, which increases their subjective well-being, i.e., their happiness (Cheng & O-Yang, 2018).

Application of Job Crafting

Research about the effectiveness of job crafting interventions in general is accumulating (e.g., Demerouti et al., 2017; Kooij et al., 2017; Sakuraya et al., 2016; van Wingerden et al., 2017), and shows overall positive effects for a variety of job crafting behaviours (de Devotto & Wechsler, 2019). Although further research is needed to gain a greater understanding about the differential effects of specific job crafting interventions for particular groups of employees, based on the initial positive results obtained from a variety of job crafting interventions, it can be concluded that job crafting interventions seem to be a valuable concept to increase the well-being of employees in the tourism industry. However, it is helpful to have sufficient organisational support (Chen et al., 2014; Cheng & O-Yang, 2018; Yadav, 2020). Organisational support can be increased by highlighting the benefits for employees as well as employers, along with the fact that job crafting can be implemented with very little cost to the organisation (Baker & Slemp, 2018). A common barrier to the implementation of job crafting is the fear from employers that employees might craft their job in such a way that neglects or prevents them from performing their core job roles. In most cases though, this worry is unfounded and worth dispelling as the positive effects of job crafting for employees and employers, including increased communication between employees and management, tend to prevail (van den Heuvel et al., 2015).

For the first two levels of job crafting – task and relational – a simple conversation with one's manager or colleagues can enable this. For example, through friendly agreement, or perhaps through clever negotiation, tasks could be swapped between colleagues to match strengths and preferences. In the same way, perhaps a swap of shifts or teams might help with avoiding working with one person or enabling working with another. Another idea might be to get creative and think of a better way to accomplish the end goal of an unpleasant task and suggest the change to management. This could be a more efficient method that saves time and/or other resources, or simply a more enjoyable method that uses the same amount of time and resources. Perhaps there exists an opportunity to train a new colleague in a task that is usually mundane. Providing connection and knowledge can give the employee a deeper purpose that can make the task more interesting (or tolerable at the very least). A final idea here is that the organisation could also be encouraged to gamify tasks to make roles more engaging with extrinsic (e.g., prizes) and/or intrinsic (e.g., fun) rewards (Robson et al., 2015). Baker and Slemp (2018, p. 5) also provide a list of job crafting activities highly pertinent to the hospitality and tourism sector, including such items as: noting the best customer experience at the end each day; volunteering for a new project; spending five minutes a day learning a new skill – for example learning to use a new presentation tool; making time for a face-to-face conversation each day that would previously have been dealt with by email; reclaiming their lunch break and setting up a lunch date with a colleague each week; spending time tidying up the electronic client notes on a customer relationship management system if they felt that these were not clear; finding out something new about a different colleague each day; doing a random act of kindness for colleagues; sending a note of thanks to a colleague once a day; investing extra time with a new employee once a day, where they felt that it mattered.

For those that are not able to job craft on these two levels, the cognitive level should not be underestimated. It can be applied to the job as a whole, or to any one single task. This level has an even greater emphasis on defining one's role from the perspective of how it impacts the world around the individual. An example here is of a doorman, opening the door for customers as they come through. This simple task could also be framed as the first and last impression that the client has of the organisation as they arrive and leave. It could also be interpreted as getting the customer's journey off to a smooth and welcome start, setting the tone for the whole experience. A second example could be of a ticket seller. They could simply be selling tickets, or they could be making a profound positive impact on visitors' holidays by providing them access to a wonderful local experience. These causes are far larger than simple actions and thus give the role greater meaning and purpose. The idea of cognitive crafting often also touches on the notion of performing an act of kindness (see below), as cognitive crafting frequently involves understanding the possible positive consequences a particular task has on others (Baker & Slemp, 2018). This can make this aspect of job crafting even more powerful.

Diagnostic tools are available to assess the degree of job crafting by employees. For example, the 15-item Job Crafting Questionnaire (JCQ) by Slemp and Vella-Brodrick (2013) assesses task crafting, relationship crafting, and cognitive crafting. The questionnaire asks employees to indicate to what extent they engage in behaviours such as: 'Introducing new work tasks that better suit my skills or interests' or 'changing the scope or types of tasks that need to be completed at work' (task crafting); 'making friends with people at work who have similar skills or interests' (relationship crafting); 'think about how your job gives your life purpose', 'remind yourself about the significance your work has for the success of your organisation' (cognitive crafting). This can both raise awareness of the extent to which job crafting is being undertaken, as well as potentially provide the starting point for any job crafting intervention.

There are also training courses available, such as the training intervention by van den Heuvel et al. (2015) which consists of one training day, four weeks of experimenting with job crafting, and concludes with a half-day reflection session. The job crafting intervention – like most job crafting interventions – is based on the Job Demands-Resources model (Bakker & Demerouti, 2007), and therefore asks participants to reflect on their work environment in relation to demands and resources. The objective is to think about how employees can shape their work environment in such a way as to increase their job resources (such as identifying further opportunities for development) and to decrease superfluous job demands (aspects of the role that could impede achievement of one's goals). A 'personal crafting plan' is created with self-chosen job crafting goals describing when and how to accomplish this. These weekly manageable steps are then carried out by participants in the proceeding four-week experimental phase, with progress reflected on at the end of each week. The intervention has been demonstrated to lead to enhanced employee well-being as well as increased job performance.

Although it is often the case that hospitality workers must follow standardised processes, leaving relatively little room for individual job crafting (Shamim et al., 2017; Yadav, 2020), Slemp and Vella-Brodrick (2103, p. 127) note that:

although employees are typically not able to redesign their jobs, there will be opportunities in the context of almost any job where employees can initiate changes to tasks, interactions, or ways they think about their work to make it more personally meaningful or enjoyable. Job crafting, then, can be applied across a variety of roles with different levels of seniority and degrees of autonomy (Berg, Wrzesniewski, & Dutton, 2010; Wrzesniewski & Dutton, 2001), and hence it is plausible that even in the most restricted and routine jobs employees are able to initiate changes to influence their work experience.

However, for those roles where job crafting opportunities are particularly limited, leisure crafting is a good alternative whereby one applies the idea of job crafting not to one's job role, but to one's leisure activities (Petrou et al., 2017). Just as with job crafting, leisure crafting equally aims to maximise employees' positive emotions, but interestingly, this positive effect from outside the workplace can carry through into the workplace thanks to the above-mentioned broaden-and-build theory (Fredrickson, 2004). This is because there can be a "spillover of flexibility, energy, renewal, and well-being back and forth between career and home" (Greenhaus & Kossek, 2014, p. 377). Indeed, it seems that 'serious leisure' (e.g., activities that promote education, personal enrichment, self-expression of individual, and self-gratification) significantly predicts subjective well-being (Lee & Hwang, 2018), and the enhancement of leisure subjective well-being has been found to lead to a strengthening of overall subjective well-being (Newman, 2014). On top of this, increased time spent on leisure has been found to increase self-efficacy at work (Kelly et al., 2020). However, it should be noted that this last effect was only found to be true if the leisure was either serious or similar to work activities/skills, but not both.

The application of leisure crafting can be done in a similar fashion to that of job crafting. In the same way, it starts with the notion that leisure crafting is not casual, but an intentional act (Stebbins, 2001). Therefore, leisure crafting can be defined as the deliberate pursuit of leisure activities which contain goal setting, human connection, learning, and personal development (Petrou & Bakker, 2016). Tourism and hospitality organisations can support employees with this in various ways. They could provide support materials, and possibly training workshops, to inform people about the concept of leisure crafting and how to apply it to one's own situation. They could also support employees with the idea of having a project outside of work that is both challenging and in line with their values and interests. It could also mean helping employees with thinking about setting serious leisure goals and about the skills they could develop through pursuit of those goals. This can enable individuals to experience a sense of mastery as they progress in those activities and therefore experience a sense of happiness. Essentially, leisure crafting will help tourism workers to experience positive emotions on a more regular basis, which can then spill over into work. Thus, Petrou et al. (2016) state that managers should encourage leisure crafting among employees, not least because the two life domains complement each other. To the authors' knowledge, no specific tools for leisure crafting are currently available, however the available job crafting interventions (see above) should be easily applicable to the concept of leisure crafting.

ACTS OF KINDNESS (AoK)

Definition of and Research around AoK

Performing acts of kindness is now recognised as an established method of increasing one's own long-term happiness (Lyubormirksy, 2010). In simple terms, helping others makes us happy – also labelled 'the helpers high' (Dossey, 2018). Indeed, Ko et al. (2019, p. 1) state: "One of the most common – and the most fruitful – ways to increase well-being is to engage in prosocial behaviour (i.e., perform acts of kindness for others)." Acts of kindness are typically defined as "voluntary, intentional behaviours that benefit another and are not motivated by external factors such as rewards or punishments" (Eisenberg, 1986, p. 63). There is now a substantial body of research around AoK that tourism and hospitality workers can draw on in order to maximise their happiness through performing such acts.

For example, research has now established that performing the same act of kindness repeatedly is less effective at increasing happiness than performing a variety of acts of kindness. This is best illustrated by a ten-week study conducted by Lyubomirsky (2010). One group of participants was asked to perform a variety of different acts of kindness whereas another group was asked to perform similar AoK over and over. The findings clearly demonstrated that the group that performed a wide variety of AoK reported being happier compared to the low variety group. Interestingly, the high variety group's happiness also lasted longer. The explanation for these results is based around the notion that performing the same act again and again can become routine and therefore a less meaningful experience than performing different acts of kindness. Similarly, Lyubomirsky also found that performing five AoK in one day has a bigger impact on people's happiness than performing five AoK spread over a week. The assumed reason for this is that, once again, the five spread over a week may not stand out as something distinctive, and therefore their effect is more likely to be blended into the normal routine.

Another interesting finding within the AoK research pertinent to tourism and hospitality workers is the fact that the size of the kindness given to someone might be less important than the act itself. This is best highlighted by a study conducted by Dunn et al. (2008) in which students were asked to spend a certain amount of money (either US$5 or US$20) on themselves or on someone else. The main findings of this study support the general happiness-increasing effect of AoK as the group that spent the money on someone else reported as being significantly happier than the group who spent it on themselves. However, with regard to the size of the AoK, this study also found that the group who spent US$20 on someone else did not report as feeling any happier than the group who only spent US$5 on someone else. This suggests that the absolute amount or size of the AoK is less important than the act of giving itself.

Recent research has also showed that simply remembering past AoK can have a positive impact on people's happiness. A study conducted by Ko et al. (2019) demonstrated this by randomly assigning participants to one of four groups: performing social behaviours, recalling social behaviours, performing and recalling social behaviours, and a control group. Results indicated that the group who only recalled their AoK did not differ in their happiness levels to the group who performed

AoK or to the group who performed and recalled their AoK. This is particularly interesting because recalling an AoK can be a very quick and free method of boosting one's happiness. It is also much easier to remember an AoK than to perform one.

Finally, research also suggests that it is useful to encourage people to see the impact of their acts of kindness on others by becoming closer to the person they are being kind to (Dunn & Norton, 2014). Generally speaking, giving to people we are very close to has the greatest happiness boosting effect. However, we can also intentionally become closer to the ones we are giving to, i.e., by seeing giving to others as an opportunity to become closer to them. For example, instead of simply giving someone a voucher to buy something nice, we can choose to go with that person to the shop and buy it with them. This will mean spending more time with that person, seeing the smile on their face, and essentially increasing the strength of the relationship. This strategy, according to Dunn and Norton (2014), can also be used when giving to strangers as this can be a good way to increase the connection with another person by, for example, taking the initiative to learn more about the receiver of the AoK. Closely aligned to the idea of connection is also the idea of witnessing the positive impact of our Aok on others as an important source for our own happiness. Therefore, research findings strongly suggest encouraging people to go out and discover the positive effects of their actions.

Application of AoK

Despite most of the literature on AoK within the tourism and hospitality sectors being focussed on the positive consequences for customers/guests, the above research clearly demonstrates that performing AoK can also have a positive impact for the workers themselves. As such, organisations can develop a clear set of guidelines for the benefit of their employees. To start with, organisations could draw up a comprehensive (though relevant) list of all potential acts of kindness that could be performed within their own organisation. Drawing on change management literature, these should be specifically developed within each organisation using an inclusive, bottom-up approach involving all members of staff, i.e., in a participative way, to increase buy-in of staff (Mullins, 2013; Yukl, 2002).

As variety of AoK is important, the list should be diverse. However, as mentioned above, the AoK do not need to be large. Little things (such as a free drink) can go a long way. This suggests that tourism workers could be given a small budget which they could use at their own discretion to perform various AoK for their guests/ clients, or even co-workers. This could include gifting guests for special occasions (e.g., a fruit basket, bottle of wine, voucher for a service, etc.).

The Ritz-Carlton Hotel Company is often cited for their exemplary gold standard service, "focused on serving "even the unexpressed wishes" of its guests, and doing so through fully-empowered "Ladies and Gentlemen" (as the employees of Ritz-Carlton proudly refer to themselves)" (Solomon, 2015). One extraordinary example of an elaborate, yet inexpensive AoK was enacted by two of their frontline employees. A family's two-year old son had lost his favourite toy, Thomas the Tank Engine, on their final day of holiday and his mother asked staff to help her look for it as the loss would be heart-breaking for him. They could not find it by the time the family left, so formulated a plan. They drove to a toy store, bought the same toy, and took pictures of its adventures around the property, including cooking in the kitchen wearing a mini chef's hat. They then posted the toy, the photos, and a handwritten note from the toy to the boy telling him about his extended holiday.

A few recommended guidelines for employees could also be included on the list of potential AoK. For example, the above-mentioned research findings on timing suggest that workers should focus on a short period of time in which to perform a large number of AoK as opposed to just one act a day spread over a longer period of time. In addition, workers can capitalise on the research showing that remembering an AoK can be an equally strong happiness booster. Therefore, organisations could find a system to encourage their employees to think back on their day and remember the various AoK they have performed. Finally, remembering that AoK have a greater happiness-increasing effect if the positive impacts can be seen suggests that workers could focus their AoK particularly on those groups of customers with whom the organisation has a more long-term relationship. For example, in a hotel this would mean focussing on those hotel guests who stay longer or those who stay on a regular basis. Equally, the idea of witnessing the impact of one's kindness on others can also be applied to colleagues. This is potentially even more powerful as it can build connectedness with stronger relationships in the workplace, which also boosts happiness (Dunn, Aknin, & Norton, 2014).

As a final point on AoK, it should be noted that the most powerful way to use a recommended list as described above is at the employees' own discretion. It is by giving them autonomy and empowering them to make the effort to decide what to give and when in response to the situations unfolding in front of their eyes that can make their work more satisfying, thus increasing happiness (Gladwell, 2008).

GRATITUDE EXERCISES

Definition of and Research around Gratitude Exercises

Gratitude exercises have been associated with improved levels of subjective happiness in a variety of populations (Al-Seheel & Noor, 2016; Dickens, 2017; Emmons & McCullough, 2003; Emmons & Stern, 2013; McCullough et al., 2004; O'Leary &

Dockray, 2015; Salces-Cubero et al., 2018; Seligman et al., 2005; Watkins et al., 2003). Peterson and Seligman (2004, p. 30) define gratitude as "being aware of and thankful for the good things that happen; taking time to express thanks". This definition seems most appropriate as it is in keeping with a broader life orientation approach (Wood et al., 2010). Seligman et al. (2005) add to this that if one expresses their appreciation, it should be done elaborately and sincerely.

Wood et al. (2010) note that there are three main types of gratitude exercises. The first is the gratitude list. This exercise involves maintaining a list of the things one is grateful for on a regular basis, often kept in the format of a diary or journal. For example, participants would write three to five specific things for which they are grateful at the end of each day. The second type of exercise is grateful contemplation. This exercise is similar to the gratitude list but focuses on thinking and reflecting on more general aspects or events in life. Finally, the third exercise is behavioural expressions of gratitude. This is often completed in the form of writing a letter of thanks to a specific individual. The letter could be simply delivered, or it could be read to the recipient in person ("the gratitude visit"). The list and letter have been the most popular strategies among scholars (Davis et al., 2016).

The effectiveness of gratitude interventions is increased when they are completed daily (Emmons & McCullough, 2003), are longer (Koydemir et al., 2020), and are combined with other positive interventions (Seligman et al., 2005). In terms of the enduring effects of gratitude exercises, Seligman et al. (2005) demonstrated that a single gratitude visit exercise boosted happiness levels for a month. Wong et al. (2016) noted that improved levels of mental health were reported 12 weeks after the gratitude exercise. However, the studies that have identified the beneficial outcomes of gratitude exercises tend to rely on self-reported measures. Seligman et al. (2005) note that this method is not sufficiently solid since people may not always clearly remember when and in what situations they were happy in the past. Nevertheless, other studies have confirmed the positive impacts by external observers, such as friends, family, or romantic partners, in terms of higher levels of positive affect (McCullough et al., 2002) or subjective well-being (Martínez-Martí et al., 2010).

Increasing awareness of the benefits of gratitude among management would be the first step in the journey towards increasing happiness among those working in the tourism and hospitality industries. For example, research shows that gratitude encourages networking (Williams & Barlett, 2014), improves people's self-esteem (Chen & Wu, 2014), reduces envy (McCullough et al., 2002), improves sleep quality (Digdon & Koble, 2011), and increases purpose and meaning at work (Dik et al., 2015). Studies have also shown that those who are more grateful make better decisions and are willing to wait for greater financial gain (DeSteno, 2009; DeSteno et al., 2014). As such, workers who are more grateful will be able to focus better on their long-term career goals. Those in their early stages of a career who are appreciated come to feel increasingly grateful to their organisations for providing a positive job experience and are less likely to leave (Ng, 2016).

Gratitude encourages collaboration along with prosocial and moral behaviour, meaning that those who received sincere expressions of gratitude were more likely to help their colleagues and also help others (DeSteno, 2009; DeWall et al., 2012; McCullough et al., 2001; McCullough et al., 2008). Therefore, encouraging gratitude among staff can help support collaboration and nurture positive working relationships. Expressing gratitude also serves as a mechanism of upstream reciprocity, or in other words, "pay it forward" (McCullough et al., 2008). This may explain why those who have been successfully mentored in the past are more likely to become a mentor in the future.

It is not uncommon for those working in frontline positions in hospitality to encounter verbal abuse and aggression (Karatepe & Ehasani, 2012). DeWall et al. (2012) refer to gratitude as one of the underlying mechanisms of resilience to aggression – their study shows that grateful people perceive provocation as less threatening and are more likely to behave in a prosocial manner, even when others behave less kindly.

Application of Gratitude Exercises

Gratitude exercises are particularly relevant to the hospitality and tourism industries for at least two reasons. First, the positive impacts of gratitude exercises have been demonstrated to be effective in many different countries, such as Spain (Martínez-Martí et al., 2010), China (Deng, et al., 2019), Turkey (Isik & Erguner-Tekinalp, 2017; Yüksel & Oğuz Duran, 2012), Hong Kong (Chan, 2013), and Malaysia (Al-Seheel & Noor, 2016; Senf & Liau, 2013). These findings appeal to the diversity of these industries given that large parts operate on a global scale (e.g., airlines, hotels, tour operators, restaurants, cruises). In addition, many of them rely on an international body of staff, even within single organisations. The second reason is that research suggests that women tend to derive greater benefits from the experience and expression of gratitude compared to men (Kaczmarek et al., 2015; Kashdan et al., 2009). Given that females can occupy up to 70% of service sector positions in some regions (Baum, 2013), gratitude exercises can be particularly relevant for these sectors.

Employing gratitude at work can be relatively straightforward. Cheng et al. (2015) demonstrated that using a simple diary exercise twice a week whereby workers were asked to count their blessings (take stock of people and things that have benefited them at work) reduced their stress and depressive symptoms significantly. This may in turn reduce turnover intentions (Huang et al., 2018; Hwang et al., 2014).

To apply all three main gratitude exercises, managers could, for example, introduce a three-week gratitude intervention program, implementing a different exercise each week. In the first week, the gratitude list could be introduced whereby employees write a daily list of three things (Three Good Things, Seligman et al., 2005) that occurred during their shift for which they are grateful. In the second week employees can be asked to think and write about the bigger things they are grateful for (events, people, or experiences). This is often referred to as counting one's blessings and is done to instil an appreciation for various people or other things in their lives. The third and last week could task the employees with writing a gratitude letter to a colleague describing what that person did and what it meant to them.

Hospitality businesses could also use gratitude exercises to form, maintain, and strengthen relationships among their staff members (Algoe & Zhaoyang, 2016; Kubacka et al., 2011). For example, a short list could be developed for each employee, where colleagues briefly jot down what they appreciate about the person. This shorter gratitude list can be pinned to their cubicle wall, personal locker, or office door. Management could also install a notice board where employees express their gratitude using post-it notes.

Organisations who are keen to assess personal levels of gratitude among their staff should consider the Gratitude Questionnaire-6 (GQ-6; McCullough et al., 2002). This robust diagnostic tool has been used extensively and shown to be reliable with Cronbach's alpha coefficients exceeding .80 in recent studies (e.g., $\alpha = .85$ in Reeves et al., 2021; $\alpha = .84$ in Barnett et al., 2021; $\alpha = .90$ in Bundick et al., 2021). The survey tool requests respondents to rate six questions from 1 (= strongly disagree) to 7 (= strongly agree). Two example items are: *"I have so much in life to be thankful for"* and *"I am grateful to a wide variety of people"*.

Tourism and hospitality organisations could also encourage their staff to install a gratefulness application on their smartphone. These applications generally rely on the same gratitude exercises as mentioned previously and there are several gratitude applications freely available for both Android and iOS operating systems. Ghandeharioun et al. (2016) evaluated the effectiveness of the 'Kind and Grateful' application which displayed inspirational content to encourage expression of gratitude. Ghandeharioun et al. demonstrated that the smartphone application led to increased practice of gratitude, increased positive emotional valence, and increased levels of subjective well-being compared to baseline.

CONCLUSION

This chapter focussed on three happiness interventions pertinent to the tourism and hospitality industries: job crafting, performing acts of kindness, and gratitude exercises. These industries are well-known for their long working hours, high levels of stress, and high standardisation – three conditions typically associated with lower happiness levels (Huang et al., 2018). Fortunately, the interventions described in this chapter can provide workers with a (partial) solution and increase well-being. The advantage is that they can be implemented despite the above-mentioned constraints and are cost-effective to implement from an organisational perspective. Although all interventions are intended to increase the happiness of tourism workers at work, research clearly shows that happiness at work has positive spillover effects into other domains of a person's life (Greenhaus & Kossek, 2014; Judge et al., 2001), and so are likely to increase global happiness levels. Therefore, given that happiness is a central aim for many individuals around the globe (Lyubomirsky, Sheldon, & Schkade, 2005), the three selected interventions should be highly appealing to most tourism workers.

Key to application of the interventions is for employees to be proactive in engaging in these intentional activities. At the same time, the success of implementing these interventions also benefits from organisational support. Indeed, it is well worth the investment from an organisational perspective as, whilst there is a direct positive effect for workers, there is also a secondary but important organisational benefit. Combining and applying multi-component positive psychology interventions has been associated with improved mental health (Hendriks et al., 2020). Therefore, employers may consider applying practices from all three of the proposed interventions (i.e., job crafting, acts of kindness, and gratitude).

Notwithstanding the importance of applying the latest research findings on job crafting, acts of kindness, and gratitude in the tourism industry, this chapter also alludes to important future research pertinent to the three interventions in question. With regards to job crafting, it is apparent that the vast amount of research focus has been on job crafting with less attention to leisure crafting. Hence, the development and evaluation of specific leisure crafting exercises, which tourism organisations can easily implement, would be an important advancement of the leisure crafting literature. With regards to acts of kindness, further research would be useful to analyse the specific budget that would help tourism workers gain the highest happiness boost while at the same time being cost-effective for the organisation. For example, is there a difference in happiness for a front of house hotel worker in whether they can autonomously spend $5 or $10 on acts of kindness in a given period of time? With regards to gratitude exercises, one important strand of future research will be in the comprehensive evaluation of digital gratitude interventions (for example, through smartphones) for tourism workers, particularly in organisations with a global and therefore dispersed workforce. Finally, research evaluating the effectiveness of the combined implementation of all three interventions in the tourism industry is lacking thus far, indicating another important area for future research.

The tourism and hospitality industries are in a vulnerable situation from the effects of the pandemic. Many of the studies around the effectiveness of the interventions examined by this chapter have been conducted in a variety of different cultural settings, making these interventions even more relevant to these predominantly globally operating industries. Furthermore, the focus has been on research with practical application to provide an effective path in the rebuilding of these industries through clear explanations and industry-specific examples. Perhaps the easy and practical nature of these three interventions will make them useful tools in restructuring the industries in a more positive and mentally healthy way.

REFERENCES

Al-Seheel, A. Y., & Noor, N. M. (2016). Effects of an Islamic based gratitude strategy on Muslim students' level of happiness. *Mental Health, Religion & Culture*, *19*(7), 686–703. doi:10.1080/13674676.2016.1229287

Algoe, S. B., & Zhaoyang, R. (2016). Positive psychology in context: Effects of expressing gratitude in ongoing relationships depend on perceptions of enactor responsiveness. *The Journal of Positive Psychology*, *11*(4), 399–415. doi:10.1080/17439760.2015.1117131 PMID:27800009

Amin, Z., & Akbar, K. P. (2013). Analysis of psychological well-being and turnover intentions of hotel employees: An empirical study. *International Journal of Innovation and Applied Studies*, *3*(3), 662–671.

Baker, R., & Slemp, G. (2018, December). *An introduction to micro job crafting: How job crafting for 12 minutes or less a day contributes to sustainable positive behaviour change*. Paper presented at the CIPD Applied Research Conference, Nottingham, UK.

Bakker, A. B., & Demerouti, E. (2007). The Job Demands-Resources model: State of the art. *Journal of Managerial Psychology*, *22*(3), 309–328. doi:10.1108/02683940710733115

Barnett, M. D., Van Vleet, S. C., & Cantu, C. (2021). Gratitude mediates perceptions of previous generations' prosocial behaviors and prosocial attitudes toward future generations. *The Journal of Positive Psychology*, *16*(1), 54–59. doi:10.1080/17439760.2019.1676459

Baum, T. (2013). *International perspectives on women and work in hotels, catering and tourism*. International Labour Office.

Bundick, M. J., Remington, K., Morton, E., & Colby, A. (2021). The contours of purpose beyond the self in midlife and later life. *Applied Developmental Science*, *25*(1), 62–68. doi:10.1080/10888691.2018.1531718

Chan, D. W. (2013). Counting blessings versus misfortunes: Positive interventions and subjective well-being of Chinese school teachers in Hong Kong. *Educational Psychology*, *33*(4), 504–519. doi:10.1080/01443410.2013.785046

Chen, C. Y., Yen, C. H., & Tsai, F. C. (2014). Job crafting and job engagement: The mediating role of person-job fit. *International Journal of Hospitality Management*, *37*(2), 21–28. doi:10.1016/j.ijhm.2013.10.006

Chen, L. H., & Wu, C. H. (2014). Gratitude enhances change in athletes' self-esteem: The moderating role of trust in coach. *Journal of Applied Sport Psychology*, *26*(3), 349–362. doi:10.1080/10413200.2014.889255

Cheng, J. C., & O-Yang, Y. (2018). Hotel employee job crafting, burnout, and satisfaction: The moderating role of perceived organizational support. *International Journal of Hospitality Management*, *72*, 78–85. doi:10.1016/j.ijhm.2018.01.005

Cheng, S.-T., Tsui, P. K., & Lam, J. H. M. (2015). Improving mental health in health care practitioners: Randomized controlled trial of a gratitude intervention. *Journal of Consulting and Clinical Psychology*, *83*(1), 177–186. doi:10.1037/a0037895 PMID:25222798

Chia, Y. M., & Chu, M. J. T. (2016). Moderating effects of presenteeism on the stress-happiness relationship of hotel employees: A note. *International Journal of Hospitality Management*, *55*, 52–56. doi:10.1016/j.ijhm.2016.02.005

Davis, D. E., Choe, E., Meyers, J., Wade, N., Varjas, K., Gifford, A. Z., Quinn, A., Hook, J. N., Tongeren, D. R., Griffin, B. J., & Worthington, E. (2016). Thankful for the little things: A meta-analysis of gratitude interventions. *Journal of Counseling Psychology*, *63*(1), 20–31. doi:10.1037/cou0000107 PMID:26575348

de Devotto, R. P., & Wechsler, S. M. (2019). Job crafting interventions: Systematic review. *Trends in Psychology*, *27*(2), 371–383. doi:10.9788/TP2019.2-06

Demerouti, E., Xanthopoulou, D., Petrou, P., & Karagkounis, C. (2017). Does job crafting assist dealing with organizational changes due to austerity measures? Two studies among Greek employees. *European Journal of Work and Organizational Psychology*, *26*(4), 574–589. doi:10.1080/1359432X.2017.1325875

Deng, Y., Xiang, R., Zhu, Y., Li, Y., Yu, S., & Liu, X. (2019). Counting blessings and sharing gratitude in a Chinese prisoner sample: Effects of gratitude-based interventions on subjective well-being and aggression. *The Journal of Positive Psychology*, *14*(3), 303–311. doi:10.1080/17439760.2018.1460687

DeSteno, D. (2009). Social emotions and intertemporal choice: 'Hot' mechanisms for building social and economic capital. *Current Directions in Psychological Science*, *18*(5), 280–284. doi:10.1111/j.1467-8721.2009.01652.x

DeSteno, D., Li, Y., Dickens, L., & Lerner, J. S. (2014). Gratitude: A tool for reducing economic impatience. *Psychological Science*, *25*(6), 1262–1267. doi:10.1177/0956797614529979 PMID:24760144

DeWall, C. N., Lambert, N. M., Pond, R. S. Jr, Kashdan, T. B., & Fincham, F. D. (2012). A grateful heart is a nonviolent heart: Cross-sectional, experience sampling, longitudinal, and experimental evidence. *Social Psychological & Personality Science*, *3*(2), 232–240. doi:10.1177/1948550611416675

Dickens, L. R. (2017). Using gratitude to promote positive change: A series of meta-analyses investigating the effectiveness of gratitude interventions. *Basic and Applied Social Psychology*, *39*(4), 193–208. doi:10.1080/01973533.2017.1323638

Digdon, N., & Koble, A. (2011). Effects of constructive worry, imagery distraction, and gratitude interventions on sleep quality: A pilot trial. *Applied Psychology. Health and Well-Being*, *3*(2), 193–206. doi:10.1111/j.1758-0854.2011.01049.x

Dik, B. J., Duffy, R. D., Allan, B. A., O'Donnell, M. B., Shim, Y., & Steger, M. F. (2015). Purpose and meaning in career development applications. *The Counseling Psychologist*, *43*(4), 558–585. doi:10.1177/0011000014546872

Dossey, L. (2018). The helper's high. *Explore (New York, N.Y.)*, *14*(6), 393–399. doi:10.1016/j.explore.2018.10.003 PMID:30424992

Dunn, E., & Norton, M. (2014). *Happy money: The science of smarter spending*. Simon & Schuster Paperbacks.

Dunn, E. W., Aknin, L. B., & Norton, M. I. (2008). Spending money on others promotes happiness. *Science*, *319*(5870), 1687–1688. doi:10.1126cience.1150952 PMID:18356530

Dunn, E. W., Aknin, L. B., & Norton, M. I. (2014). Spending money on others: Using money to benefit others pays off. *Current Directions in Psychological Science*, *23*(1), 41–47. doi:10.1177/0963721413512503

Ehrlich, C. (2018). The development of an extended goal-striving reasons framework: Evidence for its relevance in the workplace, for its theoretical difference to self-concordance and for its buffering effect on work intensity. *Journal of Positive Psychology and Wellbeing*, *2*(2), 1–23.

Ehrlich, C. (2019). The relevance of identified and intrinsic goal motivation for work engagement. *European Journal of Applied Positive Psychology*, *3*(5), 1–5.

Eisenberg, N. (1986). *Altruistic emotion, cognition and behavior*. Erlbaum.

Emmons, R. A., & McCullough, M. E. (2003). Counting blessings versus burdens: An experimental investigation of gratitude and subjective well-being in daily life. *Journal of Personality and Social Psychology*, *84*(2), 377–389. doi:10.1037/0022-3514.84.2.377 PMID:12585811

Emmons, R. A., & Stern, R. (2013). Gratitude as a Psychotherapeutic Intervention. *Journal of Clinical Psychology*, *69*(8), 846–855. doi:10.1002/jclp.22020 PMID:23775470

Fredrickson, B. L. (2004). The broaden-and-build theory of positive emotions. *Philosophical Transactions of the Royal Society of London. Series B, Biological Sciences*, *359*(1449), 1367–1378. doi:10.1098/rstb.2004.1512 PMID:15347528

Ghandeharioun, A., Azaria, A., Taylor, S., & Picard, R. W. (2016). "Kind and Grateful": A context-sensitive smartphone app utilizing inspirational content to promote gratitude. *Psychology of Well-Being*, *6*(9), 9. Advance online publication. doi:10.118613612-016-0046-2 PMID:27441170

Gladwell, M. (2008). *Outliers: The story of success*. Little, Brown and Company.

Gordon, S., Tang, C. H., Day, J., & Adler, H. (2019). Supervisor support and turnover in hotels: Does subjective well-being mediate the relationship? *International Journal of Contemporary Hospitality Management*, *31*(1), 496–512. doi:10.1108/IJCHM-10-2016-0565

Greenhaus, J. H., & Kossek, E. E. (2014). The contemporary career: A work–home perspective. *Annual Review of Organizational Psychology and Organizational Behavior*, *1*(1), 361–388. doi:10.1146/annurev-orgpsych-031413-091324

Hendriks, T., Schotanus-Dijkstra, M., Hassankhan, A., de Jong, J., & Bohlmeijer, E. (2020). The efficacy of multi-component positive psychology interventions: A systematic review and meta-analysis of randomized controlled trials. *Journal of Happiness Studies*, *21*(1), 357–390. doi:10.100710902-019-00082-1

Huang, S., van der Veen, R., & Song, Z. (2018). The impact of coping strategies on occupational stress and turnover intentions among hotel employees. *Journal of Hospitality Marketing & Management*, *27*(8), 926–945. doi:10.1080/19368623.2018.1471434

Hwang, J., Lee, J. J., Park, S., Chang, H., & Kim, S. S. (2014). The impact of occupational stress on employee's turnover intentions in the luxury hotel segment. *International Journal of Hospitality & Tourism Administration*, *15*(1), 60–77. doi:10.1080/15256480.2014.872898

Isik, S., & Erguner-Tekinalp, B. (2017). The effects of gratitude journaling on Turkish first year college students' college adjustment, life satisfaction and positive affect. *International Journal for the Advancement of Counseling*, *39*(2), 164–175. doi:10.100710447-017-9289-8

Judge, T. A., Thoreson, C. J., Bono, J. E., & Patton, G. K. (2001). The job satisfaction-job performance relationship: A qualitative and quantitative review. *Psychological Bulletin*, *127*(3), 376–407. doi:10.1037/0033-2909.127.3.376 PMID:11393302

Kaczmarek, L. D., Kashdan, T. B., Drazkowski, D., Enko, J., Kosakowski, M., Szaefer, A., & Bujacz, A. (2015). Why do people prefer gratitude journaling over gratitude letters? The influence of individual differences in motivation and personality on web-based interventions. *Personality and Individual Differences*, *75*, 1–6. doi:10.1016/j.paid.2014.11.004

Karatepe, O. M., & Ehsani, E. (2012). Work-related depression in frontline service jobs in the hospitality industry: Evidence from Iran. *Journal of Human Resources in Hospitality & Tourism*, *11*(1), 16–35. doi:10.1080/15332845.2012.621052

Kashdan, T. B., Mishra, A., Breen, W. E., & Froh, J. J. (2009). Gender differences in gratitude: Examining appraisals, narratives, the willingness to express emotions, and changes in psychological needs. *Journal of Personality*, *77*(3), 691–730. doi:10.1111/j.1467-6494.2009.00562.x PMID:20078735

Kelly, C. M., Strauss, K., Arnold, J., & Stride, C. (2020). The relationship between leisure activities and psychological resources that support a sustainable career: The role of leisure seriousness and work-leisure similarity. *Journal of Vocational Behavior*, *117*, 1–15. doi:10.1016/j.jvb.2019.103340

Ko, K., Margolis, S., Revord, J., & Lyubomirsky, S. (2019). Comparing the effects of performing and recalling acts of kindness. *The Journal of Positive Psychology*. doi:10.1080/17439760.2019.1663252

Kooij, D. T. A. M., van Woerkom, M., Wilkenloh, J., Dorenbosch, L., & Denissen, J. J. A. (2017). Job crafting towards strengths and interests: The effects of a job crafting intervention on person-job fit and the role of age. *The Journal of Applied Psychology*, *102*(6), 971–981. doi:10.1037/apl0000194 PMID:28277726

Koydemir, S., Sökmez, A. B., & Schütz, A. (2020). *A Meta-Analysis of the Effectiveness of Randomized Controlled Positive Psychological Interventions on Subjective and Psychological Well-Being*. Applied Research in Quality of Life. doi:10.100711482-019-09788-z

Kubacka, K. E., Finkenauer, C., Rusbult, C. E., & Keijsers, L. (2011). Maintaining Close Relationships: Gratitude as a Motivator and a Detector of Maintenance Behavior. *Personality and Social Psychology Bulletin*, *37*(10), 1362–1375. doi:10.1177/0146167211412196 PMID:21670220

Lee, K. J., & Hwang, S. (2018). Serious leisure qualities and subjective well-being. *The Journal of Positive Psychology*, *13*(1), 48–56. doi:10.1080/17439760.2017.1374437

Lyubomirski, S., Sheldon, K. M., & Schkade, D. (2005). Pursuing happiness: The architecture of sustainable change. *Review of General Psychology*, *9*(2), 111–131. doi:10.1037/1089-2680.9.2.111

Lyubomirsky, S. (2010). *The How of Happiness*. Piatkus Books.

Lyubomirsky, S., King, L. A., & Diener, E. (2005). The benefits of frequent positive affect: Does happiness lead to success? *Psychological Bulletin*, *131*(6), 803–855. doi:10.1037/0033-2909.131.6.803 PMID:16351326

Martinez-Marti, M. L., Avia, M. D., & Hernández-Lloreda, M. J. (2010). The effects of counting blessings on subjective well-being: A gratitude intervention in a Spanish sample. *The Spanish Journal of Psychology*, *13*(2), 886–896. doi:10.1017/S1138741600002535 PMID:20977036

McCullough, M. E., Eammons, R. A., & Tsang, J. A. (2002). The grateful disposition: A conceptual and empirical topography. *Journal of Personality and Social Psychology*, *82*(1), 112–127. doi:10.1037/0022-3514.82.1.112 PMID:11811629

McCullough, M. E., Kilpatrick, S. D., Emmons, R. A., & Larson, D. B. (2001). Is gratitude a moral affect? *Psychological Bulletin*, *127*(2), 249–266. doi:10.1037/0033-2909.127.2.249 PMID:11316013

McCullough, M. E., Kimeldorf, M. B., & Cohen, A. D. (2008). An adaptation for altruism? The social causes, social effects, and social evolution of gratitude. *Current Directions in Psychological Science*, *17*(4), 281–285. doi:10.1111/j.1467-8721.2008.00590.x

McCullough, M. E., Tsang, J.-A., & Emmons, R. A.McCullough. (2004). Gratitude in intermediate affective terrain: Links of grateful moods to individual differences and daily emotional experience. *Journal of Personality and Social Psychology*, *86*(2), 295–309. doi:10.1037/0022-3514.86.2.295 PMID:14769085

Melief, A. E. F., Schulp, J. A., & de Boer, M. R. (2020). Happiness at work: A phenomenological study of the experiences of hospitality industry employees. *Research in Hospitality Management*, *10*(1), 63–66. doi:10.1080/22243534.2020.1790212

Mullins, L. J. (2013). *Management & Organisational Behaviour* (10th ed.). Pearson.

Newman, D. B., Tay, L., & Diener, E. (2014). Leisure and subjective well-being: A model of psychological mechanisms as mediating factors. *Journal of Happiness Studies*, *15*(3), 555–578. doi:10.100710902-013-9435-x

Ng, T. W. H. (2016). Embedding employees early on: The importance of workplace respect. *Personnel Psychology*, *69*(3), 599–633. doi:10.1111/peps.12117

O'Leary, K., & Dockray, S. (2015). The Effects of Two Novel Gratitude and Mindfulness Interventions on Well-Being. *Journal of Alternative and Complementary Medicine (New York, N.Y.)*, *21*(4), 243–245. doi:10.1089/acm.2014.0119 PMID:25826108

O'Neill, J. W., & Davis, K. (2011). Work stress and well-being in the hotel industry. *International Journal of Hospitality Management*, *30*(2), 385–390. doi:10.1016/j.ijhm.2010.07.007 PMID:23794780

Park, J. & Min, H. (2020). Turnover intention in the hospitality industry: A meta-analysis. *International Journal of Hospitality Management, 90*, 1-11. doi:10.1016/j.ijhm.2020.102599

Peterson, C., & Seligman, M. E. P. (2004). *Character strengths and virtues: A handbook and classification*. Oxford University Press.

Petrou, P., & Bakker, A. B. (2016). Crafting one's leisure time in response to high job strain. *Human Relations*, *69*(2), 507–529. doi:10.1177/0018726715590453

Petrou, P., Bakker, A. B., & van den Heuvel, M. (2017). Weekly job crafting and leisure crafting: Implications for meaning-making and work engagement. *Journal of Occupational and Organizational Psychology*, *90*(2), 129–152. doi:10.1111/joop.12160

Reeves, M., Vogt, F., & Marks, E. (2021). Dispositional mindfulness, gratitude and self-compassion: Factors affecting tinnitus distress. *Mindfulness*, *12*(4), 1002–1008. Advance online publication. doi:10.100712671-020-01569-2

Robson, K., Plangger, K., Kietzmann, J., McCarthy, I., & Pitt, L. (2015). Is it all a game? Understanding the principles of gamification. *Business Horizons*, *58*(4), 411–420. doi:10.1016/j.bushor.2015.03.006

Sakuraya, A., Shimazu, A., Imamura, K., Namba, K., & Kawakami, N. (2016). Effects of a job crafting intervention program on work engagement among Japanese employees: A pretest-posttest study. *BMC Psychology*, *4*(1), 49. doi:10.118640359-016-0157-9 PMID:27776553

Salces-Cubero, I. M., Ramírez-Fernández, E., & Ortega-Martínez, A. R. (2018). Strengths in older adults: Differential effect of savoring, gratitude and optimism on well-being. *Aging & Mental Health*, *23*(8), 1017–1024. doi:10.1080/13607863.2018.1471585 PMID:29781723

Seligman, M. E. P., Steen, T. A., Park, N., & Peterson, C. (2005). Positive psychology progress: Empirical validation of interventions. *American Psychologist, 60*(5), 410-421. doi:10.1037/0003-066X.60.5.410

Senf, K., & Liau, A. K. (2013). The effects of positive interventions on happiness and depressive symptoms, with an examination of personality as a moderator. *Journal of Happiness Studies: An Interdisciplinary Forum on Subjective Well-Being*, *14*(2), 591–612. doi:10.100710902-012-9344-4

Shamim, S., Cang, S., Yu, H., & Li, Y. (2017). Examining the feasibilities of Industry 4.0 for the hospitality sector with the lens of management practice. *Energies*, *10*(4), 499. Advance online publication. doi:10.3390/en10040499

Slemp, G. R., & Vella-Brodrick, D. A. (2013). The Job Crafting Questionnaire: A new scale to measure the extent to which employees engage in job crafting. *International Journal of Wellbeing*, *3*(2), 126–146.

Solomon, M. (2015). *Ritz-Carlton President Herve Humler's leadership, culture and customer service secrets*. Retrieved 6th February 2021, from https://www.forbes.com/sites/micahsolomon/2015/04/21/ritz-carlton-president-herve-humlers-leadership-culture-and-customer-service-secrets/?sh=62d9ccc43b55

Stebbins, R. A. (2001). Serious leisure. *Society, 38*(4), 53–57. doi:10.100712115-001-1023-8

Tims, M., & Bakker, A. B. (2010). Job crafting: Towards a new model of individual job redesign. *SA Journal of Industrial Psychology, 36*(2), 1–9. doi:10.4102ajip.v36i2.841

van den Heuvel, M., Demerouti, E., & Peeters, M. C. (2015). The job crafting intervention: Effects on job resources, self-efficacy, and affective well-being. *Journal of Occupational and Organizational Psychology, 88*(3), 511–532. doi:10.1111/joop.12128

van Wingerden, J., Derks, D., & Bakker, A. B. (2017). The impact of personal resources and job crafting interventions on work engagement and performance. *Human Resource Management, 56*(1), 51–67. doi:10.1002/hrm.21758

Watkins, P. C., Woodward, K., Stone, T., & Kolts, R. L. (2003). Gratitude and happiness: Development of a measure of gratitude and relationships with subjective well-being. *Social Behavior and Personality, 31*(5), 431–452. doi:10.2224bp.2003.31.5.431

Williams, L. A., & Bartlett, M. Y. (2015). Warm thanks: Gratitude expression facilitates social affiliation in new relationships via perceived warmth. *Emotion (Washington, D.C.), 15*(1), 1–5. doi:10.1037/emo0000017 PMID:25111881

Wong, J. O., Gabana, N. T., Brown, J. W., Mcinnis, S., Toth, P., & Gilman, L. (2016). Does gratitude writing improve the mental health of psychotherapy clients? Evidence from a randomized controlled trial. *Psychotherapy Research, 28*(2), 192–202. doi:10.1080/10503307.2016.1169332 PMID:27139595

Wood, A. M., Froh, J. J., & Geraghty, A. W. A. (2010). Gratitude and well-being: A review and theoretical integration. *Clinical Psychology Review, 30*(7), 890–905. doi:10.1016/j.cpr.2010.03.005 PMID:20451313

Wrzesniewski, A., & Dutton, J. E. (2001). Crafting a job: Revisioning employees as active crafters of their work. *Academy of Management Review, 26*(2), 179–201. doi:10.5465/amr.2001.4378011

Wrzesniewski, A., Dutton, J. E., & Debebe, G. (2003). Interpersonal sensemaking and the meaning of work. In R. M. Kramer & B. M. Staw (Eds.), *Research in organizational behavior. Research in organizational behavior: An annual series of analytical essays and critical reviews* (Vol. 25, pp. 93–135). Elsevier Science Ltd. doi:10.1016/S0191-3085(03)25003-6

Wrzesniewski, A., LoBuglio, N., Dutton, J. E., & Berg, J. M. (2013). Job crafting and cultivating positive meaning and identity in work. In A. B. Bakker (Ed.), *Advances in Positive Organizational Psychology* (Vol. 1, pp. 281–302). Emerald Group Publishing Limited. doi:10.1108/S2046-410X(2013)0000001015

Yadav, A. (2020, June). *How job crafting is shaping the future of hospitality industry: Through the lens of industry 4.0.* Paper presented at the Travel and Tourism Research Association International Conference, Victoria, Canada.

Yukl, G. (2002). *Leadership in organizations* (5th ed.). Prentice Hall International Inc.

Yüksel & Oğuz Duran. (2012). Turkish Adaptation of the Gratitude Questionnaire. *Eurasian Journal of Educational Research*, *46*, 199–216.

Zhang, T., Kandampully, J., & Choi, H. Y. (2014). The role of employee wellness programme in the hospitality industry: A review of concepts, research, and practice. *Research in Hospitality Management*, *4*(1-2), 45–54. doi:10.1080/22243534.2014.11828313

Chapter 6
Contributions of Neurosciences to Studies of Well-Being in Tourism

Verônica Feder Mayer
https://orcid.org/0000-0002-7543-5215
Federal Fluminense University, Brazil

Carla Conceição Lana Fraga
Federal University of the State of Rio de Janeiro, Brazil

Leticia Cynara Santos Silva
Federal Fluminense University, Brazil

ABSTRACT

The SWB covers individuals' emotional and cognitive assessments and the conceptualization of happiness, satisfaction with one's life, and quality of life. Thus, an exploratory and descriptive study based on a bibliography survey and textual analysis was developed, and it was possible to understand that (1) tourists were the focus of the studies analyzed, so the understanding of their behaviors is essential; (2) emotion is still a gap that can be fill in research on tourism and well-being from the standpoint of the neurosciences; (3) the experience construct is significant but subjective so that the neurosciences can contribute to the understanding of the tourist experience related to SWB; (4) technological advances in neuroimaging can contribute, but this depends on adequate financial and human resources.

DOI: 10.4018/978-1-7998-7239-9.ch006

Copyright © 2021, IGI Global. Copying or distributing in print or electronic forms without written permission of IGI Global is prohibited.

INTRODUCTION

Subjective well-being (SWB) resides in the individual's experience, including positive and negative factors, and its measures generally include a global assessment of all aspects of a person's life (Diener, 1984). According to the author, SWB tries to understand what makes people happy and content with life, and it can be acquired when a particular state, for example, a goal or need, is reached. Understanding the tourism phenomenon through SWB is also understanding life satisfaction, quality of life, happiness, hedonism and eudaimonism, positive psychology, among others, through reactions, moods, and judgments of the individual (Diener, Oishi & Lucas, 2003), whether being the tourist, resident of the community or tourism and hospitality worker.

Modern neurosciences originated a new science of mind (Kandel et al., 2014). Important findings opened an avenue for its studies, like Camilo Golgi's discovery in 1873 of the neuron structure and, after it, Santiago Ramón y Cajal's discovery of neuronal doctrine (Kandel, 2020). There is a clear tendency of interdisciplinary cooperation efforts of different and complementary fields to expand the body of knowledge in neurosciences. It is possible to identify one in tourism, which has been called *neurotourism*. For some authors, *neurotourism* is a shift in the travel paradigm, and they discuss the interface between e-tourism and neuroscience (Panyik & Gonçalves, 2017). Other authors question whether *neurotourism* is a current discipline or only a future perspective (Giudici, Detorri & Caboni, 2017).

This chapter explores subjective well-being's emerging literature on tourism and reflects on neurosciences' possible contributions to advancing the studies. It was developed an exploratory and descriptive study, based on a bibliography survey and textual analysis (Salviati, 2017). The methodology adopted in this study allows us to exam how the relationship between subjective well-being, tourism, and happiness is developing in the identified scientific literature. The chapter begins with the definitions of subjective well-being, with a table with its theories and concepts most present in the literature. Following, the explanation about neurosciences and *neurotourism*, the methods, results and discussions. The final remarks show ontological and epistemological challenges within the study fields.

SUBJECTIVE WELL-BEING AND TOURISM

The field of subjective well-being (SWB) comprises the scientific analysis of how people evaluate their lives (Diener & Suh, 1997) in moments of the past, the present, and for more extended periods. It includes people's emotional responses, domain satisfactions, and global judgments of life satisfaction, the domains being the fields of work, health, marriage, leisure, among others (Diener et al., 2003). It is also a broad category of phenomena, which need to be understood by itself; that is, it is a general area of scientific interest and not just a specific construct (Diener, Suh, Lucas & Smith, 1999).

Diener (1984) describes SWB in three hallmarks. First, it is subjective because it lives within the experience of the individual; second, subjective well-being includes positive measures and negative factors; and the third one is that SWB measures include a global assessment of all aspects of a person's life. According to the same author, SWB can be acquired when a specific state is reached, whether being an objective or need. And the discovery that SWB is associated with many desirable social and psychological results, such as higher income, a more optimistic outlook, significant relationships, better physical and mental health, and greater community involvement, demonstrate the importance of adopt this perspective (Kim & Hall, 2019).

The subjective assessment of a person's life can be based on cognitive or affective bases or a combination of both (Diener, 1984; Diener et al., 1999). Cognitive assessments are satisfaction with life, interest, and involvement, as they are based on evaluative beliefs and attitudes, whereas the affective component refers to how people perceive the balance between their pleasant and unpleasant affections, such as joy and sadness, which is called a hedonic component (Chen, Fu, & Lehto, 2016; Mayer, Machado, Marques & Goncalves Nunes, 2019).

In many cases, SWB is also known as happiness because of its important historical and famous roots, as the high level of SWB suggests that individuals experience satisfaction with life, frequent joy, and relatively uncommon unpleasant emotions (Diener et al., 1999). In the psychology literature, subjective well-being is similar to the satisfaction of life, quality of life, and happiness, but happiness can be more broadly defined as subjective well-being (McCabe & Johnson, 2013). The SWB' studies also dialogue with positive psychology and the theory of well-being (Seligman, 2011); eudemonism and hedonism (Bosnjak, Brown, Lee, Yu & Sirgy, 2016; Houge Mackenzie & Hodge, 2020); quality of life (Dolnicar, Yanamandram & Cliff, 2012) and happiness itself (Nawijn, 2011; Nawijn, Marchand, Veenhoven, & Vingerhoets, 2010a). The most important theories and concepts in the area of subjective well-being are summarized in table 1.

Subjective well-being has been widely used in the field of tourism as well as in other areas of knowledge. There is relative consistency in the SWB' description among tourism scholars, mainly rooted in the quality of life (Dolnicar et al., 2012; Sirgy, 2010). Academics focus on tourists' SWB and experiences (Chen et al., 2013; Nawijn, 2011; Nawijn et al., 2013; Nawijn & Peeters, 2010b); residents' perceptions about tourism the involvement of the local community (Bimonte & Faralla, 2016; Lv & Xie, 2017; Nawijn & Mitas, 2012; Rivera, Croes & Lee, 2016; Tang & Wang,

Table 1. Theories and concepts related to subjective well-being (SWB)

Theories and Concepts	Definition	Authors
Happiness	The word happiness is used with different meanings. It is typically used as a term for various aspects of the good life in classic philosophy. Social scientists increasingly use the word "happiness" as a synonym for subjective enjoyment of life; life-satisfaction, and 'appreciation of one's life as a whole'. Psychologists formally refer to this construct as subjective well-being (SWB).	● Veenhoven (1984); ● Eid & Diener (2004); ● Nawijn & Mitas (2012).
Life satisfaction	Life satisfaction is a cognitive, judgmental process, consisting of a global assessment of a person's quality of life and is influenced by the levels of satisfaction with different life domains (e.g., satisfaction with leisure travel, family, work, social life, health, and so on). High SWB suggests that individuals experience high life satisfaction.	● Diener et al. (1999) ● Sirgy (2010) ● Gilbert & Abdullah (2004) ● Nawijn (2011) ● McCabe & Johnson (2013)
Emotions	Emotions are direct, intense reactions to events that happen in an individual's environment and experiences. Frequent joy and relatively infrequent unpleasant emotions are related to higher levels of SWB. The authentic happiness conceptualization suggests that happiness is about having positive emotions (love, interest, joy, and contentment).	● Beedie, Terry & Lane (2005) ● Diener et al. (1999) ● (Filep, 2014)
Quality of life	Quality of life (QQL) is an individual's subjective evaluation of the degree to which his or her most important needs, goals, and wishes have been fulfilled.	● Nawijn & Mitas (2012) ● Dolnicar et al. (2012)
Hedonic and eudaimonic well-being	**Hedonic** well-being is the sense of feeling good in joyful and pleasurable activities. Involves an emphasis 'placed on happiness as pleasure, enjoyment, and absence of discomfort understood as subjective affective states'. **Eudaimonic** well-being comprises dimensions such as personal growth, life proposal, self-realization, positive relationships, self-acceptance, and is commonly called 'true happiness'. Both hedonic and eudaimonic well-being are parts of the happiness construct and the SWB.	● Veenhoven (1984) ● Mcmahan & Estes (2011) ● Waterman; Schwartz & Conti (2008) ● Vada, Prentice & Hsiao (2019) ● Huta & Waterman, 2014
Flow Theory	The theory of optimal experience rests on the notion of **flow**. The flow theory states that individuals raise their levels of well-being and happiness when participating in **high-engagement** tasks, where their skills and attention are fully applied. These experiences usually involve some kind of intellectual or physical challenge, with clear goals and immediate feedback.	● Csikszentmihalyi (1988, 1990)
PERMA Theory and Positive Psychology	PERMA theory is the acronym for five elements that together enable human flourishing and well-being: **positive emotion, engagement, relationships, meaning, and accomplishment.** The theory is within the Positive Psychology, an emerging field of research committed to addressing positive issues related to human growth. This perspective integrates the hedonic and eudaimonic dimensions of SWB.	● Seligman (2011) ● Seligman & Csikszentmihalyi (2000)

2020). These perceptions influence the general views about the community's well-being, their commitment to the place, and their general satisfaction with life. Tourism scholars have also focused, in a smaller number, on the employees' SWB (Hu, Liu & Qu, 2019; Yurcu & Akinci, 2017).

Tourism researches have shown that holidays make us happier and people who take vacations seem to be a little happier than those who do not (Nawijn, 2011). In addition, studies have determined that the holidays - a relatively long rest period spent mainly in a different and more relaxing environment - helps people to psychologically disconnect from work and the mundane environment (Chen, Lehto & Cai, 2013). Individuals who take vacations are generally happier with their families, financial situation, and health than individuals who do not take them, according to Gilbert and Abdullah (2004). Specifically, it has been found that travelers' SWB is associated with their vacation experience and their satisfaction with it (Chen et al., 2016). Is has been also established that different moods, personal emotions, interests, and perspectives mean that touristic products and experiences are interpreted differently, once emotional dimensions are frequently associated with memorability (Aydin & Omuris, 2020).

Therefore, some studies try to understand positive emotions, like Fredrickson, Cohn, Coffey, Pek & Finkel (2008) brought in their research, that positive emotions momentarily broaden people's attention and thinking, enabling them to draw on higher-level connections. These resources can be *cognitive*, *psychological*, *social*, or *physical*. The same authors explained that people with these resources are more likely to meet life's challenges effectively, take advantage of its opportunities, become successful, healthy, and happy. Thus, frequent experiences of positive emotions are posited to be keys to later increases in well-being. However, the use of neuroscience methodologies between tourism studies and emotions is still under development.

Additionally, following the Positive Psychology approach, the Positive Tourism perspective emerged as the study of hedonic and eudaimonic well-being and conditions for flourishing (Filep, Laing & Csikszentmihalyi, 2017). A fulfilling tourist experience is arguably one that is characterized not just by pleasure but also by how personally meaningful tourists found their holiday activities. (Filep, 2014). A life rich in meaning incorporates hedonic aspects of pleasure, but combines it with eudaimonic ones, for instance, a search for outcomes such as self-actualization, transcendence, engagement, personal growth, and life goals (cf. King, Richards, and Stemmerich 1998). A tourism experience felt physically, spiritually and intellectually, will be remembered long after it ends creating a memorable tourism experience (Aydin & Omuris, 2020).

In sum, evidence from tourism studies support, so far, the claim that holiday trips affect happiness, induce memorable experiences and influence different levels of subject well-being and happiness. Nevertheless, despite the relevance and development of tourism research on the last years, there are still gaps on the knowledge due to methodological challenges associated with capturing and measuring complex human phenomena such as social interactions, emotions, moods and changes on levels of SWB, more specifically when trying to observe tourists' real experiences (Mayer et al. 2019). Thus, we foresee the neurosciences will open new methodological and theoretical perspectives to tourism research on these streams of research.

NEUROSCIENCES, WELL-BEING AND NEUROTOURISM

Interest in the study of the brain is not new, but neuroscience is a recent word and discipline. Modern neurosciences originated a new science of mind (Kandel et al. 2014), and it is essential to understand some facts. One of the main landmarks for neurosciences occurred in the late 19th century, more precisely by Camillo Golgi in 1873. He discovered, using a method of staining brain tissue, the structure of the neuron. Golgi's discoveries lead Santiago Ramón y Cajal to propose the four principles of what became known as neuronal doctrine: (1) the characterization of the neuron as a fundamental structural unit of the nervous system; (2) the synapse, that is, the interaction between neurons; (3) the specificity of the connection, that is, a set of circuits; (4) dynamic polarization, indicating the direction of information, which is known today as the flow of data from the brain (Kandel, 2020). These discoveries opened an avenue for modern neuroscience.

On neuroscience, Bear, Connors and Paradiso (2002) draw attention to the levels of analysis involved, namely: (a) molecular neurosciences; (b) cellular neurosciences; (c) neuroscience of systems; (d) behavioral neurosciences; (e) cognitive neurosciences. Witch shows the complexity of neuroscientific studies and the need for articulation between various disciplines - theories and methods - to meet the analysis levels' objectives. Bear, Connors and Paradiso (2002) also highlighted the importance of the scientific process: observation, replication, interpretation and verification.

Also, the advances in microscope optics were significant for the 19th century, neuroimaging advances, notably from the 1970s onwards, are essential for analyzing the brain *in vivo*. For example, tomography analyzes the brain's structure, but it is the functional image that introduces the dimension of time and allows observing the activity of the brain. Thus, the feature of functional magnetic resonance (FmRI) enables understanding regions and even circuits of the brain according to the objectives of the researcher (Kandel, 2020).

Although brain studies have historically been derived from physics, philosophy, chemistry, and other classic sciences (Rooney, 2018), nowadays, there is a clear association between neurosciences and the formation of new disciplines. For instance, social neuroscience is a discipline devoted to explaining the brain functions on social interactions, combining work in the neurosciences, social sciences, and cognitive sciences. According to social neuroscience, sociality and well-being are closely linked once humans are essentially a social species. The key features of being a society member are a sense of self, the ability to infer others' mental states, mechanisms to detect social threats, and the capacity to regulate thought, behavior, and emotion. The brain default network is crucial to promoting positive well-being and positive feelings about oneself. Much of the research on the field has focused on aspects of ill-being such as social rejection, loneliness, and depression, but future work may help identify the brain mechanisms that help people flourish (da Silva & Heatherton, 2018).

In social neurosciences, traditional approaches to assessing and measuring brain function have been electroencephalography (EEG) and functional neuroimaging (fMRI). EEG measures electrical activity in the brain and is most often used because of its excellent temporal resolution. It helps to identify the timing of neural events that reflect psychological processes. EEG has provided insights into how different activation patterns and frequency bands relate to internalized attention, positivity, and well-being. Instead, the fMRI, which measures brain activity by detecting blood flow changes, has been used by most research to locate brain functions. Essentially, fMRI can identify brain regions whose activity is correlated with mental states or behavior. As methods have evolved, researchers in social neuroscience have sought to use brain activity to predict behavioral outcomes relevant to well-being. A recent trend is examining how different brain areas work together to process information. Once the brain regions do not work in isolation, mental states require many regions' coordinated activity. Two approaches to measuring neural networks are functional connectivity (how brain regions' activity correlate) and structural connectivity (the extent to which regions are physically connected). Functional connectivity can be examined with psychophysiological interaction (PPI) and resting-state functional connectivity (rsfMRI) (da Silva & Heatherton, 2018).

With all these advances, the emergence of *neuropsychology, neuroeconomics*, and *neuromarketing* confirm the tendency of interdisciplinary cooperation efforts of different and complementary fields to expand knowledge. Likewise, it is possible to identify the formation of a new body of knowledge in tourism, which has been called *neurotourism*.

Neurotourism is a discipline combining neuroscience and tourism to enhance tourism management studies (Giudici et al., 2017). Neurotourism practices are based on cognitive neuroscience, that is, the investigation of brain operations (Mileti, Guido & Prete, 2016). For some authors, it is a shift in the travel paradigm, and they discuss the interface between e-tourism and neuroscience (Panyik & Gonçalves, 2017). Other authors question whether neurotourism is a current discipline or only a future perspective (Giudici et al., 2017). Ma et al. (2014) take a step forward by having an applied vision of neuroscience in Tourism Management, proposing the adoption of *neuromanagement*, which may significantly advance the interface between neuroscience and tourism.

It is crucial to highlight that tourism and psychology relate to one another, shaping people's motivations and behaviors towards traveling. Their reactions and perceptions are essential psychological processes. Also, travelers' behavior changes during a trip once there are many sub-decisions in touristic purchasing behavior and travel experiences. Thus, to acquire knowledge on tourist decision-making, advances developed in the neuroscience field can be usefully applied in neurotourism. More precisely, the use of neuroscience techniques and theories constitutes a significant advantage for tourism in its neurotourism perspective. For instance, a portfolio of research procedures and practices can be built with more realistic assumptions regarding human emotions and behaviors (Giudici et al., 2017).

According to Giudici et al. (2017), neurotourism research instruments and procedures are divided into two general classes: tools that measure reactions of the body and devices that measure the brain's responses (Table 2).

Therefore, as a new paradigm of study, the lenses of neurosciences and neurotourism can help to advance the knowledge concerning happiness and subjective well-being in tourism research.

METHOD

To promote the debate on the possible contributions of neurosciences to increasing the state of the art of tourism research, we evaluated the emerging literature on SWB in tourism through an exploratory and descriptive study, based on a bibliography survey and textual analysis (Salviati, 2017).

Firstly, to identify the most relevant published studies about subjective well-being and happiness in tourism, which is the main focus of this chapter, we inserted the terms "subjective well-being" AND "touris*" AND "happiness" as topics at the Web of Science Core Collection (WoS), resulting in an initial set of 40 documents[1]. WoS was chosen for this study because it is the world's largest publisher-neutral citation index and research intelligence platform, and covers more than 34,000 journals. This platform has been consistently structured since 1950 to the present and has indexed the main scientific journals, including in tourism and psychology fields (Web of Science, 2021).

Two more WoS criteria were applied to the initial sample: type of documents (articles) and category (hospitality leisure sport tourism), thus the ultimate selection of documents was defined (n = 29). It is important to mention that our first attempt to find studies combining the terms "subjective well-being" AND "touris*" AND

Table 2. Instruments and procedures in neurotourism research

Facial expressions	The human face records a wide variety of emotions and can be interpreted in two ways: visible changes in expressions and imperceptible micro-changes (for example, small muscle contractions). They can be analyzed as positive or negative emotional indicators.
Eye tracking	This device uses infrared beams to detect eye movements and pupil dilation while seeing something or someone. Eye-tracking has numerous applications in *neurotourism*. The speed and course of changes in eye movement give good markers of attention, interest, and fascination, and they reveal the direction a person's eyes take when looking at something.
Electro-dermal activity	Skin perspiration is measured using electrodes attached to the fingers, and it is a subconscious response. Signals of this tool increase with emotional stimulation of the nervous system. When a particular memory is activated, a significant spike in emotional arousal is shown. But it is impossible to identify whether it is a positive or a negative outcome.
Respiration and heart rate	This system is based on measuring the beating speed of the heart and respiration speed using sensors placed on the chest or using a finger pulse meter.
Response time	To understand how the brain processes are revealed in the unconscious behavior, it is necessary to analyze the rate of its reaction to word correlations or visual decisions. Reaction time measures give an easy and open approach to test the quality of notions.
Functional magnetic resonance imaging (fMRI)	This method empowers the exact imaging of activity anyplace in the cerebrum, measuring the blood flow. It measures the increase in oxygen levels in the blood flow in the brain and indicates when activity in specific parts of the brain increases. At the point when parts of the cerebrum start to be dynamic, blood flows to them. Since blood contains iron, this flow can be followed by a magnet that encompasses the head of the individual being examined. They're essential for testing models and theories about brain functioning.
Electroencephalography (EEG)	This brain-scanning technology uses sensors to capture the tiny electrical signals that brain activity produces. EEG measures the quality at the scalp of a few electrical fields produced by cerebrum activity. It is the most well-known technique due to its minimal effort and adaptable equipment prerequisites.
Magnetoencephalography (MEG).	This method measures minute changes in magnetic fields delivered by the cerebrum. It is a multi-million-dollar piece of hardware and requires a cool temperature close to zero for it to work. It has been used as a part of scholarly research but not yet as part of a neurotourism study.

Source: Adapted from Giudici, Detorri & Caboni, 2017.

"happiness" with "neuroscience" and "neurotourism" had no results, clear evidence of the lack of research.

Then, we used Iramuteq software, version 0.7 alpha 2, to perform the textual analysis in two ways: (a) Word Cloud: based on a simple lexical analysis, the frequency was used, highlighting those that appear most frequently in the textual corpus; (b) Similitude Analysis: based on Graph Theory and assists "in the identification of the database structure (corpus), distinguishing common parts and specificities, in addition to allowing them to be checked against existing descriptive variables" (Salviati, 2017, p.69). For the similitude analysis, Fruchterman Reingold's presentation was used, considering parameters such as communities and halos. The textual corpus was prepared from the 29 abstracts, considering formatting adaptations outlined in the Iramuteq Manual (see Salviati, 2017). An adjustment was made in the software properties, indicating the adverbs as supplementary forms for both analyzes.

TEXTUAL ANALYSES: WELL-BEING IN TOURISM RESEARCH

The methodology adopted in this study allowed us to exam how the relationship between subjective well-being, tourism, and happiness is developing in the identified scientific literature. Preliminary results show the most repeated words in this textual corpus are: tourist (n = 54), subjective (n = 49) and tourism (n = 44). Other highlights related to the study were: acronym SWB (n = 30) and happiness (n = 29).

Furthermore, of the 29 works extracted from the Web of Science Core Collection, most of them used methods like self-report/self-administrated questionnaires, two-step survey, and interviews. However, these methods contain significant limitations when the ultimate goal is to capture the reactions and emotions of people. Instruments and procedures used in neurotourism research (Table 2), such as facial expressions, eye-tracking or functional magnetic resonance imaging (fMRI), were not found. While we did not found evidence of research integrating neurosciences and well-being in tourism, we proceeded with the textual corpus analyses.

Analyzing the Word Cloud (Figure 1), it is evident that studies on subjective well-being, happiness, and tourism focus mainly on the tourist (n = 54), including constructs such as satisfaction (n =, 41), experience (n = 32), happiness (n = 29) and the SWB itself (n = 30). Therefore, there is still need to advance the research on residents and tourism workers SWB.

When comparing these results with the theories and constructs presented in Table 1, there are still gaps related to SWB, such as the study of emotions. Emotions are considered fundamental to the process of adaptation and survival of human beings since Darwin's seminal study on "The expression of the emotions in man and animals" published in 1872. Therefore, it should be one of the central themes of the studies on neurosciences and tourism. For example, on the neuroscience side, Richards et al. (2000) were already interested in the neural bases of emotions and what was conventionally treated as "affective neuroscience." On the tourism side, Moyle, Moyle, Bec and Scott (2019) showed that the frontier of tourism research is the study of emotions. These authors seek explanations in the Cube of Emotions (or Lövheim's Cube, 2012), a physiological perspective since Lövheim (2012) three-

Figure 1. Word Cloud

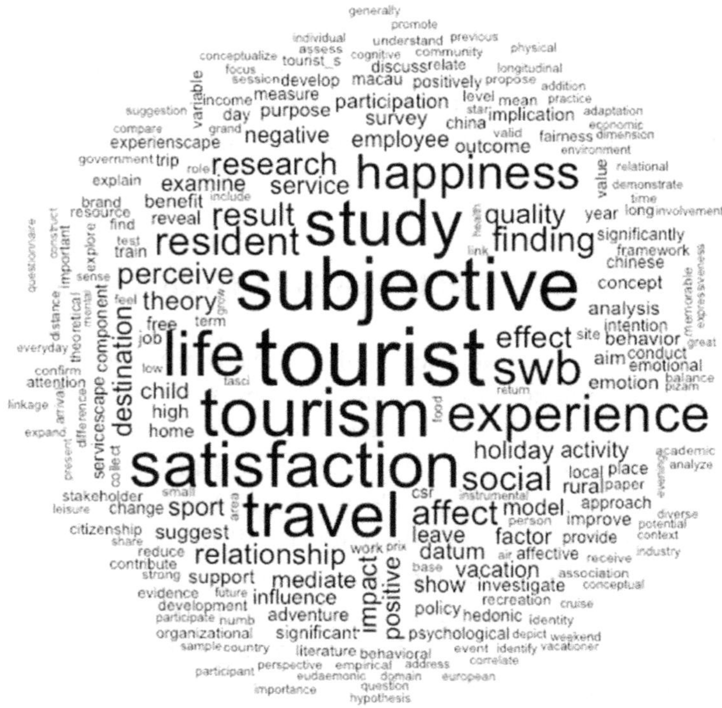

dimensional cube, that postulates about the relationship of dopamine, serotonin, noradrenaline, and emotions.

Figure 2 presents the Similitude Analysis, identifying the themes more evident in the textual corpus and the relationship among them. For this analysis, the frequency of terms equal to or greater than ten was chosen on the software. The resulting diagram shows four halos, the central halos being represented respectively by the word tourist (n = 54) and tourism (n = 44).

Similitude Analysis allows us to see that in the halo formed by the word tourist there is a direct relationship with terms such as destination (n = 14) and service (n = 12). Thus, neurotourism could contribute with aspects of neuromanagement as suggested by Ma et al. (2014). On the other hand, perceive (n = 16) and effect (n = 16) demonstrate how these dimensions need to be worked on to reach the objectives with tourists. Therefore, to study these dimensions, it is relevant to consider that the future of tourism research involves cognitive psychology and neurosciences, as pointed out by Scott (2020). Also, advances in neuroimaging may help to understand the behaviors and cognitions of tourists.

Figure 2. Similitude Analysis

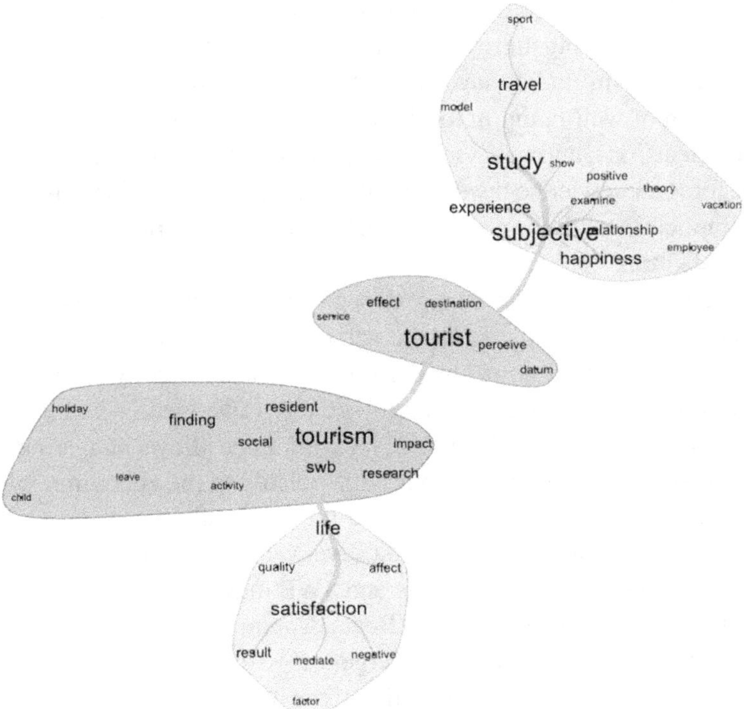

On the Similitude Analysis the acronym SWB (n = 30), although directly linked to tourism (n = 44) as a phenomenon and activity, also has a strong connection with the term tourist (n = 54) and the term subjective (n = 49), reaffirming the centrality of studies that focus on the SWB of tourists. In another halo, the acronym SWB (n = 30) connects with the term life (n = 40) and satisfaction (n = 41), demonstrating the importance that the concept of life satisfaction (Table 1) has for the relationship that the tourist establishes with the SWB. Still in Figure 2, it is evident that the subjective (n = 49) is related to both happiness (n = 29) and experience (n = 32). Therefore, to understand happiness from the perspective of the SWB, it is necessary to understand the tourist's subjectivities.

NEUROSCIENCES AND WELL-BEING IN TOURISM

The textual analyses show that emerging tourism research in well-being, although relevant, is still not applying the resources and theories developed in the neurosciences field, nor neurotourism studies have turned their focus to the subject. However, once the investigation of well-being in tourism involves psychological phenomena such as emotions, decisions, and behaviors, it is clear that the neurosciences approach is significant for the knowledge development of the field. Particularly, neurosciences may have implications on the study of social behavior, emotions, experiences, and tourists' satisfaction.

Neurosciences can collaborate to deepen the knowledge about tourists and **social behavior**. For example, from the perspective of affective neuroscience, six brain systems were identified at the core of human personality, namely: seeking, caring, playfulness, fear, anger, sadness - see Affective Neuroscience Personality Scales (ANPS) (Orri et al. 2017). Social Neuroscience allows analyzing the social interaction between individuals, as tourism is related to the encounter with the "I" and the "other", between the "visitor" and the "visited", this approach becomes essential for studies tourist and SWB (daSilva & Heatherton, 2018). Thus, future studies of tourism-related neurosciences and SWB may contribute to understanding behaviors (for the levels of analysis, see Bears, Connors and Paradiso, 2002) related to travel, observing convergences and divergences with daily life due to intrapersonal constructs, interpersonal constructs, and emotional regulation skills.

As studies on neurotourism are recent, the interface between tourism and neuroscience has a long way to go. **Emotions** are ubiquitous in tourism (Nawijn et al., 2013) and studies suggests that happiness is about having positive emotions (love, interest, joy, and contentment) (Filep, 2014). Studies such as Pearce's (2020) are highlighted in the tourism literature, which seeks to analyze links between time, emotion, and other triggers related to memory. Neurosciences could collaborate in measuring tourist's emotions and how it affects their satisfaction and influence decisions to purchase tourism and leisure services (Hosany et al., 2015).

However, it should be noted that in neuroscience studies it is already evident the need to focus more on researching **positive emotions**, as highlighted by Alexander et al. (2021, p.239):

Thus, we echo previous calls made by other researchers to counterbalance the bias toward studying negative emotions in animals and humans and continue to shift the focus toward the study of positive emotions in order to enhance our understanding of critical factors and strategies that contribute to societal happiness and wellbeing (Alexander et al. 2021, p.239).

On the tourism side, the opposite seems to be the case, that is, studies on **negative emotions** are scarce since the hegemonic narrative about tourism has been built in terms of the search for beauty, for the positive, for well-being, for happiness. However, to understand the SWB and the sense of happiness in the tourist context, adding the neuroscientific perspectives, the theoretical and conceptual framework of the SWB must be expanded and focus on studies of emotions and experiences, considering both promoters and reducers of well-being in trips and in different tourism settings (Mayer et al., 2019).

The availability of mobile devices and sensors can provide an analysis of the travel **experience** as it occurs, overcoming present methodological limitations once it is not an easy to capture the different levels of well-being during a trip (Mayer et al., 2019). For example, it is possible to capture data from mobile EEG devices (headband) and Eye-tracking glasses, crossing with georeferenced information. However, it depends on in-depth knowledge about sensors and devices and about the neuropsychology of memory.

Satisfaction seems to take center stage (see Figure 2) in studies on SWB in Tourism. Therefore, understanding the quality of life and satisfaction is essential to deepen the knowledge about the levels of well-being experienced in tourism. For instance, one of the software outputs used in studies with eye-tracking (see Table 2) is the attention maps. That is, it is possible to identify which visual projections are the most important for the observer. Hence, combining results from attention maps with other consumer behavior measures may lead to the advancement of methodological and theoretical perspectives of SWB in tourism.

Many new advances in neurosciences are developing, opening up new possibilities for neurotourism and research on well-being in tourist destinations and travel experiences. There are opportunities for tourism research, but also challenges and barriers to overcome. Thus, it will be necessary a significant investment in human and physical resources and the adoption of new epistemological perspectives to bring to tourism the latest methodological and theoretical perspectives of the neurosciences.

CONCLUSION

This chapter contributes to tourism studies' progress, especially in subjective well-being (SWB), and the debate about the contributions of neurosciences to tourism research. For this purpose, it was relevant to illuminate the discussion based on a theoretical-conceptual framework to identify the primary constructs involved in SWB studies. Besides this, the bibliographic search in the WoS database enabled identifying a robust textual corpus in the academic literature. By applying the textual analysis, it was possible to recognize the primary relationships of research about SWB, happiness, and tourism.

In light of the results, it was possible to understand that the emerging tourism research in well-being is still not adopting the resources and theories developed in the neurosciences field. This approach will possibly help us find new ways to overcome the methodological challenges involved in observing the fluctuations of well-being in real-time as the trips and experiences unfold (Mayer et al., 2019).

The tourists were the main focus of the studies analyzed, so understanding their behaviors is essential. However, social interactions and the well-being of residents and workers are still under researched and need more attention from the field. Emotion is still a gap in the studies on tourism and well-being, particularly from the standpoint of the neurosciences. Affective and social neurosciences can advance the discussions regarding the positive and negative emotions associated with happiness and well-being. Also, well-being-related constructs such as social behavior, experiences, and satisfaction may benefit from the neurosciences approach.

Technological advances in neuroimaging (e.g., functional magnetic resonance imaging - fMRI; electroencephalogram - EEG, among others), although necessary, depend on adequate financial and human resources. Giudici et al. (2017) discuss the limitations of techniques and neurological tools, that includes: a) the development of high-priced and short-time neuroimaging experiments; b) experiments confined to artificial laboratory environments; c) moral, social, and ethical abuses of neurosciences devices used for neurotourism, among others.

Future studies involving SWB, tourism, and happiness can consider different theoretical-empirical approaches, utilizing theories and methods emerging from the neurosciences to achieve epistemological and methodological advances to strengthen neurotourism. However, this depends on a series of factors. The first one is a shared vision of the importance of the neurosciences that congregate researchers and research centers. Second, multidisciplinary and interdisciplinary teams must be formed to address various theoretical and conceptual approaches based on neurosciences.

We must recognize that ontological and epistemological challenges exist when dealing with neurosciences within the tourism academy. Also, it is critical to guarantee access to resources and specialized knowledge. Therefore, research groups capable of overcoming these barriers will lead the field's future developments.

REFERENCES

Alexander, R., Arago, O. R., Bookwala, J., Cherbuin, N., Gatt, J. M., Kahrilas, I. J., Kästner, N., Lawrence, A., Lowe, L., Morrison, R. G., Mueller, S. V., Nusslock, R., Papadelis, C., Polnaszek, K. L., Richter, S. H., Silton, R. L., & Styliadis, C. (2021). The neuroscience of positive emotions and affect: Implications for cultivating happiness and wellbeing. *Neuroscience and Biobehavioral Reviews*, *121*, 220–249. doi:10.1016/j.neubiorev.2020.12.002 PMID:33307046

Aydin, D., & Omuris, E. (2020). The mediating role of meaning in life in the relationship between memorable tourism experiences and subjective well-being. *Advances in Hospitality and Tourism Research-AHTR*, *8*(2), 314–337. doi:10.30519/ahtr.656469

Bear, M. F., Connors, B. W., & Paradiso, M. A. (2002). *Neurociências: desvendando o sistema nervoso*. Artmed Editora.

Beedie, C. J., Terry, P. C., & Lane, A. M. (2005). Distinctions between emotion and mood. *Cognition and Emotion*, *19*(6), 847–878. doi:10.1080/02699930541000057

Bimonte, S., & Faralla, V. (2016). Does residents' perceived life satisfaction vary with tourist season? A two-step survey in a Mediterranean destination. *Tourism Management*, *55*, 199–208. doi:10.1016/j.tourman.2016.02.011

Bosnjak, M., Brown, C. A., Lee, D.-J., Yu, G. B., & Sirgy, M. J. (2016). Self-Expressiveness in Sport Tourism: Determinants and Consequences. *Journal of Travel Research*, *55*(1), 125–134. doi:10.1177/0047287514535845

Chen, Y., Fu, X., & Lehto, X. Y. (2016). Chinese Tourist Vacation Satisfaction and Subjective Well-being. *Applied Research in Quality of Life*, *11*(1), 49–64. doi:10.100711482-014-9354-y

Chen, Y., Lehto, X. Y., & Cai, L. (2013). Vacation and well-being: A study of chinese tourists. *Annals of Tourism Research*, *42*, 284–310. doi:10.1016/j.annals.2013.02.003

Csikszentmihalyi, M. (1988). The flow experience and its significance for human psychology. In M. Csikszentmihalyi & I. S. Csikszentmihalyi (Eds.), *Optimal experience: Psychological studies of flow in consciousness* (pp. 15–35). Cambridge University Press. doi:10.1017/CBO9780511621956.002

Csikszentmihalyi, M. (1990). *Flow: The psychology of optimal experience*. Harper & Row.

daSilva, A. W., & Heatherton, T. F. (2018). Social neuroscience of subjective well-being and life satisfaction. In J. E. Maddux (Ed.), Frontiers of social psychology. Subjective well-being and life satisfaction (p. 32–49). Routledge/Taylor & Francis Group.

Diener, E. (1984). Subjective Well-being. *Psychological Bulletin*, *75*(3), 542–575. doi:10.1037/0033-2909.95.3.542 PMID:6399758

Diener, E., Oishi, S., & Lucas, R. E. (2003). Personality, culture and subjective well-being: Emotional and cognitive evaluations of life. *Annual Review of Psychology*, *54*(1), 403–425. doi:10.1146/annurev.psych.54.101601.145056 PMID:12172000

Diener, E., & Suh, E. (1997). Measuring quality of life: Economic, social, and subjective indicators. *Social Indicators Research*, *40*(1/2), 189–216. doi:10.1023/A:1006859511756

Diener, E., Suh, E. M., Lucas, R. E., & Smith, H. E. (1999). Subjective well-being: Three decades of progress. *Psychological Bulletin*, *125*(2), 276–302. doi:10.1037/0033-2909.125.2.276

Dolnicar, S., Yanamandram, V., & Cliff, K. (2012). The contribution of vacations to quality of life. *Annals of Tourism Research*, *39*(1), 59–83. doi:10.1016/j.annals.2011.04.015

Eid, M., & Diener, E. (2004). Global judgement of subjective well-being: Situational variability and long-term stability. *Social Indicators Research*, *65*(3), 245–277. doi:10.1023/B:SOCI.0000003801.89195.bc

Filep, S. (2014). Moving Beyond Subjective Well-Being A Tourism Critique. *Journal of Hospitality & Tourism Research (Washington, D.C.)*, *38*(2), 266–274. doi:10.1177/1096348012436609

Filep, S., Laing, J., & Csikszentmihalyi, M. (2017). What Is Positive Tourism? Why Do We Need It? In S. Filep, J. Laing, & M. Csikszentmihalyi (Eds.), *Positive Tourism* (pp. 3–15). Routledge.

Fredrickson, B. L., Cohn, M. A., Coffey, K. A., Pek, J., & Finkel, S. M. (2008). Open hearts build lives: Positive emotions, induced through loving-kindness meditation, build consequential personal resources. *Journal of Personality and Social Psychology*, *95*(5), 1045–1062. doi:10.1037/a0013262 PMID:18954193

Gilbert, D., & Abdullah, J. (2004). Holidaytaking and the sense of well-being. *Annals of Tourism Research*, *31*(1), 103–121. doi:10.1016/j.annals.2003.06.001

Giudici, E., Dettori, A., & Caboni, F. (2017). Neurotourism: Futuristic perspective or today's reality? 20th Excellence. In *Services International Conference Verona*. University of Verona.

Hosany, S., Prayag, G., Deesilatham, S., Caušević, S., & Odeh, K. (2015). Measuring tourists' emotional experiences: Further validation of the destination emotion scale. *Journal of Travel Research, 54*(4), 482–495. doi:10.1177/0047287514522878

Houge Mackenzie, S., & Hodge, K. (2020). Adventure recreation and subjective well-being: a conceptual framework. *Leisure Studies, 39*(1), 26–40. doi:10.1080/02614367.2019.1577478

Hu, B., Liu, J., & Qu, H. (2019). The employee-focused outcomes of CSR participation: The mediating role of psychological needs satisfaction. *Journal of Hospitality and Tourism Management, 41*, 129–137. doi:10.1016/j.jhtm.2019.10.012

Huta, V., & Waterman, A. (2014). Eudaimonia and Its Distinction from Hedonia: Developing a Classification and Terminology for Understanding Conceptual and Operational Definitions. *Journal of Happiness Studies, Springer, 15*(6), 1425–1456. doi:10.100710902-013-9485-0

Kandel, E. R. (2020). *Mentes diferentes: O que cerébros incomuns revelam sobre nós*. Manole.

Kandel, E. R., Schwartz, J. H., Jessell, T. M., Siegelbaum, S. A., & Hudspeth, A. J. (2014). *Princípios de Neurociências*. 5o ed. Porto Alegre. Mc Graw Hill Education. *The Art of Medication*.

Kim, M. J., & Hall, C. M. (2019). A hedonic motivation model in virtual reality tourism: Comparing visitors and non-visitors. *International Journal of Information Management, 46*, 236–249. doi:10.1016/j.ijinfomgt.2018.11.016

King, L. A., Richards, J. H., & Stemmerich, E. (1998). Daily goals, life goals, and worst fears: Means, ends, and subjective well-being. *Journal of Personality, 66*(5), 713–744. doi:10.1111/1467-6494.00030 PMID:9802231

Lövheim, H. (2012). A new three-dimensional model for emotions and monoamine neurotransmitters. *Medical Hypotheses, 78*(2), 341–348. doi:10.1016/j.mehy.2011.11.016 PMID:22153577

Lv, Q., & Xie, X. (2017). Community involvement and place identity: The role of perceived values, perceived fairness, and subjective well-being. *Asia Pacific Journal of Tourism Research, 22*(9), 951–964. doi:10.1080/10941665.2017.1345772

Ma, Q. G., Hu, L. F., Pei, G. X., Ren, P. Y., & Ge, P. (2014). Applying Neuroscience to Tourism Management: A Primary Exploration of Neurotourism. *Applied Mechanics and Materials*, *670–671*, 1637–1640. . doi:10.4028/www.scientific.net/AMM.670-671.1637

Mayer, V. F., Machado, J. dos S., Marques, O., & Goncalves Nunes, J. M. (2019). Mixed feelings? Fluctuations in well-being during tourist travels. *Service Industries Journal, 40*(1–2), 158–180. doi:10.1080/02642069.2019.1600671

McCabe, S., & Johnson, S. (2013). The happiness factor in tourism: Subjective well-being and social tourism. *Annals of Tourism Research*, *41*, 42–65. doi:10.1016/j.annals.2012.12.001

McMahan, E. A., & Estes, D. (2011). Hedonic versus eudaimonic conceptions of wellbeing: Evidence of differential associations with experienced well-being. *Social Indicators Research.* Advance online publication. https://psycnet.apa.org/doi/10.1007/s11205-010-9698-0

Mileti, A., Guido, G., & Prete, I. M. (2016). Nanomarketing: A new frontier of neuromarketing. *Psychology and Marketing*, *33*(8), 664–674. doi:10.1002/mar.20907

Moyle, B. D., Moyle, C., Bec, A., & Scott, N. (2019). 'The next frontier in tourism emotion research'. *Current Issues in Tourism*, *22*(12), 1393–1399. doi:10.1080/13683500.2017.1388770

Nawijn, J. (2011). Determinants of Daily Happiness on Vacation. *Journal of Travel Research*, *50*(5), 559–566. doi:10.1177/0047287510379164

Nawijn, J., Marchand, M. A., Veenhoven, R., & Vingerhoets, A. J. (2010a). Vacationers Happier, but Most not Happier After a Holiday. *Applied Research in Quality of Life*, *5*(1), 35–47. doi:10.100711482-009-9091-9 PMID:20234864

Nawijn, J., & Mitas, O. (2012). Resident Attitudes to Tourism and Their Effect on Subjective Well-Being: The Case of Palma de Mallorca. *Journal of Travel Research*, *51*(5), 531–541. doi:10.1177/0047287511426482

Nawijn, J., & Peeters, P. M. (2010b). Travelling 'green': Is tourists' happiness at stake? *Current Issues in Tourism*, *13*(4), 381–392. doi:10.1080/13683500903215016

Orri, M., Pingault, J. B., Rouquette, A., Lalanne, C., Falissard, B., Herba, C., Côté, S. M., & Berthoz, S. (2017). Identifying affective personality profiles: A latent profile analysis of the Affective Neuroscience Personality Scales. *Scientific Reports*, *7*(1), 4548. doi:10.103841598-017-04738-x PMID:28674393

Panyik, E., & Gonçalves, J. (2017). Neurotourism: A paradigm shift in travel on the intersection of e-tourism and neuroscience. *Revista Turismo & Desenvolvimento*, (27/28), 321-323.

Pearce, P. L. (2020). Tourists' perception of time: Directions for design. *Annals of Tourism Research*, *83*, 102932. doi:10.1016/j.annals.2020.102932

Rivera, M., Croes, R., & Lee, S. H. (2016). Tourism development and happiness: A residents' perspective. *Journal of Destination Marketing & Management, 5*(1), 5–15. doi:10.1016/j.jdmm.2015.04.002

Rooney, A. (2018). *A História da Neurociência*. M.Books.

Salviati, M. E. (2017). *Manual do Aplicativo Iramuteq*. Planaltina. http://iramuteq.org/documentation/fichiers/manual-do-aplicativo-iramuteq-par-maria-elisabeth-salviati

Scott, N. (2020). Cognitive psychology and tourism – surfing the "cognitive wave": A perspective article. *Tourism Review*, *75*(1), 49–51. doi:10.1108/TR-06-2019-0217

Seligman, M. E. P. (2011). *Flourish: A visionary new understanding of happiness and well-being*. Free Press.

Seligman, M. E. P., & Csikszentmihalyi, M. (2000). Positive Psychology: An introduction. *The American Psychologist*, *55*(1), 5–14. doi:10.1037/0003-066X.55.1.5 PMID:11392865

Sirgy, M. J. (2010). Toward a Quality-of-Life Theory of Leisure Travel Satisfaction. *Journal of Travel Research*, *49*(2), 246–260. doi:10.1177/0047287509337416

Tang, J., & Wang, Y. (2020). Does tourism sports event make residents happier? - Exploring the SWB of Macau residents in the case of Macau Grand Prix. *Journal of Tourism and Cultural Change*, 1–19. Advance online publication. doi:10.1080/14766825.2019.1711103

Vada, S., Prentice, C., & Hsiao, A. (2019). The role of positive psychology in tourists' behavioural intentions. *Journal of Retailing and Consumer Services*, *51*, 293–303. doi:10.1016/j.jretconser.2019.06.015

Veenhoven, R. (1984). *Conditions of Happiness Reidel*. Springer. doi:10.1007/978-94-009-6432-7

Waterman, A. S., Schwartz, S. S., & Conti, R. (2008). The implications of two conceptions of happiness (hedonic enjoyment and eudaimonia) for the understanding of intrinsic motivation. *Journal of Happiness Studies*, *9*(1), 41–79. doi:10.100710902-006-9020-7

Yurcu, G., & Akinci, Z. (2017). influence of organizational citizenship behavior on hotel employees' job satisfaction and subjective well-being. *Advances in Hospitality and Tourism Research-AHTR*, *5*(1), 57–83.

ENDNOTE

1 The search was conducted in February of 2021.

Section 2

Chapter 7
Nautical Sports Tourism:
Improving People's Wellbeing and Recovering Tourism Destinations

Mariana Brandão Cavalheiro
https://orcid.org/0000-0001-6045-9866
Federal Fluminense University, Brazil

Verônica Feder Mayer
https://orcid.org/0000-0002-7543-5215
Federal Fluminense University, Brazil

Aline Barbosa Tinoco Luz
Federal Fluminense University, Brazil

ABSTRACT

While positive news on COVID-19 vaccines has boosted the hopes of the travel and tourism sector (T&T), the pandemic scenario has brought new concerns to tourists and different motivations to travel. For instance, safety, health, and hygiene are paramount. Moreover, as the pandemic impacted not only the physical, but also the mental health, wellness has been acknowledged as a key ingredient for the T&T recovery. The appreciation of nature surroundings and the practice of outdoor activities are the new trend in the post-COVID-19 world. As such, the nautical tourism with special interest in sports has the potential to become very popular. Thus, with the purpose to identify managerial, governmental, and academic opportunities for this sector, a literature review on nautical sports tourism was developed. The analysis revealed that there is still a long way for this literature research to reach maturity. Nevertheless, it was observed that the combination of sports and tourism in nautical environments opens a myriad of possibilities for places to enhance their public welfare.

DOI: 10.4018/978-1-7998-7239-9.ch007

Copyright © 2021, IGI Global. Copying or distributing in print or electronic forms without written permission of IGI Global is prohibited.

INTRODUCTION

The outbreak of the new coronavirus disease, officially reported by China to the World Health Organization (WHO) on 31st December 2019, has posed critical health, social and economic challenges worldwide (Mayer & Coelho, 2021). The WHO categorized COVID-19 as a pandemic on 11th March 2020 after confirming the disease spread in 114 countries(WHO, 2020). One year after that, 220 countries and territories have reported more than 165 million COVID-19 cases and more than 3,4milions of deaths (WHO, n.d.).

In the face of this global crisis, caused by a virus with a high speed of transmission and for which, at first, there was no vaccine nor treatment, governments worldwide had no other option but to close their borders, restrict people's movements and impose social distancing (Weed, 2020). Although such measures effectively slowed down the spread of COVID-19, they caused economic shocks across all industries, hitting hard the Travel and Tourism sector (T&T) (Sharma et al., 2021). By itself, in 2020, international arrivals dropped by 74% (when compared to 2019), representing an estimated loss of USD 1.3 trillion in export revenues, USD 3.4 trillion in global GDP, and around 120 million direct jobs at risk (UNWTO, 2021).

While positive news on vaccines has boosted the hopes of the T&T sector, the pandemic scenario has brought new concerns to tourists, as well as different motivations to travel. For instance, safety, health, and hygiene are paramount in this new era (WTTC/ Oliver Wyman, 2020). In addition, according to the OECD (2020), the re-start of tourism is going to be driven primarily by domestic markets and the search for low-density destinations (in terms of population and tourists) with outdoor activities opportunities.

In fact, the COVID-19 pandemic has impacted not only the physical, but also the mental health globally. The subsequent lockdowns and livelihood losses, and the constant search for mental healing, have shown the value of slower and simpler lifestyles, including tourism (Cooper & Buckley, 2021). Therefore, researchers have appointed wellness tourism as a key ingredient for the T&T sector recovery (Rogerson & Rogerson, 2021; Wen et al., 2020).

In this context, Weed (2020) observed that the interface of sports and tourism could make significant contributions to outcomes and policy goals for wellbeing, physical health, and mental health. Additionally, a study developed by the United Nations Conference on Trade and Development on COVID-19 impacts on the blue economy projected a growth in tourists preferences for "coastal and rural areas in a quest for contact with nature, open air and water" (UNCTAD, 2020). Given that, nautical sports tourism has the potential to become a popular travel interest in the post-COVID-19 world.

Aware of the potential of nautical sports tourism for the T&T sector recovery, some regions worldwide have launched development projects to be prepared and attract such tourists. For instance, in partnership with the city hall of Niteroi (located in Rio de Janeiro state, Brazil), the Federal Fluminense University is implementing a project[1] to map the offer and define the strategic path to develop nautical sports tourism in the region. Likewise, the Galician Association of Nautical activities is engaged in transforming the euro region of Galicia-Northern Portugal into a nautical educational, cultural, and sports destination[2].

Despite the fact that some researchers have made an effort to review the literature on nautical tourism (e.g. Vázquez, 2020; Vázquez et al., 2021), the specific topic of nautical sports tourism still lacks an analysis of the studies addressing this field. With the purpose to identify managerial, governmental, and academic opportunities to develop the promising sector of nautical sports tourism, literature review on the subject was developed.

NAUTICAL TOURISM

With almost two thirds of our planet being covered with water, it is no surprise that the maritime surroundings have always been a major part of humankind history, whether as a source of food, a route for transportation or a place for sports, leisure and tourism (Luck, 2007). Over the past decades, with substantial discussion on the importance of the marine environment for human life and the need for sustainable use of natural resources, the concept of Blue Economy (BE) was created as an extension of the green economy (Smith-Godfrey, 2016). More specifically, BE regards the management and optimization of natural water resources (seas, oceans, rivers), considering its environmental and social dimensions to achieve sustainable growth (Kabil et al., 2021).

According to the UNCTAD (2020), coastal and nautical tourism is the largest of all Blue economy sectors and is an essential source of jobs, investments, and incomes for many countries. For instance, in 2019, the cruise industry added over $154 billion to the global economy and created 1.8 million jobs around the world (CLIA, 2020). Likewise, only in the European Union context, recreational nautical activities created up to 234,000 jobs and generated € 28 billion annual revenue in 2016 (European Commission, 2017). Besides the direct socio-economic impacts, nautical activities offer many other opportunities to the development of the tourism sector, such as easy fit with other tourism packages, lack of seasonality, promotion of other services, and the incentives they provide to the qualification and diversification of the coastal infrastructures (González et al., 2015).

The nautical tourism can be divided into two main kinds: the Nautical Tourism based on recreational and sports activities and the Nautical Cruise Tourism. While the Recreational and Sports type is motivated by activities on or underwater, the main motivation for the cruising tourism is closely related to the ship itself, not to the water environment (Cavalheiro et al., 2021; Vázquez et al., 2021). Another substantial difference between these two segments is the issue of overtourism that can be caused by the giant cruise ships with huge capacity, not to mention the environmental impact of these vessels in terms of waste disposal (Capocchi et al., 2019).

It is worth to mentioning that, as recreational and sports nautical tourism and cruise tourism present diverse characteristics and motivations, the COVID-19 pandemic also impacted them differently. On the one hand, the coronavirus outbreak has turned the cruise segment upside down (Radic et al., 2020). In March 2020, all cruise operations were suspended worldwide, and the whole segment stood still until July when few ships resumed navigation in parts of Europe, Asia and the South Pacific (CLIA, 2020). On the other hand, the quest for outdoor experiences and contact with nature and water, after COVID-19 lockdowns suspensions, has renewed the popularity for recreational nautical activities. For instance, the boat rental market has expanded worldwide and presents an estimated growth rate of 5% per year, for the next five years (ReportLinker, 2020).

As many authors advise that the two nautical tourism types should be studied separately (Moreno & Otamendi, 2017), and the recreational nautical tourism is the one growing in popularity, the present study will address this promising segment. As such, from now on, every time we use the term nautical tourism we will be relating to the recreational and sports kind.

Although recreational nautical activity was initially and historically associated with higher social classes, nowadays such elitism can no longer be generalized (Chapapría, 2000). In fact, nautical tourism is poly-functional, comprising a wide and growing number of maritime activities (Luković, 2012). Such multifunctional nature makes it difficult for the scholar community to reach a consensus on the definition and scope of nautical tourism (Moreno & Otamendi, 2017). In addition, other terms, like maritime and marine tourism, have been used in the literature as synonyms of nautical tourism (Vázquez, 2020).

Recently, Vázquez et al. (2021) analyzed the scientific production on nautical tourism since 1986. These researchers observed that the classification of nautical tourism encompasses a wide range of categories, for example: nature of the activities (recreational, sportive, labor); modalities (sailing, surfing, scuba diving, water skiing); geographical features of the water (seas, oceans, rivers, lakes); equipment to be used and so on. Moreover, regardless of the considerable array of definitions concerning “Nautical Tourism” Vázquez et al. (2021) have found a common ground on the literature they reviewed: the water is the main element in this practice and thus, the contact between the tourist and the water is a premise.

In addition to the search for outdoor activities in contact with water, Sports have become one of the most promising sector, with people growing concern for a healthy lifestyle (Yang et al., 2020). As proposed by Weed (2021), the sports tourism is fertile soil for increasing people's well-being. Thus, we focus our study on the nautical tourism with special interest in sports.

NAUTICAL SPORTS TOURISM

Travelling to participate in sports or watching sporting events is one of the first tourism motivations recorded through human history. "As early as 900 BC the Greeks were travelling to take part in and watch the Ancient Greek Games" (Gibson, 1998, p.50). Later on, the Romans continued the travel element associated with sports (Weed & Bull, 2004). Famous for their sports spectacles that would pack entire coliseums the Roman Empire used such events to shift the focus of the populace away from their poor living conditions.

According to Standeven (1998), sports tourism needs to be understood as a two-dimensional concept: it is a cultural experience of physical activity (albeit sometimes vicariously as a spectator) integrally tied to a cultural experience of place. In the same line, Weed & Bull (2004) contributed to refining the sports tourism definition by adding people to the mix, as for many participants an important part of the leisure experience is the people with whom participation takes place. Therefore, they conceptualize it as "a social, economic and cultural phenomenon that arises from the unique interaction of activity, people and place" (Weed & Bull, 2004, p.37). Taking these two definitions into account, nautical sports tourism may be characterized as a unique interaction of people, activity and water.

With regard to the activity dimension of the sports tourism (e.g. kayaking, sailing, surfing etc), it may be formal or informal, competitive or recreational, actively or vicariously/passively participated in (Weed, 2009), as well as it might involve multi-sports or single-sport participation (De Knop, 1990). Many are the sports activities that can take place in a marine environment. In order to classify such sports, Li & Niu (2020) have proposed a framework with three categories: maritime sports tourism (parasols, sailboats, yachts etc.), coastal sports tourism (beach volleyball, Binhai bike, coastal rock climbing, island exploration, etc.) and sports tourism in the sea (swimming, scuba diving, surfing, etc.). As the foundation of nautical tourism is strongly related to the water (Vázquez et al., 2021), nautical sports tourism activities can be organized into two of these categories: maritime sports tourism and sports tourism in the sea.

Regarding the people element, they might be competitors and/or co-participants. As explained by Weed (2005):

For vicarious/passive participants, the people element is likely to be both other vicarious/passive participants (i.e., other spectators) and the active participants (i.e., competitors). Similarly, active competitors and co-participants may experience other people as active and/or vicarious/passive participants. Even activities that are sometimes participated in alone (e.g., mountaineering, running) are likely to involve other people because participants may reference their participation in terms of the subculture of the activity and thus experience a feeling of communitas. (Weed, 2005, p.233)

Furthermore, considering the interaction of activity, people and place, Weed & Bull (2004) have suggested a typology of sports tourism as depicted on table 1.

Table 1. Types of sports tourism

Type	Definition
Tourism with sports content	"tourism that involves sport as an incidental activity rather than the prime purpose of the trip"(Weed & Bull, 2004:124)
Sports participation tourism	"sport is the prime purpose of the trip"(Weed & Bull, 2004:127)
Sports training	"It comprises, quite simply, sports tourism trips where the prime purpose is sports instruction or training."(Weed & Bull, 2004:130)
Sports events	"It refers to tourism where the prime purpose of the trip is to take part in sports events, either as a participant or a spectator."(Weed & Bull, 2004: 131)
Luxury sports tourism	It is defined by "the quality of the facilities and the luxurious nature of the accommodation and attendant facilities" (Weed & Bull, 2004:133)

These five types of sports tourism can also be applied to the nautical context. For instance, several tourists of sun and beach tourism end up practicing some sort of nautical sports (W. Carvache-Franco et al., 2019), thus taking part in tourism with sports content. On the other hand, practicing a nautical sport, such as scuba diving or surfing, has been reported as some people's primary motivation for traveling (Geldenhuys et al., 2014; Scheske et al., 2019). As an example of sports training tourism, before the 2016's Olympic Games, several sailing teams have traveled to Rio de Janeiro to practice in the real setting of the games (Cavalheiro et al., 2021). Furthermore, some sailing events, such as the Volvo Ocean Race, attract millions of visitors to the host destination (J. Vieira et al., 2018). At last, the type of recreational sailing involving luxury motor yachts would fall into the luxury sports tourism (Weed & Bull, 2004).

Moreover, regarding the coronavirus pandemic impact on the types of nautical tourism, undoubtedly, the sports events were the most affected. In fact, as the COVID-19 outbreak grew, sports event hosts and administrators worldwide began to consider whether their events should be postponed or cancelled (Weed, 2020). The most remarkable example of the pandemic impact on the sports events market was the postponement of the 2020 Olympic Games, affecting not only economically by also psychologically the host community (Yamamura & Tsutsui, 2020).

In contrast, tourism with sports content and sports tourism participation might be more prized than ever before. In most countries around the world, no matter how severe the lockdown, some provision or allowance was made to allow some form of physical activity or exercise, given its importance for people's well-being (Weed, 2020). Additionally, during the confinement periods, it was also observed that the exercise and the place in which it is practiced impacted well-being, as exercising in natural open spaces was acknowledged more effective than in urban areas (Cooper & Buckley, 2021). Additionally, well-being also seems to be linked to a sense of movement, no matter how small, and to places perceptually distinct from home (Su et al., 2020). Thus, it is no surprise that the demand for sports tourism is expected to grow, especially in rural and coastal areas (UNCTAD, 2020).

Given the relevance of nautical sports tourism for improving people's wellbeing and the clear potential of this sector to recover tourism destinations worldwide, we seek to identify managerial, governmental, and academic opportunities for its development. To accomplish this purpose, we developed a literature review on the subject.

MATERIALS AND METHOD

This study concerns an exploratory analysis of the nautical sports tourism scholar, as it analyzes a topic on which there is little accumulated and systematized knowledge (Schlüter, 2003). The methods applied were a systematic literature review that is a formal, systematic, structured approach to reviewing all the relevant literature on a topic (Jiménez-García et al., 2020), followed by a bibliometric analysis, a technic used to measure the advances of science, identifying the paths taken by researchers, the productivity, and the quality researched so far (Nascimento et al., 2019).

This research is based on articles published in English until 2020, in scientific journals available on Web of Science (WoS) and/or on Scopus, the two most acknowledged international databases (E. S. Vieira & Gomes, 2009). Data was collected on March 2021. The search of documents was carried out using the terms "tourism" AND "sport" AND ("nautical" OR "marine" OR "maritime"), which should appear either in article title, abstract or keywords. The terms marine and maritime were included in the search because, as mentioned earlier, they are often used in the tourism literature as synonymous of nautical (Vázquez et al., 2021). While the query on Scopus resulted in 77 publications, on WoS we found 65 articles. After gathering all articles, and removing repeated ones (those that were present in the two databases), keeping only one version of them, we end up with an initial sample of 92 studies.

Then, a pre-analysis of the 92 articles was carried out to filter and select those articles that have actually addressed the topic of nautical sports tourism and ensure that the searched terms were not only used as examples in the abstract. After this pre-analysis, we ended up with a final sample of 55 articles.

At last, we performed an analysis of the 55 articles with the aid of three softwares: Microssoft Excel, Nvivo version 12 and VosViwer. Other than analyzing the network of keywords and the most relevant authors and journals, we have also observed the longitudinal evolution of the literature, the main destinations studied, the types of activities covered in the article sample, and the relationship between nautical sports tourism and well-being. These results are presented in the next section.

Results

Publications' Longitudinal Analysis

Given the exploratory nature of this study, we did not establish a time frame for the articles query. As such, it was possible to analyze the temporal evolution of research on nautical sports tourism. The oldest article from our sample was published in 1991. This pioneering research (Tatsuyuki-Shishido, 1991) analyzed a governmental plan for the development of Wakanoura bay, in Japan, as a marine recreation resort, namely Wakayama Marina City. One of the goals of such project was "developing the new Wakayama Marina City as a major sports center of the Prefecture by providing a large marina and a variety of other sports facilities to promote the use by youth from neighboring major cities as well as regional population" (Tatsuyuki-Shishido, 1991: 390).

Since 1991, the scholar production on nautical sports tourism has considerably grown, reaching its peak of annual publications in 2020, with 20 articles published. Thus, it is observed a soaring academic interest on this topic. Figure 1 shows the chronological evolution of scientific production on nautical sports tourism, considering the number of articles published.

Figure 1. Chronological evolution of the publication on nautical sports tourism

It is worthwhile mentioning that out of the 20 articles published in 2020, 13 were published in a special issue of the Journal of Costal Research, entitled "Sustainable Development in Coastal Regions: A Perspective of Environment, Economy, and Technology", which might bias our analyses. However, this growing trend of publications concerning nautical sports tourism was also observed in 2019, when the number of publications increases considerably (9 articles), when compared to 2018 and 2017, with 1 publication each year.

Authors and Source Analyses

The analyses of the most prominent journals publishing articles on nautical sports tourism have shown that such literature has a quite spars publication source. In fact, the 55 articles of our sample were published in 32 different journals. Hereafter, on table 2, are presented the journals containing more than one article addressing this topic.

Table 2. Number of articles published by journal

Journal	Number of Articles
Journal of Coastal Research	16
Tourism in Marine Environments	4
Marine Policy	3
Ocean and Coastal Management	2
Journal of Sustainability Science and Management	2
European Sport Management Quarterly	2

Again, it is important to acknowledge that out of the 16 articles published on the Journal of Costal Research, 13 were published in a 2020's special edition mentioned earlier.

In respect of the Authors, the 55 articles were written by 143 different researchers. In addition, we observed that most of these researchers had coauthored only one article, making it difficult to define the most relevant ones. The authors with the highest number of articles are the coauthors Mauricio Carvache-Franco, Wilmer Carvache-Franco and Ana Beatriz Hernandez-Lara, who have published 3 articles together (M. Carvache-Franco, Carvache-Franco, Carvache-Franco, & Hernández-Lara, 2020; W. Carvache-Franco et al., 2019, 2020).

Other than the number of articles, we have also analyzed the origin of the authors. Through our sample we have found authors from 26 different countries. However, we found that 75% of these authors were from only 9 countries, as depicted on figure 2. As shown hereafter, China is the origin with the higher number of authors, 25.

Textual Analyses and Keywords Trends

After a quantitative textual analysis of all 55 articles, we observed that the most repeated words are: tourism (n = 3709), marine (n = 2009), management (n = 1071) and development (n =1065). These findings suggest that most of the studies in our sample were developed with a managerial perspective. In addition, the exhaustive repetition of the word development might suggest that these studies sought to analyze the tourism, the community or the nautical activity development.

A word cloud (figure 3) with the 500 most repeated words was arranged with the Nvivo 12 software for visualization purposes.

By analyzing the keywords of these articles, we identified the most commonly used terms have been identified and the most current trends related to emerging areas of interest in nautical sports tourism are also revealed. A trend graphic (figure 4)

Figure 2. Origin of authors

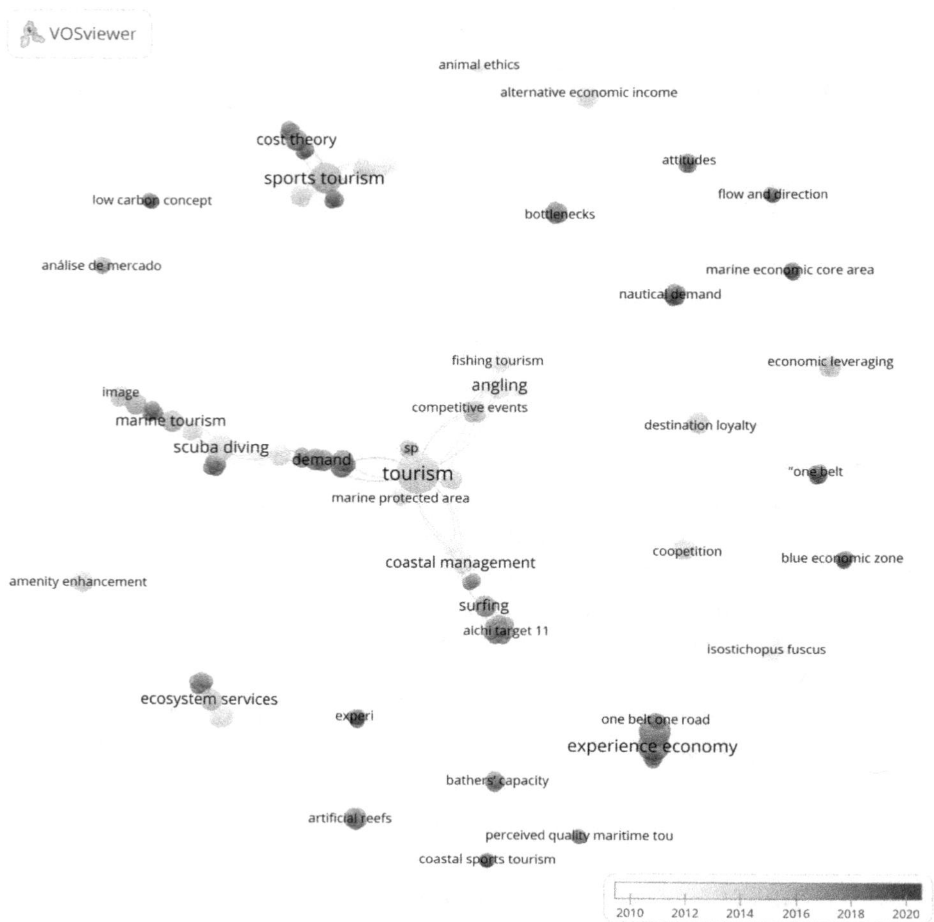

was elaborated with VOSviewer software, which uses a color scale that goes from lighter (less recent terms) to darker (more recent terms).

Figure 4 comprises analyses of 185 keywords, showing a lack of concentration in the terms used. Even though, it is possible to observe a more recent concentration in keywords related to coastal management, cost theory, competitive events, demand, blue economic zone, experience economy, sports tourism, and sustainable tourism, among others.

Taking the specific cluster of sports tourism (a group of 21 keywords) to a closer analysis, the most prominent terms used are directly related to sustainable tourism, marine ecotourism, carrying capacity, innovation (in blue and purple variations),

Figure 3. Word cloud – 500 most frequent words

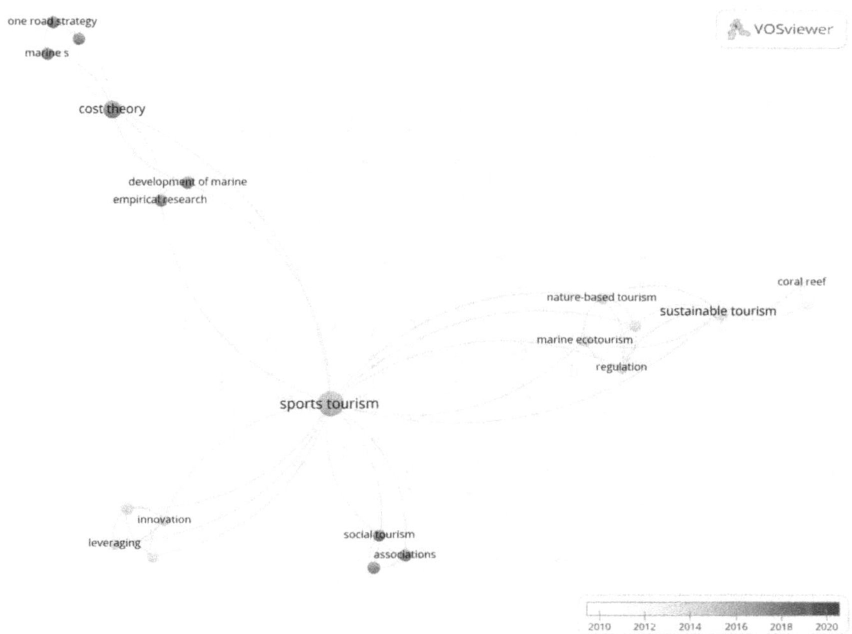

social tourism, association, development of marine and cost theory, as depicted in figure 5.

The presence of the terms "cost theory" and "one belt one road strategy" linked as a trend of research suggests an economical and managerial perspective of nautical sports tourism literature, at a local, regional, national and international levels (M. Carvache-Franco, Carvache-Franco, Carvache-Franco, Hernández-Lara, et al., 2020; Z. Li & Niu, 2020). On the other hand, however, the prominent keywords "social tourism" and "association" linked, demonstrates the great complexity of touristic system and the concern of the researches regarding sustainable development.

Types of Sports (Activities)

As explained earlier, the nautical sports tourism may be characterized as a unique interaction of people, activity and water. Hence, one of the topics of interest of our analysis was the type of nautical activities already explored by the literature. Among the 55 articles, 20 addressed a specific sport, as portrayed in table 3. Moreover, fishing was the activity most researched, with 7 articles.

Figure 4. Trends in keywords

Although other articles did not address a specific sport, they mentioned sundry activities that can be performed in the nautical environment. All these activities are exposed on the word cloud below (figure 6).

In fact, in figure 6, it is possible to observe activities belonging to both Li and Niu (2020)'s categories: maritime sports tourism (e.g., sailing, kayaking, banana boating, jet skiing) and sports tourism in the sea (e.g., snorkeling, swimming, surfing, body boarding).

Contexts Analysis

Though the analysis, we observed that most of the articles were theoretical-empirical studies applied to a specific city or region. Over the 55 publications, 45 addressed a specific destination. Furthermore, places from 23 different countries were studied by these articles. The country most researched was China. In Table 4 we compile the countries contemplated by these studies and the number of articles that addressed them.

Figure 5. Trends of keywords related to sports tourism

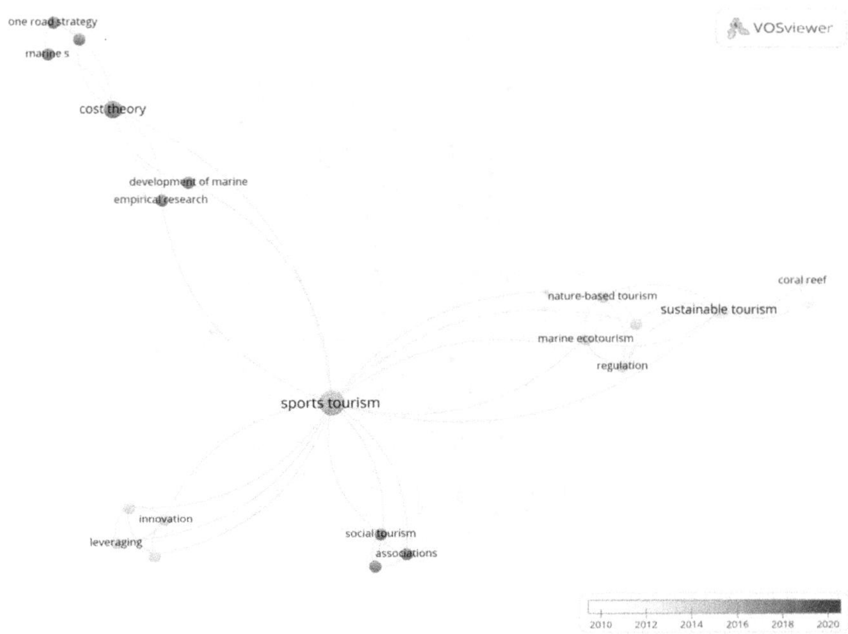

Well-Being and the Nautical Sports Tourism Literature

At last, given the importance of wellness tourism to the T&T sector recovery after the COVID-19 pandemic (Wen et al., 2020), we have also analyzed how the studies have addressed the relationship between nautical sports tourism and well-being.

Table 3. Articles addressing a specific activity

Sport	Articles	Sum of articles
Fishing	(Bull, 2007; Freire et al., 2016; Pawson et al., 2008; Schuhbauer & Koch, 2013; Szylvian, 2004; Waldo & Paulrud, 2012; Wood et al., 2013)	**7**
Scuba Diving	(Anuar et al., 2020; Cater, 2008; Davis & Tisdell, 1995; Geldenhuys et al., 2014; Seaman & Depper, 2019)	**5**
Surfing	(Hume et al., 2019; Ponting & O'Brien, 2014; Rendle & Rodwell, 2014; Scheske et al., 2019)	**4**
Sailing	(Jones & Navarro, 2018; Philippe, 2020; Rahman et al., 2020)	**3**
Canoeing	(Correia & Sebastião, 2017)	**1**

Figure 6 Nautical activities word cloud

Actually, we observed that this field is a fertile soil for the growth of well-being expressions, not only for tourists but also for the hosting community. From this sample, it was possible to capture important ones, as described below.

- **Tourists' physical well-being.** Due to sports' immanent value regarding humans' need for health improvement and quality of life, more people integrate leisure, entertainment and physical activities in a daily basis.
- **Tourists' mental and spiritual well-being.** Aside from the benefits regarding leisure activities, it is possible to find even more challenging, sensorial, communal, nature-connected experiences boosting tourists' positive emotions in the nautical sports tourism context.
- **Economic well-being.** The nautical sports tourism has a huge potential to increase investments for infrastructure and logistics, to create job opportunities, to generate and support other related service industries and in addition, to break the homogenized marine tourism industry.
- **Citizen well-being.** The host communities, in which nautical sports tourism happens (if well developed), brings a strong sense of belonging to individuals and a maritime cultural sense, raising a better understanding of the environment and its conservation.

Table 4. Number of Articles by Country

Number of articles	Country Studied
8	China
5	Ecuador
4	Portugal
3	Croatia, France, USA
2	Australia, Malasia, Spain
1	Brazil, Fiji, Iran, Japan, Malta, Mexico, New Zeland, Peru, Chile, South Africa, Marroco, Sweden, Taiwan, Uk, Uruguai

Table 5 shows an example of references from our sample that tackle each perspective of well-being expressions presented.

Table 5. Perspectives of nautical sports tourism well-being by author

Perspective	Author
Tourists' physical well-being	(Hu & Yang, 2020; Yu, 2020; J. Zhang, 2020; R. Zhang, 2020; Zhao & Lv, 2020)
Tourists' mental and spiritual well-being	(Hu & Yang, 2020; Z. Li & Niu, 2020; Scheske et al., 2019; Shao & Sun, 2020; Wang, 2020; Yang et al., 2020; Yu, 2020; J. Zhang, 2020; R. Zhang, 2020; Zhao & Lv, 2020)
Economic well-being	(H. Li et al., 2020; Lillebø et al., 2019; Liu et al., 2020; Marušić et al., 2020; Rahman et al., 2020; Schuhbauer & Koch, 2013; Shao & Sun, 2020; Wang, 2020; Yang et al., 2020; Yu, 2020; J. Zhang, 2020; R. Zhang, 2020; Z. Zhang & Ding, 2020; Zhao & Lv, 2020)
Citizen well-being	(Lillebø et al., 2019; Marušić et al., 2020; Scheske et al., 2019; Schuhbauer & Koch, 2013; Z. Zhang & Ding, 2020)

DISCUSSION

The results presented here have shown that the publications on nautical sports tourism are quite sparse in terms of authors, source and focus. Such scattered nature suggests that this literature is not mature enough so as to have a collaborative scholar community engaged in studying this field. Nonetheless, we have observed that the academic interest on the topic has consistently displayed an upward trend, especially over the last two years, with 29 articles published in 2019 and 2020.

In fact, these most recent articles have often acknowledged the growing interest on the nautical sports tourism, not only by researchers but also by tourists. For instance Hu and Yang (2020) highlight that the nautical sports tourism is developing rapidly under the dual pursuit of leisure tourism and sports fitness. Likewise, Z. Li and Niu (2020) argue that marine sports tourism has become a new tourism hotspot, which is willing to help tourists experience a new way of life, including experiential, entertaining, and challenging. Moreover, Zhao and Lv (2020) defines nautical sports tourism as a new sunrise industry, as sports tourists deeply love marine sports.

Furthermore, the authors also justify the increasing interest on nautical sports tourism due to the many benefits that this sector might bring. As observed by Marušić et al. (2020), this sector plays a significant role in job creation and growth in the blue economy, and it has substantial qualities for sustainable socio-economic development. Indeed, throughout the literature review, we came across various positive impacts that may result from a well-developed nautical sports tourism initiative such as: enhancement of destination competitiveness (Wang, 2020), improvement of regional brand image (Shao & Sun, 2020), qualification and diversification of the coastal infrastructures (Lam González et al., 2015), fostering social interaction (Pereira et al., 2015), enhancement of environmental concerns (Philippe, 2020), among others.

In addition to positive impacts, some studies have also highlighted the challenges and critical issues for the adequate development of nautical tourism with special interest in sports. For example, Liu et al. (2020) called attention to the need of skilled labor. According to these authors, the human resources of marine sports tourism industry are relatively scarce, and the professional standard of both quality and level, of managers and practitioners, is low, which leads to weak environmental and market awareness. Additionally, other concerns such as the dependence on suitable climate conditions (Shao & Sun, 2020), the possible threats it may inflict to local folks and culture (H. Li et al., 2020), the need to ascertain destinations carrying capacity (Basterretxea-Iribar et al., 2019), and the challenge of organic developing the marine tourism economy and marine ecological environment (Liu et al., 2019) are also mentioned.

In respect of the scopes of the studies analyzed, we observed a concentration of contextual and case studies. Essentially, most of this literature addressed development initiatives and happened or potential impacts of nautical sports tourism in a specific region. As explained by Walsham (1995), any phenomena in the social sciences should be viewed as "tendencies", which are valuable in explanations of past data but are not wholly predictive for future situations. Therefore, despite the current literature unquestionable relevance, there is still a lack of specific theory development and framework propositions in the field of nautical sports tourism.

Regarding the nautical sports activities, although the articles from our sample specifically addressed only 5 types of activities, a big variety of them are mentioned in both Li and Niu (2020)'s categories (maritime sports tourism and sports tourism in the sea). Still, none of the articles analyzed proposed a comparison between these sports nor appointed those that attracted more tourists. In addition, the Weed & and Bull (2004)'s typology of sports tourism is not addressed nor mentioned by this articles.

At last, notwithstanding a growing acknowledgment of the importance of physical, mental and subjective well-being as a motivation to travel, among the articles reviewed, none have specifically approached the nautical sports tourism as a kind of wellness tourism. In fact, as mentioned in the results section, the sample analyzed here has a managerial bias, more dedicated to exploring the economic and environmental effects of this sector. Nonetheless, over the articles, we could identify some of the potential well-being repercussions for tourists and the hosting community, mentioned by the authors as an outcome of nautical sports tourism.

CONCLUSION

This study was developed to identify managerial, governmental, and academic opportunities related to the nautical sports tourism sector, aiming to rebuild and prepare destinations to resume tourist activity sustainably. Our appreciation of the literature, combined with the rising public interest in well-being due to the COVID-19 pandemic, leaves no doubt that the nautical sports tourism is a remarkable field with many development opportunities. The benefits involve improving tourists' physical well-being, tourists' mental and spiritual well-being, destinations' economic well-being, and citizens' well-being.

Thus, we envision that the combination of sports and tourism in marine and coastal environments opens a myriad of possibilities, with many sub segments of activities, for places willing to enhance their competitiveness and public welfare, taking advantage of their distinctive nautical culture. Notably, those destinations with a solid nautical sports culture, including a history of outstanding athletes, well-known competitors and events, will have the chance to lead the trend, if the right investments are made in human resource development and nautical infrastructure. In addition, the concern with the environment and the host community engagement are paramount for developing a successful nautical sports tourism destination.

Our analysis also revealed that there is still a long way for the nautical sports tourism literature research to reach maturity. The studies are contextual, still fragmented and disperse, and there is a lack of models that can be replicated, which prevents comparisons among destinations and practices from being made. We also observed that the existing literature fails to integrate the findings and theoretical perspectives of nautical tourism and sports tourism. Additionally, none of the studies in our sample directly analyzed the consequences of the nautical sports tourism on the subjective well-being of tourists, residents, and workers.

The contributions made in this chapter are not without its limitations. The literature analyzed here encompassed articles, written in English, from academic journals, available in the Scopus and Web of Science databases. Future researches can extend the search of studies addressing nautical sports tourism to other bases and sources, such as books, book chapters and conference proceedings. Moreover, the literature in other languages should also be explored (e.g. French and Spanish) given the tradition of some European countries on nautical activities.

REFERENCES

Anuar, M. A. H., Salleh, N. H. M., & Jeevan, J. (2020). Developing dive site risk assessment model (DSRAM) to enhance tourism safety and sustainability in Perhentian Island. *Journal of Sustainability Science and Management, 15*(1), 125–135.

Basterretxea-Iribar, I., Sotés, I., & Maruri, M. (2019). Managing bathers' capacity at overcrowded beaches: A case on the Spanish North Atlantic coast. *Tourism Management, 71*, 453–465. doi:10.1016/j.tourman.2018.10.016

Bull, A. (2007). Oceans and coastal tourism: Integrating tourism into bioeconomic models. *Tourism in Marine Environments, 4*(1), 29–39. doi:10.3727/154427307784835688

Capocchi, A., Vallone, C., Pierotti, M., & Amaduzzi, A. (2019). Overtourism: A literature review to assess implications and future perspectives. *Sustainability (Switzerland), 11*(12), 3303. Advance online publication. doi:10.3390u11123303

Carvache-Franco, M., Carvache-Franco, W., Carvache-Franco, O., & Hernández-Lara, A. B. (2020). Analysis of Push and Pull Motivations and the Intentions to Return and Recommend a Coastal or Marine Destination. *Journal of Coastal Research, 36*(6), 1313–1322. doi:10.2112/JCOASTRES-D-20-00006.1

Carvache-Franco, M., Carvache-Franco, W., Carvache-Franco, O., Hernández-Lara, A. B., & Buele, C. V. (2020). Segmentation, motivation, and sociodemographic aspects of tourist demand in a coastal marine destination: A case study in Manta (Ecuador). *Current Issues in Tourism, 23*(10), 1234–1247. doi:10.1080/13683500.2019.1600476

Carvache-Franco, W., Carvache-Franco, M., Carvache-Franco, O., & Hernández-Lara, A. B. (2019). Segmentation of foreign tourist demand in a coastal marine destination: The case of Montañita, Ecuador. *Ocean and Coastal Management, 167*(1), 236–244. doi:10.1016/j.ocecoaman.2018.10.035

Carvache-Franco, W., Carvache-Franco, M., & Hernández-Lara, A. B. (2020). From motivation to segmentation in coastal and marine destinations: A study from the Galapagos Islands, Ecuador. *Current Issues in Tourism*, *0*(0), 1–17. doi:10.1080/13683500.2020.1811651

Cater, C. I. (2008). The life aquatic: Scuba diving and the experiential imperative. *Tourism in Marine Environments*, *5*(4), 233–244. doi:10.3727/154427308788714777

Cavalheiro, M. B., Luz, A. B. T., & Mayer, V. F. (2021). *Turismo Náutico em Niterói -Levantamento Documental*. http://turismonautico.uff.br/2021/05/07/relatorio-do-levantamento-documental-sobre-esportes-nauticos-em-niteroi-e-sobre-iniciativas-de-turismo-nautico-no-brasil/

Chapapría, V. E. (2000). El planteamiento de infraestructuras para el turismo náutico. *Cuadernos de Turismo*, *6*, 29–44.

CLIA. (2020). *State Of The Cruise Industry Outlook 2021*. https://cruising.org/en/news-and-research/research/2020/december/state-of-the-cruise-industry-outlook-2021

Cooper, M., & Buckley, R. (2021). Tourist Mental Health Drives Destination Choice, Marketing, and Matching. *Journal of Travel Research*. Advance online publication. doi:10.1177/00472875211011548

Correia, A. B. da C., & Sebastião, S. P. (2017). Os desportos náuticos e o turismo em Portugal. O potencial dos centros de treino de remo. *Revista Investigaciones Turísticas*, *13*(13), 114–139. doi:10.14198/INTURI2017.13.06

Davis, D., & Tisdell, C. (1995). Recreational scuba-diving and carrying capacity in marine protected areas. *Ocean and Coastal Management*, *26*(1), 19–40. doi:10.1016/0964-5691(95)00004-L

De Knop, P. (1990). Sport for All and Active Tourism. *World Leisure & Recreation*, *32*(3), 30–36. doi:10.1080/10261133.1990.10559120

do Nascimento, C. A. X., Meneghatti, M. R., Hsu, P. L., & da Silva, D. (2019). Revisiting Co-creation - Bibliometric Study. *Marketing & Tourism Review*, *4*(1), 1–25.

European Commission. (2017). Commission Staff Working Document on Nautical tourism. In EPSC. doi:10.2495/RAV110541

Freire, K. M. F., Tubino, R. A., Monteiro-Neto, C., Andrade-Tubino, M. F., Belruss, C. G., Tomás, A. R. G., Tutui, S. L. S., Castro, P. M. G., Maruyama, L. S., Catella, A. C., Crepaldi, D. V., Daniel, C. R. A., Machado, M. L., Mendonça, J. T., Moro, P. S., Motta, F. S., Ramires, M., Silva, M. H. C., & Vieira, J. P. (2016). Brazilian recreational fisheries: Current status, challenges and future direction. *Fisheries Management and Ecology*, *23*(3–4), 276–290. doi:10.1111/fme.12171

Geldenhuys, L.-L., Van der Merwe, P., & Slabbert, E. (2014). Who is the scuba diver that visits sodwana bay and why? *S.A. Journal for Research in Sport Physical Education and Recreation*, *36*(2), 91–104. https://www.researchgate.net/publication/274376878_Who_is_the_scuba_diver_who_visits_Sodwana_Bay_and_why

Gibson, H. J. (1998). Sport Tourism: A Critical Analysis of Research. *Sport Management Review*, *1*(1), 45–76. doi:10.1016/S1441-3523(98)70099-3

González, L., Yen, E., González, L., Carmelo, J., & Ledesma, D. L. (2015). Highlights of consumption and satisfaction in nautical tourism. A comparative study of visitors to the Canary Islands and Morocco. *Gestion y Ambiente*, *18*(1), 129–145.

Hu, D., & Yang, Y. (2020). The Development of Marine Sports Tourism Industry Based on Low-carbon Economy. *Journal of Coastal Research*, *112*(sp1), 97–99. doi:10.2112/JCR-SI112-028.1

Hume, T. M., Mulcahy, N., & Mead, S. T. (2019). An Overview of Changing Usage and Management Issues in New Zealand's Surf Zone Environment. *Journal of Coastal Research*, *87*(sp1), 1–12. doi:10.2112/SI87-001.1

Jiménez-García, M., Ruiz-Chico, J., Peña-Sánchez, A. R., & López-Sánchez, J. A. (2020). A bibliometric analysis of sports tourism and sustainability (2002-2019). *Sustainability (Switzerland)*, *12*(7), 1–18. doi:10.3390u12072840

Jones, A., & Navarro, C. (2018). Events and the blue economy: Sailing events as alternative pathways for tourism futures – the case of Malta Andrew. *International Journal of Event and Festival Management*, *9*(2), 204–222. doi:10.1108/IJEFM-09-2017-0055

Kabil, M., Priatmoko, S., Magda, R., & Dávid, L. D. (2021). Blue Economy and Coastal Tourism: A Comprehensive Visualization Bibliometric Analysis. *Sustainability*, *13*(7), 3650. doi:10.3390u13073650

Lam González, Y. E., De León Ledesma, J., & León González, C. J. (2015). European nautical tourists: Exploring destination image perceptions. *Tourism and Hospitality Management*, *21*(1), 33–49. doi:10.20867/thm.21.1.3

Li, H., Wang, G., & Niu, Y. (2020). The Development Strategy of Marine Sports Tourism Based on Cost Theory under the Belt and Road Strategy. *Journal of Coastal Research, 112*(sp1), 112–115. doi:10.2112/JCR-SI112-033.1

Li, Z., & Niu, Y. (2020). Development Strategy of Marine Sports Tourism: A Perspective of Cost Theory. *Journal of Coastal Research, 107*(sp1), 17–20. doi:10.2112/JCR-SI107-005.1

Lillebø, A. I., Teixeira, H., Morgado, M., Martínez-López, J., Marhubi, A., Delacámara, G., Strosser, P., & Nogueira, A. J. A. (2019). Ecosystem-based management planning across aquatic realms at the Ria de Aveiro Natura 2000 territory. *The Science of the Total Environment, 650*, 1898–1912. doi:10.1016/j.scitotenv.2018.09.317 PMID:30286356

Liu, Z., Li, H., & Cao, Q. (2019). The Construction of a Marine Tourism Image Based on Grounded Theory and the Construction of Sports Thinking. *Journal of Coastal Research, 98*(sp1), 267–270. doi:10.2112/SI98-066.1

Liu, Z., Li, H., & Cao, Q. (2020). Grey Prediction and Development Path of the Fusion of Marine Sports and Tourism. *Journal of Coastal Research, 107*(sp1), 153–156. doi:10.2112/JCR-SI107-039.1

Luck, M. (2007, Feb.). Nautical tourism development: opportunities and threats. *Nautical Tourism: Concepts and Issues,* 3–13.

Luković, T. (2012). Nautical Tourism and Its Function in the Economic Development of Europe. In M. Kasimoglu (Ed.), *Visions for Global Tourism Industry - Creating and Sustaining Competitive Strategies*. InTech. http://www.intechopen.com/books/visions-for-global-tourism- industry-creating-and-sustaining-competitive-strategies/nautical-tourism-in-the-function-of-the-economic-development-of-europe

Marušić, E., Šoda, J., & Krčum, M. (2020). The Three-Parameter Classification Model of Seasonal Fluctuations in the Croatian Nautical Port System. *Sustainability, 12*(12), 5079. doi:10.3390u12125079

Mayer, V. F., & Coelho, M. de F. (2021). Interrupted dreams: Memories and emotions of travel experiences during Covid-19 breakthrough. *RBTUR, 15*(1), 1–20.

Moreno, M. J., & Otamendi, F. J. (2017). Fostering Nautical Tourism in the Balearic Islands. *Sustainability, 9*(12), 1–20. doi:10.3390u9122215

OECD. (2020). Rebuilding tourism for the future: COVID-19 policy responses and recovery. In *OECD Economic Outlook* (Vol. 2020, Issue 2). https://www.oecd.org/coronavirus/policy-responses/rebuilding-tourism-for-the-future-covid-19-policy-responses-and-recovery-bced9859/

Pawson, M. G., Glenn, H., & Padda, G. (2008). The definition of marine recreational fishing in Europe. *Marine Policy*, *32*(3), 339–350. doi:10.1016/j.marpol.2007.07.001

Pereira, E. C. S., Mascarenhas, M. V. M., Flores, A. J. G., & Pires, G. M. V. S. (2015). Nautical small-scale sports events portfolio: A strategic leveraging approach. *European Sport Management Quarterly*, *15*(1), 27–47. doi:10.1080/16184742.2015.1007883

Philippe, M. (2020). Social and Associative Sports Tourism in France: The Glénans and the National Union of Outdoor Sports Centres (UCPA). *The International Journal of the History of Sport*, *37*(5–6), 433–450. doi:10.1080/09523367.2020.1754197

Ponting, J., & O'Brien, D. (2014). Liberalizing Nirvana: An analysis of the consequences of common pool resource deregulation for the sustainability of Fiji's surf tourism industry. *Journal of Sustainable Tourism*, *22*(3), 384–402. doi:10.1080/09669582.2013.819879

Radic, A., Law, R., Lück, M., Kang, H., Ariza-Montes, A., Arjona-Fuentes, J. M., & Han, H. (2020). Apocalypse now or overreaction to coronavirus: The global cruise tourism industry crisis. *Sustainability (Switzerland)*, *12*(17), 1–19. doi:10.3390u12176968

Rahman, N. S. F. A., Bazaruddin, N. H., Shah, S. F. S. S. J., Said, M. H., & Othman, M. K. (2020). The sustainability of nautical sports tourism in terengganu using a cost benefit analysis. *Journal of Sustainability Science and Management*, *15*(1), 148–161.

Rendle, E. J., & Rodwell, L. D. (2014). Artificial surf reefs: A preliminary assessment of the potential to enhance a coastal economy. *Marine Policy*, *45*, 349–358. doi:10.1016/j.marpol.2013.09.004

ReportLinker. (2020). *The Global Boat Rental Market is expected to grow from USD 17,620.57 Million in 2019 to USD 23,160.11 Million by the end of 2025 at a Compound Annual Growth Rate (CAGR) of 4.66% Market.* https://www.globenewswire.com/news-release/2020/07/03/2057347/0/en/The-Global-Boat-Rental-Market-is-expected-to-grow-from-USD-17-620-57-Million-in-2019-to-USD-23-160-11-Million-by-the-end-of-2025-at-a-Compound-Annual-Growth-Rate-CAGR-of-4-66.html

Rogerson, C., & Rogerson, J. (2021). COVID-19 and Changing Tourism Demand: Research Review and Policy Implications for South Africa. *African Journal of Hospitality, Tourism and Leisure, 10* (1), 1–21. doi:10.46222/ajhtl.19770720-83

Scheske, C., Arroyo Rodriguez, M., Buttazzoni, J. E., Strong-Cvetich, N., Gelcich, S., Monteferri, B., Rodríguez, L. F., & Ruiz, M. (2019). Surfing and marine conservation: Exploring surf-break protection as IUCN protected area categories and other effective area-based conservation measures. *Aquatic Conservation, 29*(S2), 195–211. doi:10.1002/aqc.3054

Schlüter, R. G. (2003). Metodologia da pesquisa em turismo e hotelaria. *Aleph (Jerusalem).*

Schuhbauer, A., & Koch, V. (2013). Assessment of recreational fishery in the Galapagos Marine Reserve: Failures and opportunities. *Fisheries Research, 144*, 103–110. doi:10.1016/j.fishres.2013.01.012

Seaman, A. N., & Depper, G. L. (2019). Visiting scuttled ships: An examination of the important elements of the wreck diving experience. *Tourism in Marine Environments, 14*(1), 31–44. doi:10.3727/154427319X15567670161919

Shao, Y., & Sun, Y. (2020). SWOT Analysis of Coastal Sports Tourism. *Journal of Coastal Research, 112*(sp1), 103–105. doi:10.2112/JCR-SI112-030.1

Sharma, G. D., Thomas, A., & Paul, J. (2021). Reviving tourism industry post-COVID-19: A resilience-based framework. *Tourism Management Perspectives, 37*, 1–11. doi:10.1016/j.tmp.2020.100786 PMID:33391988

Smith-Godfrey, S. (2016). Defining the blue economy. *Maritime Affairs, 12*(1), 58–64. doi:10.1080/09733159.2016.1175131

Standeven, J. (1998). Sport tourism: Joint marketing - A starting point for beneficial synergies. *Journal of Vacation Marketing, 4*(1), 39–51. doi:10.1177/135676679800400104

Su, L., Tang, B., & Nawijn, J. (2020). Eudaimonic and hedonic well-being pattern changes: Intensity and activity. *Annals of Tourism Research, 84*(July), 103008. doi:10.1016/j.annals.2020.103008

Szylvian, K. M. (2004). Transforming Lake Michigan into the "World's Greatest Fishing Hole": The Environmental Politics of Michigan's Great Lakes Sport Fishing, 1965-1985. *Environmental History, 9*(1), 102–127. doi:10.2307/3985947

Tatsuyuki-Shishido. (1991). Wakayama Marine City project in Wakanoura Bay—Utilization of the coastal zone through construction of artificial island. *Marine Pollution Bulletin*, *23*(C), 389–395. doi:10.1016/0025-326X(91)90706-X

UNCTAD. (2020). *The COVID-19 Pandemic and the Blue Economy: New challenges and prospects for recovery and resilience*. https://unctad.org/system/files/official-document/ditctedinf2020d2_en.pdf

UNWTO. (2021). *Recommendations for the Transition to a Green Travel and Tourism Economy*. UNWTO.

Vázquez, R. M. (2020). Nautical Tourism: A Bibliometric Analysis. *Journal of Spatial and Organizational Dynamics*, *8*(4), 320–330.

Vázquez, R. M., García, J. M., & Valenciano, J. D. P. (2021). Analysis and Trends of Global Research on Nautical, Maritime and Marine Tourism. *Journal of Marine Science and Engineering*, *9*(1), 93. doi:10.3390/jmse9010093

Vieira, E. S., & Gomes, J. A. N. F. (2009). A comparison of Scopus and Web of science for a typical university. *Scientometrics*, *81*(2), 587–600. doi:10.100711192-009-2178-0

Vieira, J., Ardigó, C. M., & Behling, H. P. (2018). Impactos da Volvo Ocean Race – Itajaí Stopover : análise pós- evento da percepção dos residentes da cidade de Itajaí (SC) Impacts of Volvo Ocean Race - Itajaí Stopover : a post-event analysis of the perception of the residents of the city of Itajaí (SC. *Revista Brasileira de Pesquisa em Turismo*, *12*(3), 172–196. doi:10.7784/rbtur.v12i3.1437

Waldo, S., & Paulrud, A. (2012). Obstacles to Developing Recreational Fishing Enterprises in Sweden. *Scandinavian Journal of Hospitality and Tourism*, *12*(2), 121–139. doi:10.1080/15022250.2011.633254

Walsham, G. (1995). Interpretive case studies in IS research: Nature and method. *European Journal of Information Systems*, *4*(2), 74–81. doi:10.1057/ejis.1995.9

Wang, H. (2020). The Development of Marine Sports Tourism in the Context of the Experience Economy. *Journal of Coastal Research*, *112*(sp1), 84–86. doi:10.2112/JCR-SI112-024.1

Weed, M. (2005). Sports Tourism Theory and Method—Concepts, Issues and Epistemologies. *European Sport Management Quarterly*, *5*(3), 229–242. doi:10.1080/16184740500190587

Weed, M. (2009). Progress in sports tourism research? A meta-review and exploration of futures. *Tourism Management*, *30*(5), 615–628. doi:10.1016/j.tourman.2009.02.002

Weed, M. (2020). The role of the interface of sport and tourism in the response to the COVID-19 pandemic. *Journal of Sport & Tourism*, *24*(2), 79–92. doi:10.1080/14775085.2020.1794351

Weed, M., & Bull, C. (2004). *Sports tourism: Participants, policy and providers*. Elsevier Butterworth-Heinemann., doi:10.4324/9780080474441

Wen, J., Kozak, M., Yang, S., & Liu, F. (2020). COVID-19: Potential effects on Chinese citizens' lifestyle and travel. *Tourism Review*, *76*(1), 74–87. doi:10.1108/TR-03-2020-0110

WHO. (2020). *WHO Director-General's opening remarks at the media briefing on COVID-19 - 11 March 2020*. World Health Organisation. https://www.who.int/director-general/speeches/detail/who-director-general-s-opening-remarks-at-the-media-briefing-on-covid-19---11-march-2020

WHO. (n.d.). *Coronavirus disease (COVID-19)*. World Health Organisation.

Wood, A. L., Butler, J. R. A., Sheaves, M., & Wani, J. (2013). Sport fisheries: Opportunities and challenges for diversifying coastal livelihoods in the Pacific. *Marine Policy*, *42*, 305–314. doi:10.1016/j.marpol.2013.03.005

Wyman, O. (2020). *To Recovery & Beyond: The Future of Travel & Tourism in the Wake of COVID-19*. https://wttc.org/Research/To-Recovery-Beyond

Yamamura, E., & Tsutsui, Y. (2020). The impact of postponing 2020 tokyo olympics on the happiness of O-MO-TE-NA-SHI Workers in tourism: A consequence of COVID-19. *Sustainability (Switzerland)*, *12*(19), 1–16. doi:10.3390u12198168

Yang, Y., Gong, Y., & Li, B. (2020). The Development of Marine Sports Tourism Based on Experience Economy. *Journal of Coastal Research*, *112*(sp1), 106–108. doi:10.2112/JCR-SI112-031.1

Yu, Y. (2020). Research on Current Situation and Development of Marine Leisure Sports in Shandong Peninsula. *Journal of Coastal Research*, *115*(sp1), 123–126. doi:10.2112/JCR-SI115-037.1

Zhang, J. (2020). The Development of Marine Sports Tourism Resources and Product Portfolio under the Background of the Belt and Road. *Journal of Coastal Research*, *112*(sp1), 81–83. doi:10.2112/JCR-SI112-023.1

Zhang, R. (2020). Research on the Development Strategy of China's Marine Sports Tourism from the Perspective of Experience Economy. *Journal of Coastal Research*, *115*(sp1), 75–77. doi:10.2112/JCR-SI115-022.1

Zhang, Z., & Ding, Z. (2020). The systematic spatial characteristics of sports tourism destinations in the core area of marine economy. *Journal of Coastal Research, 112*(sp1), 109–111. doi:10.2112/JCR-SI112-032.1

Zhao, P., & Lv, J. (2020). An Empirical Study on the Development of Marine Sports Tourism in Northern China. *Journal of Coastal Research, 106*(sp1), 105–108. doi:10.2112/SI106-027.1

ENDNOTES

[1] Project: Mapping and promoting nautical tourism in Niteroi. More information about the project can be found at http://turismonautico.uff.br/. Assessed in May 10th, 2021.

[2] More information about the project can be found at http://aganplus.org/. Assessed in May 10th, 2021.

Chapter 8
Pilgrimage Routes to Happiness:
Comparing the Camino de Santiago and Via Francigena

Sabrina Latusi
https://orcid.org/0000-0002-3937-1883
University of Parma, Italy

Massimiliano Fissore
University of Parma, Italy

ABSTRACT

The chapter investigates the topic of happiness and quality of life in the slow-tourism context, with specific reference to the pilgrimage tourism phenomenon. The Camino de Santiago and the Via Francigena are analysed through explorative research with the aim of understanding the effects of pilgrimage on individuals' levels of well-being and its implications for peoples' lives. The results reveal several differences in motivations and benefits obtained according to the pilgrimage road undertaken, with a more existential connotation tied to personal happiness in the Camino de Santiago pilgrim group and a more experiential connotation tied to wellness in the Via Francigena pilgrim group. These findings provide new insights into contemporary pilgrimages in a comparative perspective.

DOI: 10.4018/978-1-7998-7239-9.ch008

Copyright © 2021, IGI Global. Copying or distributing in print or electronic forms without written permission of IGI Global is prohibited.

INTRODUCTION

In recent years, quality of life and well-being have become important areas of investigation in tourism research (Jepson et al., 2019). Indeed, there has been a shift in the value of tourism towards more subjective elements, such as sustainability, wellness, well-being, and quality of life (Perdue et al., 2010). In particular, sustainability and personal/social well-being have become driving forces underpinning new forms of tourism, including slow tourism (Moore, 2012; Oh et al., 2016). Slow tourism is an umbrella term encompassing various tourism types (Serdane, 2020), such as responsible tourism (Timms & Conway, 2012) and ethical tourism (Clancy, 2015). Even pilgrimage tourism can be traced back to slow tourism (e.g. Kato & Progano, 2017; Lemmi, 2020; Olsen & Wilkinson, 2016), and this connection is promoting renewed interest in a phenomenon traditionally analysed in the stream of religious tourism (e.g. Raj et al., 2015; Terzidou et al., 2018), heritage tourism (e.g. Timothy & Boyd, 2006), or cultural tourism (e.g. Rinschede, 1992). As part of the slow-tourism model, pilgrimage tourism "offers a more sustainable tourism product that is less alienated (and alienating), more culturally sensitive, authentic and a better-paced experience for hosts and tourists alike" (Timms & Conway, 2012, p. 398). In essence, pilgrimage tourism can be considered part of a humanistic philosophy, fostering and developing the well-being of tourism destinations' stakeholders (Latusi & Fissore, 2021).

Although pilgrimage was the first form of tourism to come into existence hundreds of years ago (Timothy & Olsen, 2006), nowadays it can be considered as a relatively new phenomenon, given the changes experienced over the past decades and its ongoing process of transformation (Collins-Kreiner, 2020a). The modern idea of pilgrimage encompasses both religious and secular journeys, and since the 2000s researchers have discussed this holistic phenomenon in the context of spiritual rather than religious motivations and actions (Collins-Kreiner, 2016, 2020a, 2020b). The recent literature has pointed out that pilgrims are engaged in a quest for a variety of experiences, including enlightenment, knowledge, improved spiritual and physical well-being, challenge, and social bonds (e.g. Collins-Kreiner, 2020a, 2020b; Kim & Kim, 2019; Liutikas, 2015). Therefore, pilgrimage can be defined as a journey of becoming, a mobility for the search for deeper or higher meaning that includes an element of existential transformation and self-discovery (Collins-Kreiner, 2020a; Nilsson & Tesfahuney, 2016).

Despite these achievements, there is scarce research addressing the topic of pilgrims' travel motivations and inner experiences (Amaro et al., 2018; Blackwell, 2014; Damari & Mansfeld, 2016). The few studies that exist tend to consider a specific route (mainly the Camino de Santiago) and pilgrims as a homogenous group (Amaro et al., 2018) rather than comparing different experiences, individuals, and contexts. Above all, the relationship between pilgrimage tourism, happiness, and quality of life is still a matter of investigation. The only work that attempts to grasp this relationship is a study by Kim et al. (2016) that specifically refers to the values exhibited by travellers along the Camino de Santiago pilgrimage route.

The purpose of the present chapter is to fill this gap. Specifically, it explores the effects of pilgrimage on individuals' levels of well-being and its implications for peoples' lives. The following central research questions are posed:

- What are the main subjective motivations to undertake a pilgrimage?
- What does the pilgrimage experience provide to pilgrims in terms of emotions and personal values?

Pilgrims' inner driving forces and pilgrims' emotions and acquired personal values are explored for different pilgrimage routes (specifically, the Camino de Santiago and the Via Francigena) and travelling companions (the relational aspect), following a phenomenological approach. The results contribute to the tourism literature by providing new insights regarding the contemporary pilgrim from the perspective of subjective motivations and inner experiences driving to different levels of happiness and well-being. The study also responds to Amaro et al.'s (2018) call for research on pilgrimage on routes other than the Camino de Santiago. In particular, it sheds light on how motivations for pilgrimage and pilgrims' inner experiences can vary from one route to another. The findings also provide insights into the religion–tourism nexus, a research field that calls for a comparative approach (Collins-Kreiner, 2020b).

Understanding motivations for pilgrimage and pilgrims' inner experiences is of paramount importance to meet the needs of pilgrims during their journey and stay, contributing to their subjective well-being and happiness. Therefore, the findings of this study provide useful prospective insights to several stakeholders (e.g. destination management organizations, accommodation owners, tour operators, etc.) to improve their value proposition. The results also identify alternatives for improvement of the Camino de Santiago and the Via Francigena, complementing the management-side analysis by Lucarno (2016).

The chapter is organised as follows. First, an introductory theoretical discussion is offered regarding the main topics under consideration. Next, the methodology is presented, including the research context of the Camino de Santiago and the Via Francigena. The opinions of pilgrims on journey motivations and inner achievements are then analysed. The chapter concludes by underlining the contribution of the main empirical findings to existing knowledge and summarising emerging trends.

BACKGROUND

In the tourism and destination management fields, well-being and quality of life have become important areas of investigation (Jepson et al., 2019). Research has shown that tourists' experiences and activities contribute to positive affect in a range of life domains, having significant effects on both tourists' overall life satisfaction and the well-being of residents of host communities (e.g. Uysal et al., 2016).

Studies devoted to investigating tourists' subjective well-being have shown the influence exerted by accommodations and physical structures (Ali & Amin, 2014; Ali et al., 2018), the length of the trip (De Bloom et al., 2013), the contact with nature and the type of natural environment (Farber & Hall, 2007; Kim et al., 2015). These latter aspects mainly influence individual's emotions and psychological state (Hadavi, 2017), emphasizing the increasing centrality of the outdoor activities. Indeed, according to the psychological studies of Csikszentmihalyi (1988, 1990), people become happier and more satisfied by participating in high-engagement tasks, either intellectual or physical, characterized by an instant perception of goals and feedback. This produces the 'flow', an experience that is so gratifying that people are fully absorbed in the moment.

A great contribution in attaining a better understanding of the variations in tourists' well-being comes from PERMA theory (Seligman, 2011). According to the theory, five elements may account for variations in well-being experienced by individuals: positive emotion, engagement, relationships, meaning, and accomplishment. Therefore, people flourishing is a combination of positive emotions, such as happiness, personal commitment in activities, good relationships, finding meaningful purposes and being satisfied by reaching personal targets. In the field of tourism, the theory proved its relevance in explaining the variations in well-being during a trip. Specifically, the study by Mayer et al. (2019) revealed that tourists' well-being can be promoted by: moments of engagement during the trip, prompted by access to new knowledge and experiencing physical recreation and amusements; positive relationships; the sense of meaning and purpose of the trip activities; the sense of personal accomplishment achieved through the trip; physical structures and natural environments, such as sites that offer aesthetic value or comfortable conditions that generate well-being, or through exposure to weather that tourists consider to be pleasant or even contact with nature in general. Therefore, according to Mayer et al., positive feelings (joy, satisfaction, excitement, etc.) can be stimulated by activities and events that promote subjective well-being (engagement, positive relationships, meaning, physical and natural environments, and accomplishment) and many of these activities and factors can be traced in new forms of tourism.

In particular, personal/social well-being and sustainability have become driving forces underpinning slow tourism (Moore, 2012; Oh et al., 2016). The search for slowness in a world with a heavily stressed daily life supports this particular concept of tourism, that promotes quality of experiences and environmental consciousness. Slow tourism is based on a concept of "tourist quality" (Ozdemir & Celebi, 2018) in opposition to the mainstream mass tourism and its negative ecological impact. While motivation linked to sustainability and environment seems to be a good point in choosing slow tourism, it is the experiential benefits dimension that mainly drives people to slow tourism (Caffyn, 2012; Dickinson et al., 2010; Lumsdon & McGrath, 2011, McGrath & Sharpley, 2016).

The slow-tourism phenomenon has its roots in the sociocultural context of slow movement (Fullagar et al., 2012), with its call for a re-evaluation of life priorities (Heitmann et al., 2011) that promotes simplicity, mindfulness, and embodied experiences to facilitate renewal, revitalization, and self-enrichment (Howard, 2012). Slow tourism is linked to the concept of being within a place and live it with a deeper sense of experience (Germann Molz, 2009). It celebrates the diversity of places (Guiver & McGrath, 2017) and promotes real and meaningful connections between the tourist and the place (Caffyn, 2012) in the attempt to make every tourist part of the local community (Robinson & Cho, 2012). At least for the time of the holiday, slow tourists and host communities try to share their identities and paths (Guiver & McGrath, 2017). Engaging with local people; preserving cultural traditions and natural resources; making meaningful connections with heritage, places, and environment; and enjoying the journey as much as the destinations are priorities that are embedded in the slow-tourism phenomenon (e.g. Gibson et al., 2012). In essence, slow tourism embraces and amplifies people flourishing and well-being along with the quality of the experiences (Mayer et al., 2006; Heitmann et al., 2011; Ekinci, 2014).

Pilgrimage tourism is then part of the slow-tourism model (e.g. Lemmi, 2020; Olsen & Wilkinson, 2016). Walking is central to the meaning of a pilgrimage journey and "the associated slowness facilitates more locally engaged, mindful and lower-impact tourism, and allows for more attention to people and place" (Kato & Progano, 2017, p. 246). At the same time, walking promotes introspection, and its pace, linked to the enjoyment of time, translates in a both physical and internal slow mobility (Lopez et al., 2017).

Despite the emerging view that pilgrim movements may have favourable impacts on subjective well-being and on societies contributing to the development of sustainable forms of tourism, pilgrimage has been traditionally analysed in the research stream of religious tourism (e.g. Raj et al., 2015; Terzidou et al., 2018), heritage tourism (e.g. Timothy & Boyd, 2006), or cultural tourism (e.g. Rinschede, 1992). There is consensus that traditional pilgrimage was conducted with a strong religious motivation (Shinde, 2007). It is one of the oldest forms of expressing worship and penance for sins, of seeking care in the face of disease and problems, and of pursuing religious enlightenment (Digance, 2003; Jackowski & Smith, 1992; Smith, 1992; Vukonic, 1996). Indeed, traditional pilgrimage was defined as "a journey resulting from religious causes, externally to a holy site, and internally for spiritual purposes and internal understanding" (Barber, 1993, p. 1).

However, more recent studies have shown that the contemporary pilgrim is not necessarily motivated by religion (Stefko et al., 2015). For instance, Oviedo et al. (2014) found that spiritual growth, sensations-seeking, and seeking life direction were more important than religious motivations. In a recent study, Amaro et al. (2018) found that pilgrims who had walked the Camino de Santiago were mostly motivated by spiritual aspects, the search for new life experiences, and the opportunity to be immersed in a natural context. These findings support Reader's (2007) argument that the growth in the number of pilgrims is due to a spiritual revolution rather than the revival of religious sentiments: pilgrims take the journey as a personal quest for a change or a better situation in life (Blom et al., 2016), for reflection and guidance (Nilsson, 2018), or for internal or mental renewal (Lopez et al., 2017). It comes to light the need to break up with the pace of daily routines and to appreciate the temporal extension typical of the slow tourism perspective, with the right of really savouring each activity (Germann Molz, 2009; Lumsdon & McGrath, 2011). These arguments echoe with the notion of flow and the theory of optimal experience (Csikszentmihalyi, 1988, 1990).

The recent literature has thus discussed pilgrimage as a holistic phenomenon with religious and secular foundations (Collins-Kreiner, 2016), pointing out that pilgrims are engaged in a quest for a variety of experiences, including enlightenment, knowledge, personal advancement, improved spiritual and physical well-being, challenge, and social relationships (e.g. Collins-Kreiner, 2020a, 2020b; Kim & Kim, 2019; Liutikas, 2015). In essence, contemporary pilgrimage carries a much wider connotation beyond religion, including health, wellness, and self-improvement (Kato & Progano, 2017). This connotation has translated to a renewed definition of pilgrimage as "a mobility for the search for meaning that contains an element of transformation that is often deep and enduring" (Collins-Kreiner, 2020a, p. 145).

Despite these findings, there is scarce research addressing the topic of pilgrims' travel motivations (Blackwell, 2014; Amaro et al., 2018). The review undertaken by Amaro et al. (2018) about studies addressing pilgrims' motivations to undertake a pilgrimage pointed out that the majority of these studies concentrated on pilgrims who had walked the Camino de Santiago and considered pilgrims as a homogenous group. Collins-Kreiner (2020b), too, has highlighted the lack of comparative research in her comprehensive review in the field of religion and tourism.

Alongside motivations, understanding the experiences of pilgrims through their perceptions is a vital focus for the comprehension of the pilgrimage phenomenon in contemporary settings (Kim et al., 2019). Nevertheless, the experiential aspects of modern pilgrimage have been largely overlooked (Damari & Mansfeld, 2016). As a consequence, the emotional dimension—an essential concept in the context of pilgrimage because emotions relate to the affective process of experience (Noy, 2004)—has been neglected. Furthermore, the inner benefits and personal values that pilgrims perceive to acquire and fulfil through the pilgrimage experience remain unexplored topics, except for the work by Kim et al. (2016). The authors examined the values acquired by pilgrims through undertaking the Camino de Santiago pilgrimage route and highlighted that progressing through the pilgrimage journey benefits personal values, the most significant of which was personal happiness, followed by social bonds and cross-cultural understanding. Fun and enjoyment, together with self-satisfaction and achievement, did not emerge as salient values that pilgrims acquire through the experience of walking the Camino. Their findings are consistent with previous research that suggests that happiness could be achieved by experiencing refreshing conditions for the body, mind, and spirit (Gesler, 1996); health and wellness benefits (Devereux & Carnegie, 2006; Courtney, 2013); and various emotions, such as solidarity, reflection, and religious faith (Lopez, 2013).

Despite the contribution made by Kim et al. (2016), the inner impacts of pilgrimage experiences in terms of fulfilment of multi-dimensional personal values are still a matter of investigation and, since experiences may vary on different occasions (Kim et al., 2019), deserve to be addressed with reference to other routes. With the aim of capturing the relationship between pilgrimage tourism, happiness, and quality of life, the present study further explores pilgrims' inner driving forces and acquired personal values for different pilgrimage routes. Specifically, the two most traditional historic pilgrimage routes in Europe (i.e. the Camino de Santiago and the Via Francigena) are considered, as detailed in the following section.

PILGRIMS' INNER DRIVING FORCES AND ACQUIRED PERSONAL VALUES: COMPARING THE CAMINO DE SANTIAGO AND VIA FRANCIGENA

The Research Context

And it is worth noting that there are three separate terms for people who travel to honour the Supreme Being: they are called palmers if they travel to the Holy Land, where they often carry the palm; they are called pilgrims if they travel to the home of Galicia, since the tomb of Saint James was farther from his homeland than that of any other apostle; they are called romers if they travel to Rome—the place where those I am calling pilgrims were headed.

With these words, Dante Alighieri (1265–1321), in his *Vita Nova* (1896/1294), wrote about the three most important pilgrimage routes in the world—the travel to Jerusalem, the Camino de Santiago, and the Via Francigena—showing that, already in the Middle Ages, these paths, known as *peregrinationes maiores*, were established, renowned, and one of the main motivations of movements of people in Europe and the Mediterranean lands.

El Camino de Santiago, known in English as the Way of Saint James, is a religious and spiritual itinerary composed of a series of routes starting from different parts of Europe. The most well-known and travelled is Camino Francés, running from Saint-Jean-Pied-de-Port to Roncesvalles and then to Santiago de Compostela, the pilgrims' destination where the cathedral holds the remains of Saint James the Great. This route, leading to the north-west of Spain, in Galicia, became an international attraction at the beginning of the 10th century, with its apogee during the Middle Ages. Later, it experienced several fluctuating periods, until 1985, when UNESCO declared the city of Santiago de Compostela a World Heritage site, and the landmark was extended to the whole pilgrimage route in 1993 (Amaro et al., 2018). Today's Camino has become a polysemic route in which religious purposes are flanked by cultural, spiritual, sport, and touristic dimensions. The final destination is not always as important as the walking experience undertaken by thousands of pilgrims every year. These travellers come from around the world, with high proportions of Spanish and Italian people (Correia et al., 2017; Oficina del Peregrino, 2019; Santos, 2002).

The Via Francigena, or Via Romea Francigena, as Dante Alighieri referred to it, begins at the English Canterbury Cathedral and ends in the Eternal City, siege of the Papacy and Christianity: Rome. However, it continues beyond Rome, up to Apulia, where centuries ago crusaders and pilgrims embarked for the Holy Land (D'Atti & Cinti, 2014). This pilgrimage route, crossing Europe from the north to the south, was created in the ninth century as a route for those wishing to visit the Holy See and the tombs of the apostles, Peter and Paul. As the Way of Saint James, it also includes a number of variants, but the most well-known and travelled is the Sigeric's Itinerary, inheriting its name from the Archbishop of Canterbury who went to Rome in 990 AD to receive from the Pope the pallium, a symbol of his jurisdictional authority.

The Italian Via Francigena is approximately 1,000 km, starting in Italy from the Great Saint Bernard Pass, in the Pennine Alps, and crossing through Piedmont, the Po Valley, Apennines, and Tuscany, and then heading to Rome in Lazio. Like the Camino de Santiago, the Via Francigena knew golden times during the Middle Ages, followed by a decline in popularity in later centuries when religious atonement lost its importance. Today's Via Francigena was rediscovered during the Jubilee in 2000, when the route was re-evaluated in its significance and attractiveness, focusing especially on artistic and architectonic heritage and landscapes (Lucarno, 2016).

In fact, beyond its original purpose and according to the religious tourism vocation of developing accommodation facilities and maintaining the memory of religious tradition (de Souza et al., 2018), this route has been reconsidered to preserve its highly symbolic cultural value, which leads to the preservation and promotions of territorial development and local traditions (Fistola & La Rocca, 2018).

Both pilgrimage routes show annual growth—more for the Way of Saint James than for the Via Francigena (Kim et al., 2016; Gazzola et al., 2020)—despite a marked variation in pilgrims' purposes. These itineraries, historically attended mainly on foot for penitential reasons, are now integrated in slow-tourism and well-being trend proposals. While the predecessors undertook such travels to feel physical fatigue and deprivation before achieving the joy of redemption and absolution from their sins, most modern pilgrims seem uninterested in ancient and precise rules and utilise alternative modes of travel, such as bicycles, that are offered to those who opt for immersive experiences in spiritual routes enjoyed from other perspectives.

The Camino de Santiago and the Via Francigena have in common a historic, cultural, and religious importance. The two pilgrimage routes share an ancient origin as religious itineraries and a strong cultural and heritage imprint. Both received formal and institutional recognition from the Council of Europe as European Cultural Routes, and both are linear routes: they originated to connect territories and cross landscapes, and they have nodes that contain historic infrastructures to welcome travellers, for example, guesthouses and hospitals. Both highlight the material wealth of cultural heritage and the intangible items represented by the tradition of pilgrimage. Due to these similarities, the two pilgrimage routes were selected as the subjects of the present investigation. These similarities allow to shed light on how motivations for pilgrimage and pilgrims' inner experiences can vary from one route to another.

Research Design, Data Collection, and Analysis

The research followed a phenomenological approach, the most appropriate to understand the complex motives for contemporary pilgrimage tourism and pilgrims' inner experiences (Nilsson & Tesfahuney, 2016). Thirty in-depth, semi-structured interviews with pilgrims were conducted face-to-face during July and August 2020 in separate sessions lasting 30 minutes each, on average. The selection of interviewees considered a balance among genders, ages, geographical provenances, and presence or absence of travelling companions. An adjunctive selection criterion was the singularity and exclusivity of the pilgrimage route undertaken; this was done to avoid contamination and confusion between the narratives and experiences of different journeys. Therefore, respondents were recruited among people who had walked just one of the two pilgrimage routes: 15 travellers on the Camino de Santiago and 15 on the Via Francigena. Participants were informed about the research privacy policies and the confidentiality of their personal information and consented to participate.

The researchers proceeded by proposing the same set of open-ended questions to everyone. The interviews were digitally recorded and then transcribed with a view to conducting two text analyses: the first entailed the thematic analysis of data with a systematic read and reread to sum up core themes (Boyatzis, 1998; Fereday & Muir-Cochrane, 2006), followed by a rationalization of key elements; the second was conducted with T-Lab software, employing statistical and lexical methods to highlight the relevance and frequency of words and topics. T-Lab software was used to highlight both the frequency of lemmas used by pilgrims and the associations among them in relation to keywords concerning the main themes of motivations and inner experiences. Some grouping with internal homogeneity and external heterogeneity became necessary to avoid fragmentation resulting from the use of synonyms (Lancia, 2012).

For each topic, two radial diagrams (one for each route) provided the visual representation of the degree of association among words considered significant by the software, applying an adjunctive tool of the Chi-square test ($p < .05$) to consider only statistically strong associations. The distance between keywords and other lemmas is proportional to the strength of association: the shorter the distance, the greater the correlation.

Findings

To answer the research questions, the analysis was carried out in three sections dealing with: the general motivations for undertaking a pilgrimage today, the ex-ante specific motives for undertaking the Camino de Santiago/Via Francigena route, and the ex-post inner experience of the pilgrims. The first two were used to answer the first research sub-question because they are focused on the inner driving forces inducing people to become pilgrims and to undertake the chosen pilgrimage route; the last fulfilled the second research sub-question highlighting the inheritance dimension brought home by pilgrims in terms of satisfaction, happiness, and well-being.

With reference to the general motivations for contemporary pilgrimage tourism, converging elements were found. As shown in Figure 1, both groups of respondents stressed topics related to time. In particular, lemmas, such as *time*, *moments*, and *years*, highlight two temporal aspects (lemmas are in italics). The first is the meditation theme: people externalized their intention to become pilgrims as something pre-meditated, or thought about (*to think*), that lasts for years. The decision to undertake a pilgrimage is something that takes time, a reflection that requires a longer decision-making process prior to engaging in this form of tourism; it is not an instinctive or spur-of-the-moment decision. The second temporal aspect is people's desire to pursue this type of tourism in order to take a breath, to depart from their normal, everyday routine. These elements emerged strongly for the Santiago pilgrims group but less so for the Francigena group. In fact, lemmas associated with the key phrase "pilgrimage motivations" are almost opposite for the two groups of pilgrims: Camino de Santiago pilgrims tend to consider pilgrimage as an element of disruption from the daily life (*to leave*), with a strong focus on internal elements (*spiritual*, *experience*), while Via Francigena pilgrims think about pilgrimage as something *to discover, to walk*, to perform with others (*we*) and for meeting *people*.

Figure 1. Motivations for contemporary pilgrimage tourism
Source: Author's own elaboration

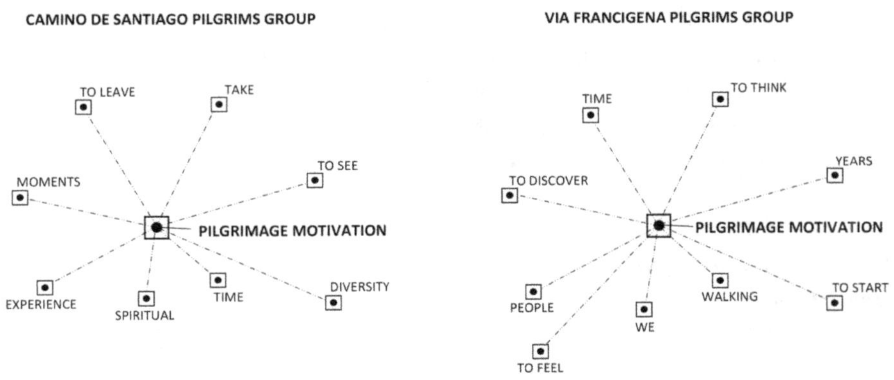

For respondents in the Camino de Santiago group, pilgrimage is a spiritual experience, an introspection conducted by people who wish to analyse the world through new perspectives: it is about *to leave* and *to see*, where the former may be preparatory for the latter. Numerous interviewees saw pilgrimage as a way to escape from a frustrating daily life. The following comments exemplify this:

For some time, I was thinking my life didn't . . . that it was not my way! My girlfriend broke up with me and sports didn't give me the adrenaline of yore. (RCS1)

Why do people become pilgrims? I don't know, it happens and stops. . . . [T]o rediscover yourself, to leave delusions behind, sometimes for hope that the future can be better. . . . I suppose that the real motivation is the exigence of inner freedom. (RCS12)

For interviewees in the Via Francigena group, contemporary pilgrimage tourism seemed to be much more oriented towards a natural but not particularly challenging adventure. They used lemmas such as *discover*, *begin*, and *walking*, in adjunction to a relational factor highlighted by *people* and *we*. A particularly eloquent comment on these two aspects was reported by one respondent:

I believe it's very important and good, the contact with nature. I also spoke about that with an environmental association, but also for meeting people. Because, often, people you meet are really beautiful inside, they tell you incredible stories. (RVF7)

To deepen the understanding of pilgrims' inner driving forces, respondents were asked to clarify the specific motives for undertaking the pilgrimage route they chose. The text analysis results highlight a strong incidence of words like *people*, *culture*, *beauty*, and *landscape*. Taking "Camino de Santiago/Via Francigena reasons" as the keyword linking all the other lemmas related to this main focus of analysis, the radial diagrams shown in Figure 2 illustrate another interesting difference between the two pilgrim groups in terms of conception and motivation.

Figure 2. Motivations for undertaking the Camino de Santiago/Via Francigena pilgrimage route
Source: Author's own elaboration

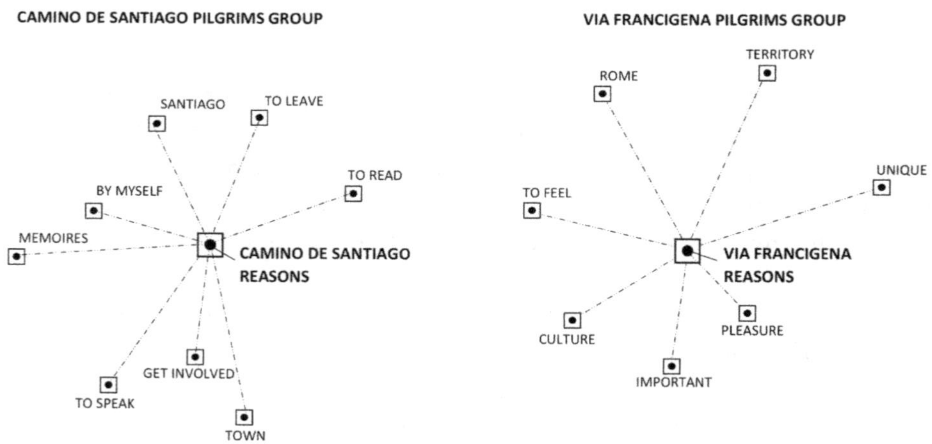

For pilgrims who decided to follow the Camino de Santiago, the mental aspect again emerged: words like *by myself* (or *alone*), *memories*, and *to leave* emphasise the urge for finding time for themselves. *Santiago*, *Town*, and *get involved* describe the final goals of the pilgrimage journey: being fully involved in the experience of walking the Camino to the final destination of Santiago de Compostela. *To speak* and *to read* are two curious reasons that were shared: they refer to the decisional phase. Many pilgrims were attracted to the Way of Saint James by speaking with other people who had travelled this road, and by reading about this path either online or in books. A specific book cited was *The Pilgrimage* (in Portuguese, *O Diário de Um Mago*) by Paulo Coelho. It was considered very inspiring by some interviewees who mentioned it, for example:

It was something I've always wanted to do. When I was 18, I received Paulo Coelho's book, The Pilgrimage. By reading it, the wish to undertake the Camino de Santiago arose in me. But after that, I graduated, I started working, and I got engaged. Years passed and effectively I never did this walk, but the Camino still was calling to me, from Paulo Coelho's book where he describes it as a magical path. Then, last year, in September, I planned to take a holiday with my friends, but in July, while I was cleaning the bookshelf, suddenly, the book falls into my hands. For months I felt bad and then I told myself that maybe it was the moment to undertake the Camino de Santiago! (RCS2)

Pilgrims who decided to undertake the Via Francigena seemed to emphasise much more the environmental and cultural elements of the trip. Explaining the reasons for their choice, respondents mentioned *Rome*, *territory*, *culture*, *unique*, *pleasure*, *to feel*, and *important*. The inner driving forces seem to be unrelated to the internal need to break routine and the frenetic pace of life, and are instead connected to cultural and leisure elements. A true slow-tourism conception, pilgrims looked for something that merged cultural aspects—highlighted by the numerous references to churches, museums, and castles cited in the corpus analysed—and environmental ones. People also sought unique places and landscapes, highlighting the aesthetical attractions rather than the spiritual ones. They seem to have been looking for a feeling of peace more through the enjoyment of landscapes and cultural atmospheres, than through a mental or spiritual challenge, as were respondents in the Camino de Santiago group. They also looked for food, an element that does not emerge from the radials but was sometimes cited by pilgrims, showing how the Italian pilgrimage route is seen more as a cultural route rather than a religious or spiritual one. The motivations that pushed people to undertake the Via Francigena are therefore much more linked to the slow-tourism concept in which the focus is on *pleasure*, something more relaxing rather than taxing.

This orientation is likely due to two main reasons. The first is Italy itself. Italy is renowned for its significant cultural heritage, delicious food, and breath-taking landscapes. This may have become a driving force, aided also by the Italian geography that naturally converges at the heart of the Mediterranean Sea, showing the truth in the proverb, "All routes lead to Rome", with the Via Francigena being the most famous one. The second possible explanation is related to promotion and mental associations. The Camino de Santiago is much better known around the world for pilgrimages, with an organized host system that has been built up over time. On the contrary, despite its ancient roots, the Via Francigena has only been rediscovered in the last few decades and still needs to find both a distinctive identity and a hosting system in which pilgrims are treated as such, and not as normal tourists.

The last analysis conducted on the interviews focused on the inner impacts of the pilgrimage experience. It embraces the "ex-post" feelings to examine more deeply pilgrims' emotions and discover which positive values they gained. Using the specific pilgrimage route undertaken by respondents as a keyword, Figure 3 illustrates the main strong connections, presenting a discussion of the pilgrims' inner experiences.

Figure 3. Pilgrims' inner experiences for the Camino de Santiago/Via Francigena pilgrimage route
Source: Author's own elaboration

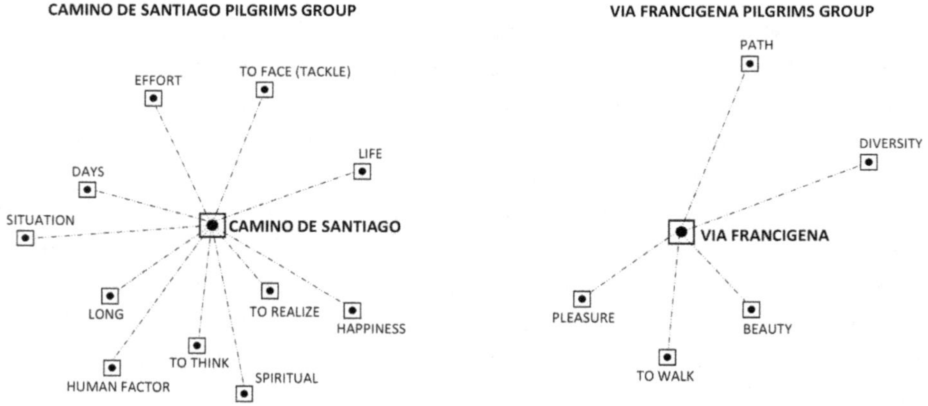

The differences between the two routes immediately catch the eye. In the Camino de Santiago group there is a series of words that emphasise length and physical task, such as *to face* (or *tackle*), *days*, *long*, and *situation*. These terms highlight the classical conception of pilgrimage, in which people undertake the journey to purify themselves in mind and body. Though in past centuries this purification was oriented more towards a religious or Catholic perspective, today it tends to have a wider framework in terms of spiritual and meditative dimensions. This is confirmed by another series of linked lemmas: *to think*, *life*, and *spiritual*. People still walk the ancient roads of the Way of Saint James with the same intimate desire of leaving behind the grudges and sufferings of their predecessors despite changing times and secularism. They walk and grow mentally to become new people. They are aware of such an inner transformation, as is grasped in the *happiness* and *to realize* lemmas. Many pilgrims realise how such experiences allow them to see everything from a different point of view, with reference not only to the dimension of the self, but also to the relational aspects, as the phrase *human factor* proposes. People undertaking the Way of Saint James try to find a balance between time for themselves and socialization, considering the importance of spontaneous and genuine social relations. Two respondents underlined these elements as the base of their happiness and experienced values:

During the Camino de Santiago, I've also seen a strong willingness towards solidarity, and this made me happy: it gave me hope. (RCS9)

In Roncesvalles I knew my travelling companions, and there I learnt my first big lesson taught by the Camino de Santiago: happiness is real only if shared. I realize that facing a pilgrimage alone, on foot, is wonderful, but at some point you need to share joy and effort with someone else. (RCS7)

The Via Francigena shows different patterns. Lemmas associated with analysing pilgrims' inner experience are *path*, *diversity*, *beauty*, *to walk*, and *pleasure*. What is highlighted in terms of the personal values of respondents is quite different from the Way of Saint James. The words used are softer, less concerned with mental aspects, and more focused on a natural experience. The walking element remains as the core aspect of the pilgrimage journey, but the desire for purification and transformation is not highlighted, giving way to the beauty offered by nature, landscapes, and the diverse art and architectural works of Italy. The Via Francigena remains an authentic pilgrimage route, but it also assumes a distinct and modern touch. If the primordial form of slow tourism can be considered the classic religious pilgrimage, now in the Via Francigena a contemporary conception has been added, encompassing the experiential dimension. In fact, among the words used by pilgrims to describe their inner experiences, *pleasure* emerges, but *happiness* does not. This difference in wording between the two pilgrim groups seems at first to be slight. However, the word *happiness* is representative of an inner spiritual dimension: it is an emotion strongly related to the personal growth achieved through walking the Camino and includes the joy of overcoming the difficulties in reaching the goal. The word *pleasure* clearly falls in a wellness dimension: pilgrims undertaking the Via Francigena route experience benefits that extend to health and well-being, enjoying the spiritual, cultural, natural, and gastronomic dimensions of the journey. Some respondents commented about this topic as follows:

When I started Via Francigena I was overweight, and I started to go on a diet. Before, I used to drink five coffees a day; now, drinking only two and waking up early in the morning is no longer a problem, so the pilgrimage did a lot of miracles. (RVF3)

You walk through . . . how can I say, I've been in Tuscany at least 30 or 40 times in my life but walking there on foot is totally different; you enjoy small towns, small paths . . . This, in my opinion, is part of the Italian beauty. (RVF6)

The human dimension, landscape, finding yourself at the end of any leg, friendship, hugs, greetings, speaking with my travelling companions, small hamlets, rows of cypress trees, water that quenched me, bread I ate, the journey in my world, tears I spilled, words remembered . . . you see, they are small things we have lost. (RVF2)

Returning to the central research questions, the study shows the multifaceted complexity of pilgrims' motivations and inner experiences accompanying all the stages of the chosen pilgrimage route. Pilgrims' inner experiences are multifaceted according to the different forms of pilgrimage tourism and routes, converging more towards feelings of personal happiness, in which the emotional aspect has a strong relevance, as in the case of the Camino de Santiago, or towards feelings of wellness, in which the contemporary experiential features drive people's sensations towards a more cultural and aesthetic dimension, as in the case of the Via Francigena. Therefore, pilgrimage can contribute to people flourishing according to a specific combination of positive emotions, personal commitment in the activity, good relationships, finding meaningful purposes and reaching personal targets. Such a combination tends to vary according to the different pilgrimage route undertaken.

SOLUTIONS AND RECOMMENDATIONS

The present research explored happiness and quality of life in pilgrimage tourism, clarifying the differences between the Camino de Santiago and the Via Francigena both in terms of individuals' motivations and the implications for people's lives from a comparative perspective. The implications deriving from the results emerged from face-to-face interviews with pilgrims.

First, deciding to undertake a pilgrimage is a long and pre-meditated process, passing through different information channels and involving weighty decision-making. This particular aspect partially collides with other studies that blur the differences between pilgrims and tourists (e.g. Collins-Kreiner, 2010a, 2010b; Timothy & Olsen, 2006) and shows that pilgrims, at least in the early phases of the decision-making process, have their own specific identity.

Second, the pilgrimage route undertaken is not only a matter of cost-effectiveness or geographical proximity, but also varies according to pilgrims' motivations. This element suggests the presence of various clusters of pilgrims converging on different routes according to their personal motivations and attitudes. From another perspective, it emerges that the Camino de Santiago and the Via Francigena highlight a multifaceted and specific context of pilgrims' motivations. Thus, the study identifies a variety of motivations of pilgrims walking the same path, as already noted by other studies (e.g. Amaro et al., 2018). Meanwhile, comparing the two pilgrimage routes, the findings highlight different sets of motivations of pilgrims undertaking each of the two itineraries. In sum, intra-heterogeneity and inter-heterogeneity emerge.

Third, pilgrims' inner experiences differ in emotional aspects and personal values according to the chosen route, although related to the "lifestyle revolution" of slow tourism philosophy (Honoré, 2005). Happiness and pleasure clearly arise as different feelings representing different inner experiences that emerge by walking the different routes. Pilgrimage promotes very intense, unrepeatable inner experiences involving both the recalibration of life perspectives, as previous studies highlighted (e.g. Lopez et al., 2017), and unique sensations. If both pilgrimage routes exhibit signs of growing secularism in terms of experience, confirming extant studies (e.g. Kim et al., 2016), pilgrims' feelings seem to show a stronger coherence with the classical pilgrimage archetype in the Camino de Santiago case, while distinctive and contemporary experiential concerns tend to overcome spirituality and religiosity in Via Francigena. In sum, pilgrims undertaking different routes reach the goal of slow tourism, identifying the different meaning of each place and discovering their identity and soul (Shang et al., 2020).

FUTURE RESEARCH DIRECTIONS

This work paves the way to new and necessary comparative research projects among pilgrimage itineraries, considering the multifaceted aspects of pilgrims and pilgrimage. Broadening studies to other religious or spiritual routes even outside of Europe may be very useful, in particular, Holy Land and India routes that are highly renowned in the context of pilgrimage and tourism. Adding and confronting spiritual tourism with Train of Memory or other emotional/psychological journeys concerning Shoah or similar phenomena may also shed new light on different perspectives of happiness and well-being in terms of self-consciousness and self-enrichment deriving from particularly touching experiences.

CONCLUSION

Pilgrimage tourism implies a journey along a tangible and intangible road in which the slowness of walking allows more conscious engagement with life and locality (environment, culture, and people). Its growth creates opportunities for sustainable development by creating jobs, fostering infrastructure and investment, and stimulating local culture, crafts, and food. Policymakers and managers who want to promote pilgrimage tourism need to realize that pilgrims' motivations are different, just as the perceived values (end-states) acquired through undertaking a pilgrimage are different. Therefore, it is necessary to create tourism products and services that appeal to these diverse needs and experience outcomes.

There are limited studies about motivations for pilgrimage and pilgrims' inner experiences (Amaro et al., 2018; Blackwell, 2014; Damari & Mansfeld, 2016), and understanding the antecedents (motives) and consequences (benefits obtained and perceived by pilgrims) of pilgrimage experiences is of paramount importance for promotional and development purposes.

From a managerial point of view, the Camino de Santiago and Via Francigena need to be managed just like any product in the tourism context. The findings of this study offer a better understanding of contemporary pilgrims walking along these pilgrimage roads. Therefore, they provide useful insight for stakeholders to devise suitable development strategies.

Ultimately, each pilgrim who participated in the study was asked to discuss the motivations and emotions they experienced. Several differences emerged in motivations and benefits obtained according to the road they had undertaken, with a more *existential* connotation tied to personal happiness in the Camino de Santiago pilgrims group and a more *experiential* connotation tied to well-being in the Via Francigena pilgrims group. Nevertheless, each pilgrim is in search of something that surely exists, but cannot be seen: "invisible energies". Regardless of whether they derive from religious, spiritual, cultural, or eco-natural dimensions, these invisible energies become visible when the world is explored in a slow way that contributes to happiness and wellness.

REFERENCES

Ali, F., & Amin, M. (2014). The influence of physical environment on emotions, customer satisfaction and behavioural intentions in Chinese resort hotel industry. *Journal for Global Business Advancement*, *7*(3), 249–266. doi:10.1504/JGBA.2014.064109

Ali, F., Kim, W. G., Li, J., & Jeon, H. M. (2018). Make it delightful: Customers' experience, satisfaction and loyalty in Malaysian theme parks. *Journal of Destination Marketing & Management*, *7*, 1–11. doi:10.1016/j.jdmm.2016.05.003

Alighieri, D. (1896). Vita Nova. Houghton, Mifflin and Company. (original work published 1294)

Amaro, S., Antunes, A., & Henriques, C. (2018). A closer look at Santiago de Compostela's pilgrims through the lens of motivations. *Tourism Management*, *64*, 271–280. doi:10.1016/j.tourman.2017.09.007

Barber, R. (1993). *Pilgrimages*. The Boydell Press.

Blackwell, R. (2014). Motivation for pilgrimage: Using theory to explore motivations. *Scripta Instituti Donneriani Aboensis*, *22*, 24–37. doi:10.30674cripta.67360

Blom, T., Nilsson, M., & Santos, X. (2016). The way to Santiago beyond Santiago. Fisterra and the pilgrimage's post-secular meaning. *European Journal of Tourism Research*, *12*, 133–146.

Boyatzis, R. E. (1998). Transforming Qualitative Information: Thematic Analysis and Code Development. *Sage (Atlanta, Ga.)*.

Caffyn, A. (2012). Advocating and implementing slow tourism. *Tourism Recreation Research*, *37*(1), 77–80. doi:10.1080/02508281.2012.11081690

Clancy, M. (2015). What's at stake in the slow movement? Consumption, responsibility and citizenship. *The Association of American Geographers Annual Meeting*.

Collins-Kreiner, N. (2010a). Researching pilgrimage: Continuity and transformations. *Annals of Tourism Research*, *37*(2), 440–456. doi:10.1016/j.annals.2009.10.016

Collins-Kreiner, N. (2010b). The geography of pilgrimage and tourism: Transformations and implications for applied geography. *Applied Geography (Sevenoaks, England)*, *30*(1), 153–164. doi:10.1016/j.apgeog.2009.02.001

Collins-Kreiner, N. (2016). Life cycle of concepts: The case of pilgrimage tourism. *Tourism Geographies*, *18*(3), 322–334. doi:10.1080/14616688.2016.1155077

Collins-Kreiner, N. (2020a). Pilgrimage tourism-past, present and future rejuvenation: A perspective article. *Tourism Review*, *75*(1), 145–148. doi:10.1108/TR-04-2019-0130

Collins-Kreiner, N. (2020b). A review of research into religion and tourism Launching the Annals of Tourism Research Curated Collection on religion and tourism. *Annals of Tourism Research*, *•••*, 82.

Correia, M. I. G., Lopez, L., Gonzalez, R. C. L., & Santos, X. M. (2017). The Challenges of the First European Cultural Itinerary: The Way to St. James. *AlmaTourism - Journal of Tourism. Culture and Territorial Development*, *8*(6), 1–19.

Courtney, M. (2015). A review of the literature exploring the occupation of contemporary pilgrimage. *Journal of Occupational Science*, *22*(2), 170–182. doi: 10.1080/14427591.2013.764816

Csikszentmihalyi, M. (1988). The flow experience and its significance for human psychology. In M. Csikszentmihalyi & I. S. Csikszentmihalyi (Eds.), *Optimal experience: Psychological studies of flow in consciousness* (pp. 15–35). Cambridge University Press. doi:10.1017/CBO9780511621956.002

Csikszentmihalyi, M. (1990). *Flow: The psychology of optimal experience*. Harper & Row.

D'Atti, M., & Cinti, F. (2014). *Guida alla Via Francigena: Oltre 900 Chilometri a Piedi sulle Strade del Pellegrinaggio verso Roma*. Terre di mezzo.

Damari, C., & Mansfeld, Y. (2016). Reflections on pilgrims' identity, role and interplay with the pilgrimage environment. *Current Issues in Tourism*, *19*(3), 199–222. doi: 10.1080/13683500.2014.957660

De Bloom, J., Geurts, S. A., & Kompier, M. A. (2013). Vacation (after-) effects on employee health and well-being, and the role of vacation activities, experiences and sleep. *Journal of Happiness Studies*, *14*(2), 613–633. doi:10.100710902-012-9345-3

Devereux, C., & Carnegie, E. (2006). Pilgrimage: Journeying beyond self. *Tourism Recreation Research*, *31*(1), 47–56. doi:10.1080/02508281.2006.11081246

Dickinson, J. E., Robbins, D., & Lumsdon, L. M. (2010). Holiday travel discourses and climate change. *Journal of Transport Geography*, *18*(3), 482–489. doi:10.1016/j.jtrangeo.2010.01.006

Digance, J. (2003). Pilgrimage at contested sites, tourism management. *Annals of Tourism Research*, *30*(1), 143–159. doi:10.1016/S0160-7383(02)00028-2

Ekinci, M. B. (2014). The Cittaslow philosophy in the context of sustainable tourism development: The case of Turkey. *Tourism Management*, *41*, 178–189. doi:10.1016/j.tourman.2013.08.013

Farber, M. E., & Hall, T. E. (2007). Emotion and environment: Visitors' extraordinary experiences along the Dalton Highway in Alaska. *Journal of Leisure Research*, *39*(2), 248–270. doi:10.1080/00222216.2007.11950107

Fereday, J., & Muir-Cochrane, E. (2006). Demonstrating rigor using thematic analysis: A hybrid approach of inductive and deductive coding and theme development. *International Journal of Qualitative Methods*, *5*(1), 80–92. doi:10.1177/160940690600500107

Fistola, R., & La Rocca, R. A. (2018). Slow Mobility and Cultural Tourism. Walking on Historical Paths. In R. Papa, R. Fistola, & C. Gargiulo (Eds.), *Smart Planning: Sustainability and Mobility in the Age of Change* (pp. 301–322). Springer. doi:10.1007/978-3-319-77682-8_18

Fullagar, S., Markwell, K., & Wilson, E. (2012). Starting slow: Thinking through slow mobilities and experiences. In S. Fullagar, K. Markwell, & E. Wilson (Eds.), *Slow tourism: Experiences and mobilities* (pp. 1–8). Channel View Publications. doi:10.21832/9781845412821-003

Gazzola, P., Grechi, D., Romanelli, M., & Pollice, F. (2020). Il turismo religioso: La via Francigena, una opportunità per l'Italia. *Economia Aziendale Online*, *11*(3), 341–358.

Germann Molz, J. (2009). Representing pace in tourism mobilities: Staycations, slow travel and the amazing race. *Journal of Tourism and Cultural Change*, *7*(4), 270–286. doi:10.1080/14766820903464242

Gesler, W. (1996). Lourdes: Healing in a place of pilgrimage. *Health & Place*, *2*(2), 95–105. doi:10.1016/1353-8292(96)00004-4

Gibson, D., Pratt, S., & Movono, A. (2012). Tribe tourism: A case study of the Tribewanted project in Vorovoro, Fiji. In S. Fullagar, K. Markwell, & E. Wilson (Eds.), *Slow Tourism: Experiences and mobilities* (pp. 185–200). Channel View Publications. doi:10.21832/9781845412821-016

Guiver, J., & McGrath, P. (2017). Slow Tourism: Exploração de discursos. *Dos Algarves: A Multidisciplinary e-Journal*, (27), 11-34.

Hadavi, S. (2017). Direct and indirect effects of the physical aspects of the environment on mental well-being. *Environment and Behavior*, *49*(10), 1071–1104. doi:10.1177/0013916516679876

Heitmann, S., Robinson, P., & Povey, G. (2011). Slow food, slow cities and slow tourism. In P. Robinson, S. Heitmann, & P. Dieke (Eds.), *Research Themes for Tourism* (pp. 114–127). CABI., doi:10.1079/9781845936846.0114

Honoré, C. (2005). *In Praise of slow: How a worldwide movement is challenging the cult of speed*. Orion Books Ltd.

Howard, C. (2012). Speeding up and slowing down: Pilgrimage and slow travel through time. In S. Fullagar, K. Markwell, & E. Wilson (Eds.), *Slow Tourism: Experiences and Mobilities* (pp. 11–24). Channel View Publications.

Jackowski, A., & Smith, V. L. (1992). Polish pilgrim-tourists. *Annals of Tourism Research*, *19*(1), 92–106. doi:10.1016/0160-7383(92)90109-3

Jepson, A., Stadler, R., & Spencer, N. (2019). Making positive family memories together and improving quality-of-life through thick sociality and bonding at local community festivals and events. *Tourism Management*, *75*, 34–50. doi:10.1016/j.tourman.2019.05.001

Kato, K., & Progano, R. N. (2017). Spiritual (walking) tourism as a foundation for sustainable destination development: Kumano-kodo pilgrimage, Wakayama, Japan. *Tourism Management Perspectives*, *24*, 243–251. doi:10.1016/j.tmp.2017.07.017

Kim, B., & Kim, S. (2019). Hierarchical value map of religious tourists visiting the Vatican City/Rome. *Tourism Geographies*, *21*(3), 529–550. doi:10.1080/14616688.2018.1449237

Kim, B., Kim, S., & King, B. (2019). Religious tourism studies: Evolution, progress, and future prospects. *Tourism Recreation Research*, *45*(2), 185–203. doi:10.1080/02508281.2019.1664084

Kim, B., Kim, S. S., & King, B. (2016). The sacred and the profane: Identifying pilgrim traveler value orientations using means-end theory. *Tourism Management*, *56*, 142–155. doi:10.1016/j.tourman.2016.04.003

Kim, H., Lee, S., Uysal, M., Kim, J., & Ahn, K. (2015). Nature-based tourism: Motivation and subjective well-being. *Journal of Travel & Tourism Marketing*, *32*(sup1), S76–S96. doi:10.1080/10548408.2014.997958

Lancia, F. (2012). *The Logic of the T- LAB Tools Explained*. www.tlab.it/en/toolsexplained.php

Latusi, S., & Fissore, M. (2021). The Seven Sisters of Piedmont: a case study of potentially worthwhile cooperation. In M. Della Lucia & E. Giudici (Eds.), *Humanistic Management and Sustainable Tourism: Human, Social and Environmental Challenges*. Routledge. doi:10.4324/9781003108955-3

Lemmi, E. (2020). Heritage and new communication technologies: development perspectives on the basis of the Via Francigena experience. In *Conference Preceedings of Heritage, tourism and Hospitality international Conference HTHIC* (pp. 43–63). USI-Università della Svizzera Italiana.

Liutikas, D. (2015). In search of miracles: Pilgrimage to the miraculous places. *Tourism Review*, *70*(3), 197–213. doi:10.1108/TR-08-2013-0046

Lopez, L. (2013). How Long Does the Pilgrimage Tourism Experience to Santiago de Compostela Last? *International Journal of Religious Tourism and Pilgrimage*, *1*, 1–14.

Lopez, L., González, R. C. L., & Fernández, B. M. C. (2017). Spiritual tourism on the way of Saint James the current situation. *Tourism Management Perspectives*, *24*, 225–234. doi:10.1016/j.tmp.2017.07.015

Lucarno, G. (2016). The Camino de Santiago de Compostela (Spain) and the Via Francigena (Italy): A comparison between two important historic pilgrimage routes in Europe. *International Journal of Religious Tourism and Pilgrimage*, *4*(7), 48–58.

Lumsdon, L. M., & McGrath, P. (2011). Developing a conceptual framework for slow travel: A grounded theory approach. *Journal of Sustainable Tourism*, *19*(3), 265–279. doi:10.1080/09669582.2010.519438

Mayer, H., & Knox, P. L. (2006). Slow cities: Sustainable places in a fast world. *Journal of Urban Affairs*, *28*(4), 321–334. doi:10.1111/j.1467-9906.2006.00298.x

Mayer, V. F., Machado, J. D. S., Marques, O., & Nunes, J. M. G. (2019). Mixed feelings?: Fluctuations in well-being during tourist travels. *Service Industries Journal*, *40*(1-2), 158–180. doi:10.1080/02642069.2019.1600671

McGrath, P., & Sharpley, R. (2016). Slow travel and slow tourism: New concept or new label? In M. Clancy (Ed.), *Slow tourism, food and cities: Pace and the search for the 'good life'*. Routledge.

Moore, K. (2012). On the periphery of pleasure: Hedonics, eudaimonics and slow travel. In S. Fullagar, K. Markwell, & E. Wilson (Eds.), *Slow Tourism: Experiences and Mobilities* (pp. 25–35). Channel View Publications. doi:10.21832/9781845412821-005

Nilsson, M. (2018). Wanderers in the shadow of the sacred myth: Pilgrims in the 21st century. *Social & Cultural Geography*, *19*(1), 21–38. doi:10.1080/14649365.2016.1249398

Nilsson, M., & Tesfahuney, M. (2016). Performing the "post-secular" in Santiago de Compostela. *Annals of Tourism Research*, *57*, 18–30. doi:10.1016/j.annals.2015.11.001

Noy, C. (2004). This trip really changed me: Backpackers' narratives of self-change. *Annals of Tourism Research*, *31*(1), 78–102. doi:10.1016/j.annals.2003.08.004

Oficina del Peregrino. (2019). *Informe estadístico Año 2019*. Publisher.

Oh, H., Assaf, A. G., & Baloglu, S. (2016). Motivations and goals of slow tourism. *Journal of Travel Research, 55*(2), 205–219. doi:10.1177/0047287514546228

Olsen, D. H., & Wilkinson, G. (2016). Are fast pilgrims true pilgrims? The Shikoku pilgrimage. *Annals of Tourism Research, 61*, 228–230. doi:10.1016/j.annals.2016.08.004

Oviedo, L., de Courcier, S., & Farias, M. (2014). Rise of pilgrims on the Camino to Santiago: Sign of change or religious revival? *Review of Religious Research, 56*(3), 433–442. doi:10.100713644-013-0131-4

Ozdemir, G., & Celebi, D. (2018). Exploring dimensions of slow tourism motivation. *Anatolia, 29*(4), 540–552. doi:10.1080/13032917.2018.1460854

Perdue, R., Tyrrel, T., & Uysal, M. (2010). Understanding the value of tourism: a conceptual divergence. In D. Pearce & R. Butler (Eds.), *Tourism Research: A 20:20 Vision* (pp. 123–134). Goodfellow Publishers.

Raj, R., Griffin, K., & Blackwell, R. (2015). Motivations for religious tourism, pilgrimage, festivals and events. In R. Raj & K. Griffin (Eds.), *Religious Tourism and Pilgrimage Management: An International Perspective* (pp. 103–117). CABI. doi:10.1079/9781780645230.0103

Reader, I. (2007). Pilgrimage growth in the modern world: Meanings and implications. *Religion, 37*(3), 210–229. doi:10.1016/j.religion.2007.06.009

Rinschede, G. (1992). Forms of religious tourism. *Annals of Tourism Research, 19*(1), 51–67. doi:10.1016/0160-7383(92)90106-Y

Santos, X. M. (2002). Pilgrimage and Tourism at Santiago de Compostela. *Tourism Recreation Research, 27*(2), 41–50. doi:10.1080/02508281.2002.11081219

Seligman, M. E. P. (2011). *Flourish: A visionary new understanding of happiness and well-being*. Free Press.

Serdane, Z. (2020). Slow philosophy in tourism development in Latvia: The supply side perspective. *Tourism Planning & Development, 17*(3), 295–312. doi:10.1080/21568316.2019.1650103

Shang, W., Qiao, G., & Chen, N. (2020). Tourist experience of slow tourism: From authenticity to place attachment–a mixed-method study based on the case of slow city in China. *Asia Pacific Journal of Tourism Research, 25*(2), 170–188. doi:10.1080/10941665.2019.1683047

Shinde, K. A. (2007). Case study 6: Visiting sacred sites in India: Religious tourism or pilgrimage. In R. Raj & N. D. Morpeth (Eds.), *Religious Tourism and Pilgrimage Management: An International Perspective* (pp. 184–197). CABI. doi:10.1079/9781845932251.0184

Smith, V. L. (1992). Introduction: The quest in guest. *Annals of Tourism Research, 19*(1), 1–17. doi:10.1016/0160-7383(92)90103-V

Souza de, J. M. R., Brito, A. S., & Perinotto, A. R. C. (2018). Tradition and religious tourism: The Passion of Christ in Parnaíba-Piauí-Brazil. *Turismo e Sociedade, 11*(2), 261–288.

Stefko, R., Kiralova, A., & Mudrík, M. (2015). Strategic marketing communication in pilgrimage tourism. *Procedia: Social and Behavioral Sciences, 175*, 423–430. doi:10.1016/j.sbspro.2015.01.1219

Terzidou, M., Scarles, C., & Saunders, M. N. (2018). The complexities of religious tourism motivations: Sacred places, vows and visions. *Annals of Tourism Research, 70*, 54–65. doi:10.1016/j.annals.2018.02.011

Timms, B. F., & Conway, D. (2012). Slow tourism at the Caribbean's geographical margins. *Tourism Geographies, 14*(3), 396–418. doi:10.1080/14616688.2011.610112

Timothy, D., & Olsen, D. (2006). Management issues for religious heritage attractions. In D. Timothy & D. Olsen (Eds.), *Tourism, Religion and Spiritual Journeys* (pp. 120–134). Routledge. doi:10.4324/9780203001073-18

Timothy, D. J., & Boyd, S. W. (2006). Heritage tourism in the 21st Century: Valued traditions and new perspectives. *Journal of Heritage Tourism, 1*(1), 1–16. doi:10.1080/17438730608668462

Uysal, M., Sirgy, M. J., Woo, E., & Kim, H. L. (2016). Quality of life (QOL) and well-being research in tourism. *Tourism Management, 53*, 244–261. doi:10.1016/j.tourman.2015.07.013

Vukonic, B. (1996). *Tourism and Religion*. Elsevier Science Ltd.

ADDITIONAL READING

Badone, E., & Roseman, S. (Eds.). (2004). *Intersecting Journeys: The Anthropology of Pilgrimage and Tourism*. University of Illinois Press.

Bravi, M. (2019). Luoghi sacri e turismo religioso. spunti di riflessione dalla santa sede. *Documenti geografici*, (1), 187-194.

Corinto, G. L. (2017). Food and gastronomic tourism for developing rural areas around the via Francigena in Tuscany. *Almatourism-Journal of Tourism. Culture and Territorial Development*, *8*(6), 106–122.

Pavione, E., Pezzetti, R., & Gazzola, P. (2017). The Role of "Slow Territories" in the Development of Sustainable Tourism. *European Scientific Journal*, *13*, 238–248.

Scidurlo, P., & D'Angeli, G. (2020). Vie Francigene for All? *Almatourism-Journal of Tourism. Culture and Territorial Development*, *10*(20), 69–91.

Sesana, G. (2006). *Pellegrini e Turisti. L'Evoluzione del Viaggio Religioso*. Hoepli Editore.

Smith, M. K., & Diekmann, A. (2017). Tourism and wellbeing. *Annals of Tourism Research*, *66*, 1–13. doi:10.1016/j.annals.2017.05.006

KEY TERMS AND DEFINITIONS

Cultural Route: A route that is physically determined and characterized by having its own historical dynamics, showing movements of people.

Destination Management: The coordinated management of all elements that make up a destination, including the attractions, amenities, access, marketing, and pricing.

Motivation: The process that initiates, guides, and maintains goal-oriented behaviours. It is the driving force behind human actions.

Personal Values: The things that are important and desirable to an individual and that guide decisions.

Pilgrimage: A journey to a shrine or other location of importance to a person's beliefs and faith.

Slow Tourism: Travel for a prolonged period of time at a slow pace, allowing the tourist to have a deep, authentic, and cultural experience.

Spiritual Tourism: Subjective and individualistic travel for spiritual betterment and self-discovery.

Chapter 9
Wellness Tourism:
Conceptual Analysis and Trends

Leticia Cynara Santos Silva
Federal Fluminense University, Brazil

Verônica Feder Mayer
https://orcid.org/0000-0002-7543-5215
Federal Fluminense University, Brazil

ABSTRACT

Crisis, social discontent, and dissatisfaction among people lead to a greater interest in relaxing the soul, body, and mind as a refuge within tourism. Thus, there are changes and new trends in international health tourism's demand and supply conditions. Even before the deep crisis caused by the new coronavirus pandemic in 2020, people increasingly turned to health and wellness activities while traveling, especially using tourist activities to achieve health. Since consumption patterns are changing among travelers, such as the growing popularity of wellness tourism, new forms of tourism may stimulate future tourism activities, and the changes are expected to push businesses to reevaluate their service designs and distribution channels. When the recovery of the tourism sector is under discussion, it is worth questioning how wellness trends could help tourism and what the perspectives are for the future. Therefore, this chapter makes recommendations based on new tourists' demands and new processes and protocols for the operation of wellness tourism services.

DOI: 10.4018/978-1-7998-7239-9.ch009

Copyright © 2021, IGI Global. Copying or distributing in print or electronic forms without written permission of IGI Global is prohibited.

INTRODUCTION

The world is undergoing profound changes in people's attitudes, behavior, and economic life, especially in recent years (Ahmed et al., 2020; Faulkner et al., 2020; Wen et al., 2021). Facing crisis, economic decline, social discontent, and dissatisfaction among people lead to a greater interest in relaxing the soul, body, and mind as a refuge within tourism (Koncul, 2012; Cai et al., 2020). Thus, there are changes and new trends in international health tourism's demand and supply conditions (Bushueva, 2020; Csirmaz & Pető, 2015; Lee et al., 2020; Wen et al., 2021).

Even before the deep crisis caused by the new coronavirus pandemic in 2020, people increasingly turned to health and wellness activities while traveling, especially using tourist activities to achieve health (Chia & Liao, 2020; Csirmaz & Pető, 2015; Thal & Hudson, 2019). The new ways of life in society have increased the concern of individuals with well-being and body care, widening tourist motivations to explore health tourism (Cunha, 2006).

Consumption patterns are changing among travelers, such as the growing popularity of wellness tourism. New forms of tourism may stimulate future tourism activities, and the changes are expected to push businesses to reevaluate their service designs and distribution channels (Wen et al., 2021). When the recovery of the tourism sector is under discussion, it is worth questioning how wellness trends could help tourism and what the perspectives are for the future. Therefore, this chapter makes recommendations for wellness tourism business sector, according to the emerging tourism service management practices. The recommendations are based on new tourists' demands and new processes and protocols for the operation of wellness tourism services.

HEALTH TOURISM AND ITS CONTRIBUTIONS TO SOCIETY

The *'health tourism'* has many definitions and terms, such as health tourism, medical tourism, hospital and clinical tourism, wellness tourism, and sometimes medical travel. They are all used to describe the relationship between health and tourism (Quintela et al., 2016). Goodrich and Goodrich (1987, p. 217) define it as:

An attempt on the part of a tourist facility to attract tourists by deliberately promoting its health-care services and facilities, in addition to its regular amenities. These health care services may include medical examinations by qualified doctors and nurses at the resort or hotel, special diets, acupuncture, trans-vital injections, vitamin complex intakes, special medical treatments for various diseases such as arthritis, and herbal medicines.

The term has come to be more widely used internationally with the notion that health nowadays focuses simultaneously on prevention and curing illness. The World Tourism Organization (UNWTO, 2018, p. 9) defines Health Tourism as:

Health tourism covers those types of tourism which have as a primary motivation, the contribution to physical, mental and/or spiritual health through medical and wellness-based activities which increase the capacity of individuals to satisfy their own needs and function better as individuals in their environment and society.

However, travel for health reasons is nothing new (Hall, 2011). Although health tourism (and its terms) is a relatively new segment within the global travel and tourism industry, the practice of health-focused travel is not new (P. S. & Abraham, 2017). This market explores an ancient format of using different types of wellness services in one place. Medicine and disease treatment date back to the dawn of civilization (Scliar, 2007) and is one of the most ancient forms of tourism. Its' roots can be found around 5000 BC with the Indian Ayurvedic practices, with beauty and cosmetic rituals that women used around 3000 BC in the Ancient Egypt Civilization, and also with the Chinese medical practices around 1000 BC (Quintela et al., 2016). The search in tourism for treatments to promote and maintain health, prevention, and cure of diseases or even the performance of aesthetic procedures grows, evolves, and becomes a revenue-generating activity, motivated by the will of society to stay healthy (Paixao, 2007).

Goeldner (1989) recognized five components of health-related tourism market segments such as: i. leisure tourism; ii. outdoor recreation, adventure tourism, sports tourism and wellness tourism; iii. health tourism and wellness tourism; iv. Spa tourism and wellness tourism; v. medical and dental tourism (Hall, 2011). This approach proposes a framework of relationships between the subcategories of health tourism (Quintela et al., 2016). In this sense, we can observe different points of view and perspectives in health tourism, as shown in Table 1.

Health tourism is located in an extensive context ranging from wellness tourism to hospital-clinical tourism. Other concepts such as medical tourism, wellness tourism, and medical travel also share conceptual similarity with health tourism. Together, wellness and medical tourism could be considered the main subcategories of health tourism, and the designation of 'health tourism' can be used "as a comprehensive umbrella term that subsumes wellness and medical tourism" (Voigt et al., 2011, p. 16).

In the 1990s, traditional health tourism emerged as wellness-fitness services, and it became popular among consumers (Csirmaz & Pető, 2015). A broad spectrum of health and wellness tourism has developed in recent years (see Figure 1). Tourists' motivation can be leisure and entertainment, focusing on sports and fitness; disease

Table 1. Point of views and perspectives of health tourism

Point of views	**the therapeutic point of view –** is linked with medical tourism, including surgery and/or treatment to cure and prevent diseases.	(Chia & Liao, 2020; P. S. & Abraham, 2017; Tapia et al., 2020).
	The recreational point of view – is known as wellness or well-being tourism, focusing on relaxation, leisure, and escape from the routine.	(Bushueva, 2020; Csirmaz & Pető, 2015; Li, 2020).
Perspectives	**The consumer perspective** – health tourism is traveling for the maintenance, enhancement, or restoration of physical and mental well-being.	(Lee et al., 2020; Lubowiecki-Vikuk et al., 2021).
	The supplier perspective – health tourism can be defined as traveling to destinations that provide facilities and healthcare services explicitly and in addition to their regular tourist amenities.	(Chi et al., 2020; Csirmaz & Pető, 2015; Loke, 2020).

Source: adapted from Quintela et al., 2016.

treatment or cosmetic surgery trips; relaxation in thalassotherapy centers; meditation and yoga focusing on a spiritual level, etc.

Smith and Puczkó (2009) suggest the spectrum of health tourism, that gives a detailed overview of health tourism, including physical healing, beauty treatments, leisure/ entertainment, life and work balance, relaxation and rest, and psychological and spiritual. All examples show concern for the body, mind, and spirit. In some travel cases, more than one spectrum is involved.

Figure 1 helps to understand why the terms and spaces used in health tourism may be interchangeable. A spectrum of wellness destinations and attractions shows the wide range of products, services, activities, and typologies of wellness tourism. The holidays which take place within these environments would, in some cases, be completely different from one another. For example, there is little connection between a spa that deals specifically with physical or medical problems using water treatments and a spiritual retreat that focuses on meditation for the mind and soul (Smith & Puczkó, 2009).

WELLNESS TOURISM

Wellness is supposed to create harmony in mental, physical, spiritual or biological health in general (P. S. & Abraham, 2017). It relates more to changing lifestyle or doing something healthy than curing a specific disease (P. S. & Abraham, 2017). The concept of wellness is the co-product of the awareness of health, well-being and happiness (Koncul, 2012). The National Wellness Institute defines wellness as an

Figure 1. Spectrum of health tourism
Source: *adapted from Smith and Puczkó (2009, p. 84)*

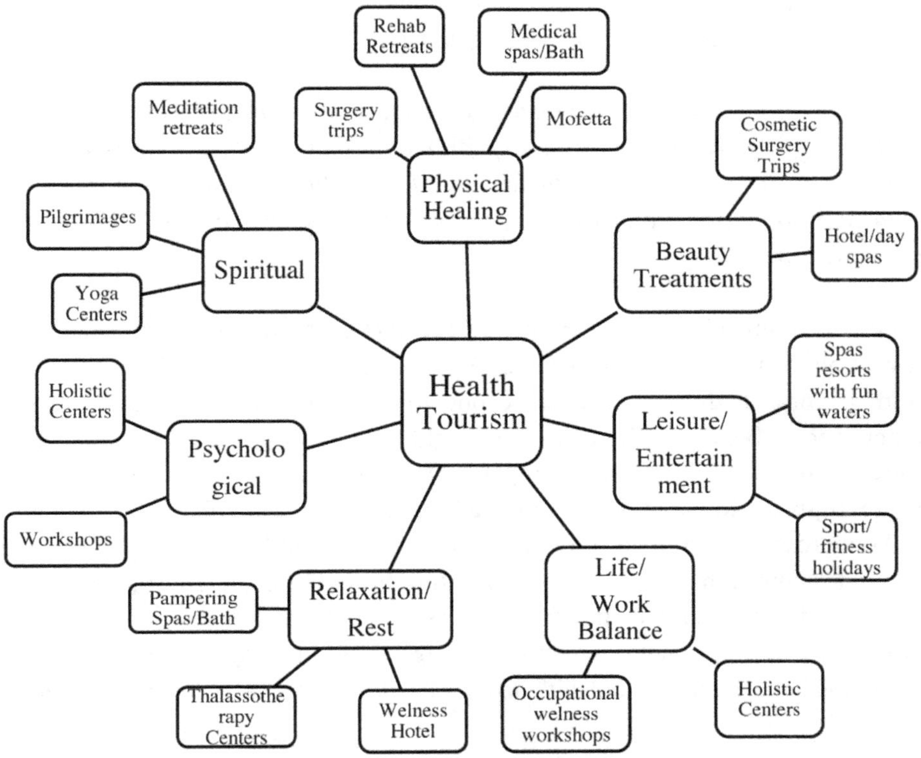

"active process through which people become aware of, and make choices toward, a more successful existence". Essentially, wellness is the pathway to achieve health (Dillette et al., 2020).

The English designation, proposed in 1961 by Dr. Halbert Dunn, seeks to cover the expression *Wellnes* resulting from the combination of the word *Wellbeing* (well-being) with the word *fitness* (physical fitness). Thus, the activities developed by *Wellness* seek to ensure a holistic conception to keep the body and mind healthy, that is, it sees the person as a whole and does not treat a disease (Cunha, 2006). Wellness advocates positive and optimistic outlook, and a holistic and balanced approach to life (Dillette et al., 2020).

The concept of wellness also includes the following domains: physical, social, intellectual, spiritual, emotional, and occupational satisfaction (National Wellness Institute, 2020). And there are two focal concerns of wellness: the fulfillment of the potential of an individual physically, psychologically, socially, spiritually, and economically; and the achievement of expectations in the family, community, place of worship, workplace, among others (Smith et al., 2006).

One of the key sectors of the wellness economy is wellness tourism. More specifically, wellness tourism is defined as:

the sum of all the relationships resulting from a journey by people whose motive, in whole or in part, is to maintain or promote their health and well-being, and who stay at least one night at a facility that is specifically designed to enable and enhance people's physical, psychological, spiritual and/or social well-being (Voigt et al., 2011, p. 17).

One distinguishing characteristic of wellness tourism is that its prime aim is prevention (Mueller and Kaufmann, 2001). Aspects such as adopting a lifestyle in a lifelong approach, the notion of self-responsibility, and the exploration of a person's potential for a better quality of life are highlighted. The search for quality of life motivates many tourism activities, particularly the ones associated with outdoor leisure, rejuvenation, and relaxation in a holistic approach (Voigt et al., 2011).

An increasing number of tourists seem to be looking for meaning and hedonic experiences through their vacation and leisure time (Cai et al., 2020), aligned with societal and consumer demands for greater social and environmental responsibility (Voigt et al., 2011). Then, "wellness tourism" should be engaged with health and well-being benefits associated with different styles and types of experience (Dillette et al., 2020).

According to Quintela et al. (2016), wellness tourism's holistic perspective can be divided into five additional types of tourism activities and experiences: spiritual and religious tourism; yoga tourism; outdoor, sports and adventure; volunteer tourism; and new age tourism (Figure 2). Holistic tourism is also known as a state of health that involves the physical, the mind, the spirit, and, more recently, the environment (Dillette et al., 2020). Thus, the concept of holistic tourism suggests wholeness and integration, implying that all dimensions of wellness could be included in the development of new tourism products or services (Kelly and Smith, 2008).

Therefore, wellness tourism encompasses a positive and holistic understanding of health that incorporates physical, psychological, social, and spiritual experiences undertaken by tourists whose primary motive is to maintain or improve their health and well-being (Voigt et al., 2011). The search for a more harmonious life, such as exercising, healthy eating, practicing hobbies, sleeping, providing individual well-being and balance, opens opportunities for many tourism products, services, and destinations. In fact, different people resort to rest centers, such as spas, in search of getting out of the routine, find harmony, inner peace, relaxation, and contact with nature (Viegas Fernandes & Viegas Fernandes, 2011). In addition, doing something

Figure 2. Holistic representation of wellness tourism
Source: *Quintela et al., 2016.*

with pleasure and desire produces endorphins, which helps to decrease mental illnesses, especially in pandemic circumstances, when there is a neuropsychiatric linkage between the outbreak of acute respiratory infections and mental disorders (Shah et al., 2020).

There are prominent approaches in emerging studies of wellness tourism examining the importance of sustainability, from the public and private sector perspective; and the consumer behavior perspective. Regarding the sustainability perspective, the benefits of the tourism sector in contributing to sustainable development goals are highlighted. Nevertheless, not all tourism providers are aware of health and wellness issues. Knowing and taking ownership of these issues makes it more feasible to provide health and wellness tourism sustainable products and services. Also, problems related to the importance of sustainability are debated in wellness tourism, explicitly referring to methods for minimizing its impacts. As the environment is close to the set of services in the health and wellness tourism sector, quality is a major requirement in services, facilities, protocols, and professional training (Alexis-Thomas, 2020; Lubowiecki-Vikuk et al., 2021; Su et al., 2018).

The recent literature stresses the necessity of understanding consumer behavior. Wellness tourism covers an extensive range of tourism activities and contexts. However, wellness tourists are not a homogenous group in terms of needs, expectations, and behavior. There are cultural differences, for instance, on price perception, brand love, revisit intention, and recommendation intention (Han et al., 2020; Bushueva, 2020). Wellness types of attributes reveal the preferences of different consumer profiles (Park et al., 2020):

1. physical attributes - cardiovascular and health-enhancing activities, including Technogym, Pilates and treadmill;
2. beauty care attributes - massage, rehabilitation, aromatherapy, remedies and facial cares;
3. nutrition and diet wellness attributes - diets, toxin-free, and detox;
4. mental and education attributes - reading, art and seminars;
5. rest and meditation attributes - rest, relaxation, yoga and mindfulness;
6. social and environmental wellness attributes - scenic, view, natural, garden and trail.

The positive relationship between quality and satisfaction is also a concern related to wellness tourists (Loke, 2020). Studies have revealed the need for medical and health practitioners to showcase high-quality and best practices for all their target markets (Soltani et al., 2020). Furthermore, the quality of treatment is an essential attribute for potential medical tourists, especially for outbound medical tourism. Additionally, medical patients now prefer to travel short distances to get medical treatment (Chia & Liao, 2020).

Finally, the literature emphasizes that the consumption patterns are changing among travelers due to COVID-19, with the growing popularity of wellness tourism. Also, new forms of tourism, including slow tourism and smart tourism, may stimulate future tourism activities and push businesses to reevaluate their service designs and distribution channels (Wen et al., 2021). Furthermore, long-stay tourism has been increasingly considered a potential tool for rural communities seeking social and economic revitalization (Lyu et al., 2021).

NEW TRENDS IN WELLNESS TOURISM MANAGEMENT

The COVID-19 pandemic has negatively impacted not only the economy but also peoples' physical and mental health. Hence, post-pandemic recovery plans must include products that focus on mental healing. In such a situation, the pandemic also presents a transformative moment or opportunity to change the world (Hall et al., 2020). A study presented at the Wellness Tourism Association's November 2020 annual meeting identifies "de-stressing," "relaxation," and "self-healing" as the top triggers in terms of what academics believe motivates consumer purchases of wellness-related travel services. Tourist attractions, accessibility, location, and link with nature and excellent wellness tourism facilities are supporting factors while selecting a destination (P. S. & Abraham, 2017).

However, new forms and practices in tourism are being studied and developed that may impact wellness tourism service providers. Standards for services management, with new practices and protocols, significantly affect the operation of tourist companies worldwide. According to Coelho and Mayer (2020), four major dimensions are related to the management of tourist services in a post-COVID scenario.

1. **Process and technology management**, encompassing redesign of processes to reduce or replace personal interactions (such as interactive machines, robots, artificial intelligence, online services, among others); new mobility controls and flows of people, materials, equipment, and information; new hygiene and biosafety protocols; and new standards for queue management and service processes.
2. **Facility and capacity management**, including the redesign of facilities and equipment to reduce crowds; new cost estimations for space usage in physical facilities, transportation, and attractions; innovations for cleaning and sanitizing facilities; new uses for idle physical facilities.
3. **People management**, including monitoring employees' health (from prevention to treatment); training in new procedures; and the adoption of remote work, when possible.
4. **Customer interactions management**, covering the redesign of experience and service encounters; monitoring the behavior and controlling the health of travelers through technological resources, digital documents and big data; the education of tourists to adopt new hygiene and safety practices; new business practices with a transition to virtual environments.

Given this, it is clear the urge to develop new practices and service protocols for the wellness tourism sector. Health is the central theme on the world scene, so situations related to social encounters and mobility are being increasingly observed. Businesses must redefine their service capacity based on the security they offer to customers, especially in regions where vaccination against COVID-19 remains slow. The adoption of measures and tests to monitor the health of employees has been adopted in many companies, and there is also a need for training professionals to meet new demands and processes. In addition, new ways of managing interactions with customers are required once the wellness tourism sector is characterized by a high degree of interactivity between people. The services must adapt, with the necessary care to guarantee the health and well-being of all those involved (Coelho & Mayer, 2020).

Regarding the reassessment of other tourism products through health tourism, the pandemic will possibly mean that: i. tourism is more about health, where standard tourist products (not belonging to the health tourism segment) will be required to use more health-related service techniques - in many ways, this will possibly dilute the limits of the so-called health and wellness tourism products; and ii. health is more than tourism since COVID-19, it seems normal to consider highly technical variables specific to health in tourism (Hernández et al., 2020)

CONCLUSION

Tourism, in general, can contribute to improving people's quality of life and health (Dolnicar et al., 2012, 2012; Filep, 2014). As demonstrated again with the pandemic situation, there is a strong relationship between health and tourism. Tourism experiences associated with the pursuit of well-being presents potential to expand, since the pandemic raised awareness of the importance of wellness – a state of physical, mental, and social wellbeing - in people's lives.

Given the trend of moving toward a wellness lifestyle, the consensus among entrepreneurs, investors, government, and the public sector will increase, particularly in large cities (Lee et al., 2020). Businesses that fail to provide environments and services related to well-being will undoubtedly lose a lot in the choice of the current consumer. Furthermore, this premise is not limited to areas of the gym, swimming pool, or sauna. The demand for well-being has gone much further, and those who are not in line with this expectation will certainly lose commercial appeal over time.

In sum, we envision that health and wellness tourism can have an essential role in the recovery of tourism service providers and destinations. The health and wellness trends identified open new opportunities for the future. Public and private sectors should contribute to the sustainable development of the field, investing in methods for minimizing its impacts. Furthermore, understanding consumer behavior is paramount since wellness tourism covers an extensive range of tourism activities and contexts, with a high degree of interaction among people during the service encounters. In addition, emerging tourism service management standards will require that businesses in this sector establish new practices and innovations based on quality and health principles, observing the new processes and protocols for wellness tourism services, and responding successfully to the new tourists' demands and behaviors.

REFERENCES

Ahmed, M. Z., Ahmed, O., Aibao, Z., Hanbin, S., Siyu, L., & Ahmad, A. (2020). Epidemic of COVID-19 in China and associated Psychological Problems. *Asian Journal of Psychiatry*, *51*, 102092. doi:10.1016/j.ajp.2020.102092 PMID:32315963

Alexis-Thomas, C. (2020). An examination of issues related to tourism and health and well-being as a sustainable development goal by tourism providers in Tobago. *Worldwide Hospitality And Tourism Themes, 12*(3), 293–303. doi:10.1108/WHATT-02-2020-0006

Bushueva, I. V. (2020). Trends of Wellness Tourism Development in Russia. *Turismo-Estudos E Praticas, 4*.

Cai, Y., Ma, J., & Lee, Y.-S. (2020). How do Chinese travelers experience the Arctic? Insights from a hedonic and eudaimonic perspective. *Scandinavian Journal Of Hospitality And Tourism, 20*(2), 144–165. doi:10.1080/15022250.2020.1744185

Chi, C. G., Chi, O. H., & Ouyang, Z. (2020). Wellness hotel: Conceptualization, scale development, and validation. *International Journal of Hospitality Management*, *89*, 102404. Advance online publication. doi:10.1016/j.ijhm.2019.102404

Chia, K. W., & Liao, Y. M. (2020). An Exploratory Study of Factors Influencing Chinese Outbound Medical Tourism. *Journal of China Tourism Research*, 1–19. Advance online publication. doi:10.1080/19388160.2020.1780177

Coelho, M. F., & Mayer, V. F. (2020). Gestão de Serviços Pós-Covid: O que se pode aprender com o setor de turismo e viagens? *Revista Eletrônica Gestão & Sociedade*, *14*(39), 3698-3706. Retrieved from: https://ges.emnuvens.com.br/gestaoesociedade/article/view/3306

Csirmaz, É., & Pető, K. (2015). International Trends in Recreational and Wellness Tourism. *Procedia Economics and Finance*, *32*(15), 755–762. doi:10.1016/S2212-5671(15)01458-6

Cunha, L. (2006). Turismo de Saúde – conceitos e mercados. *Revista Lusófona de Humanidades e Tecnologias: estudos e ensaios*, *10*, 79-84. Retrieved from: https://recil.grupolusofona.pt/handle/10437/2577

Dillette, A. K., Douglas, A. C., & Andrzejewski, C. (2020). Dimensions of holistic wellness as a result of international wellness tourism experiences. *Current Issues in Tourism*, *24*(6), 794–810. doi:10.1080/13683500.2020.1746247

Dolnicar, S., Yanamandram, V., & Cliff, K. (2012). The contribution of vacations to quality of life. *Annals of Tourism Research, 39*(1), 59–83. doi:10.1016/j.annals.2011.04.015

Faulkner, J., O'Brien, W. J., McGrane, B., Wadsworth, D., Batten, J., Askew, C. D., Badenhorst, C., Byrd, E., Coulter, M., Draper, N., Elliot, C., Fryer, S., Hamlin, M. J., Jakeman, J., Mackintosh, K. A., McNarry, M. A., Mitchelmore, A., Murphy, J., Ryan-Stewart, H., . . . Lambrick, D. (2020). Physical activity, mental health and well-being of adults during initial COVID-19 containment strategies: A multi-country cross-sectional analysis. *Journal of Science and Medicine in Sport.*

Filep, S. (2014). Moving Beyond Subjective Well-Being A Tourism Critique. *Journal of Hospitality & Tourism Research (Washington, D.C.), 38*(2), 266–274. doi:10.1177/1096348012436609

Goeldner, C. (1989). 39th Congress AIEST: English Workshop Summary. *Revue de Tourisme, 44*(4), 6-7.

Goodricj, J. N., & Goodrich, G. E. (1987, September). Health-care Tourism. *Tourism Management, 8*(3), 217–222. doi:10.1016/0261-5177(87)90053-7

Hall, C. M. (2011). Health and medical tourism: a kill or cure for global public health? *Tourism Review, 66*(1-2), 4-15.

Han, H., Kiatkawsin, K., Koo, B., & Kim, W. (2020). Thai wellness tourism and quality: Comparison between Chinese and American visitors' behaviors. *Asia Pacific Journal of Tourism Research, 25*(4), 424–440. doi:10.1080/10941665.2020.1737551

Hernández, C., León, C., & Moreno-González, A. A. (2020). Revalorización del producto turístico a través del turismo de salud. Turismo pos-COVID-19. Reflexiones, retos y oportunidades.

Kelly, C., & Smith, M. K. (2008). Holistic tourism: Integrating body, mind and spirit. In J. Jafari, R. Bushell, & P. Sheldon (Eds.), *Wellness and Tourism: Mind, Body, Spirit, Place*. Innovation and Tourism Book Series.

Koncul, N. (2012). Wellness: A New Mode of tourism. *Economic Research-Ekonomska Istraživanja, 25*(2), 525–534. doi:10.1080/1331677X.2012.11517521

Lee, L. Y.-S., Lam, K. Y.-C., & Lam, M. Y. C. (2020). Urban wellness: the space-out moment. *Journal Of Tourism Futures, 6*(3), 247–250. doi:10.1108/JTF-10-2019-0111

Li, C.-L. (2020). Quality of life: The perspective of urban park recreation in three Asian cities. *Journal of Outdoor Recreation and Tourism, 29*, 100260. doi:10.1016/j.jort.2019.100260

Loke, Z. (2020). Investigation Of Medical- And Wellness Tourists Of A Hungarian Spa To Explore Relationships Between Service Quality, Customer Satisfaction And Loyalty. *Deturope-The Central European Journal Of Regional Development And Tourism*, *12*(1), 102–118.

Lubowiecki-Vikuk, A., Dąbrowska, A., & Machnik, A. (2021). Responsible consumer and lifestyle: Sustainability insights. *Sustainable Production and Consumption, 25*, 91–101.

Lyu, J., Huang, H., & Mao, Z. (2021). Middle-aged and older adults' preferences for long-stay tourism in rural China. *Journal of Destination Marketing & Management*, *19*, 100552. doi:10.1016/j.jdmm.2020.100552

Malkina-Pykh, I. G., & Pykh, Y. A. (2014). Subjective well-being and personality: Implications for wellness tourism. *WIT Transactions on Ecology and the Environment*, *187*, 51–64. doi:10.2495/ST140051

National Wellness Institute. (2020). *The six dimensions of wellness*. Retrieved from https://cdn.ymaws.com/members.nationalwellness.org/resource/resmgr/pdfs/sixdimensionsfactsheet.pdf

P. S., S., & Abraham, S. (2017). Wellness Tourism - an Analysis of Tourist Preferences and Perceptions in Puducherry. *Journal of Hospitality Application & Research*, *12*(2), 61–79. https://search.ebscohost.com/login.aspx?direct=true&db=buh&AN=128135305&site=ehost-live&scope=site

Paixao, D. L. D. (2007). Thermae et Ludus: O início do turismo de saúde no Brasil e no mundo. *Turismo em Análise*, *18*(2), 133–147. doi:10.11606/issn.1984-4867.v18i2p133-147

Park, H., Lee, M., & Back, K.-J. (2020). Exploring the roles of hotel wellness attributes in customer satisfaction and dissatisfaction: Application of Kano model through mixed methods. *International Journal of Contemporary Hospitality Management*, *33*(1), 263–285. doi:10.1108/IJCHM-05-2020-0442

Quintela, J. A., Costa, C., & Correia, A. (2016). Health, wellness and medical tourism – a conceptual approach. *Enlightening Tourism: A Pathmaking Journal*, 1-18. Retrieved from: https://dialnet.unirioja.es/servlet/articulo?codigo=5559522

Scliar, M. (2007). História do Conceito de Saude. *Physis (Rio de Janeiro, Brazil)*, *17*(1), 29–41. doi:10.1590/S0103-73312007000100003

Shah, K., Kamrai, D., Mekala, H., Mann, B., Desai, K., & Patel, R. S. (2020). Focus on Mental Health During the Coronavirus (COVID-19) Pandemic: Applying Learnings from the Past Outbreaks. *Cureus*, *12*(3), e7405. doi:10.7759/cureus.7405 PMID:32337131

Sirgy, M. J., Kruger, P. S., Lee, D.-J., & Yu, G. B. (2011). How Does a Travel Trip Affect Tourists' Life Satisfaction? *Journal of Travel Research*, *50*(3), 261–275. doi:10.1177/0047287510362784

Smith, Kwok, Tang, & Nutbeam. (2006). WHO Health Promotion Glossary. Health Promotion International, 21(4), 340–345. doi:10.1093/heapro/dal033

Smith, M., & Puczkó, L. (2009). *Health and Wellness Tourism*. Butterworth-Heinemann– Elsevier.

Soltani, S., Ghorbanian, M., & Tham, A. (2020). South to South Medical Tourists, the Liminality of Iran? *Journal of Quality Assurance in Hospitality & Tourism*, *22*(1), 55–81. doi:10.1080/1528008X.2020.1747588

Tapia, J., Dieste, M., Royo, E., & Calvo, E. (2020). Factors affecting the choice of medical tourism destination: Spain as a host country. *Journal of Quality Assurance in Hospitality & Tourism*, 1–24. Advance online publication. doi:10.1080/1528008X.2020.1848749

Thal, K. I., & Hudson, S. (2019). A Conceptual Model of Wellness Destination Characteristics That Contribute to Psychological Well-Being. *Journal of Hospitality & Tourism Research (Washington, D.C.)*, *43*(1), 41–57. doi:10.1177/1096348017704498

Viegas Fernandes, J., & Viegas Fernandes, F. M. (2011). *Turismo de saúde e bem-estar no mundo: ética, excelência, segurança e sustentabilidade*. SENAC.

Voigt, C., Brown, G., & Howat, G. (2011). Wellness tourists: in search of transformation. *Tourism Review, 66*(1/2), 16-30.

Wen, J., Kozak, M., Yang, S., & Liu, F. (2021). COVID-19: Potential effects on Chinese citizens' lifestyle and travel. *Tourism Review*, *76*(1), 74–87. doi:10.1108/TR-03-2020-0110

World Tourism Organization. (2018). *UNWTO/ETC Launch report and health tourism*. Retrieved from: https://www.e-unwto.org/doi/book/10.18111/9789284420209

Yu, G. B., Sirgy, M. J., & Bosnjak, M. (2020). The Effects of Holiday Leisure Travel on Subjective Well-Being: The Moderating Role of Experience Sharing. *Journal of Travel Research*. Advance online publication. doi:10.1177/0047287520966381

Chapter 10
Mixed Realities as the New Reality in Tourism:
The Benefits of New Technologies for Visitors, Workers, and the Industry at Large

Maria Francisca Casado-Claro
https://orcid.org/0000-0002-4395-9620
Universidad Europea de Madrid, Spain

Marina Mattera
Universidad Europea de Madrid, Spain

ABSTRACT

This chapter proposes a comprehensive approach to understand not only how augmented reality and virtual reality operate within the tourism industry, but also how mixed reality can contribute to enhance the visitor experience and how tourism organizations can move beyond traditional communication and physical experiences into a new type of tourism approach that helps them stay relevant in the long term, as well as in the mid-term. Since the tourism industry is amongst the most affected by the COVID-19 pandemic, changes are essential to ensure an adequate adaptation to the 'new normal'. Technology enables various tourist organizations to generate greater value creation and opens possibilities to be able to extend the visit beyond physical presence, to ensure the safety of workers and visitors, to improve processes, and to make them more competitive overall. If this is carried out in collaboration with all stakeholders, one destination can generate a solid network to promote itself and become competitive during travel restrictions and in preparation for a post-pandemic new normal.

DOI: 10.4018/978-1-7998-7239-9.ch010

Copyright © 2021, IGI Global. Copying or distributing in print or electronic forms without written permission of IGI Global is prohibited.

INTRODUCTION

Tourism promotion organizations and destination management organizations (DMOs) who want to attract travelers to their destinations face harsher competition than ever. Therefore, if they want to improve their stakes of succeeding and stay relevant in the long term, they should not only update the methods they use to create value for visitors, but also the way they communicate and deliver that value. In today's attention economy, tourist destinations compete against one another to attract the interest of their audiences, as a preliminary step to transforming that interest into a visit (Casado-Claro, 2017), but they also must compete with a myriad of other actors that publicize their content online (natural and cultural attractions, resorts, influencers, etc.).

During the past decades, the increase of digitalization has been affecting humankind, daily lives, and especially leisure activities. In 2020, with the SARS-COV-2 pandemic affecting all countries around the world, this process saw an exponential increase as the many mobility restrictions called for a virtual replacement of all activities wherever it was possible. In this context, tourism was one of the most affected industries, as visitors could not travel due to their own fears and countries' travel restrictions.

However, this context inspired stakeholders in this sector to become even more creative and provide alternative solutions. To this end, new technologies have been incorporated to create experiences that would meet the requirements of consumers and the government's restrictions, such as virtual tour experiences. As some regions and countries eased their restrictions on travelling, as in the case of the Schengen Area (Europe) in the summer of 2020, digital elements became a key tool to reduce uncertainty. Technological elements provided more information in a highly interactive way, limiting contact with other individuals, or delivering updates in real time, among many other features.

Therefore, new technologies were able to contribute to reducing stress, in both visitors and workers. Through these novel methodologies it was possible to make the tourist experience at the destination more enjoyable. The inception of 5G networks and the constraints of travel in the times of COVID-19 have enhanced their application and added a new dimension to their use: it is not only about enjoying the travel experience, but also about making it safer.

The use of new technologies, such as Augmented Reality (AR) and Virtual Reality (VR), in the travel, hospitality and tourism industries promotes happiness and subjective well-being, not only by reducing various stress factors such as the fear of getting lost in an unfamiliar place, or the discomfort of communicating in a foreign language, but also by generating added value for visitor attractions and tourism-related businesses (Williams and Slak.Valek, 2019; Correira-Loureiro,

Guerreiro and Ali, 2020; Pai et al., 2020). Furthermore, in the context of the COVID-19 pandemic, these technologies enhance safety, by helping avoid physical contact in certain travel-related situations, such as when checking-in in a flight or hotel, or when visiting museums.

Following on the research conducted under the MIECAT project funded by the Erasmus+ Programme, this chapter further enlarges on the results of applying VR, AR or a combination of these elements known as Mixed Reality (MR) in the tourism industry. Specifically, the earlier studies that were undertaken under the project focused on the creation of value added in cultural heritage in Europe thanks to a variety of initiatives, including mixed realities. Following up on the elements identified through those analyses, the present study considers the impact of those technologies in the tourism industry considering the COVID-19 crisis.

In this chapter, AR, VR and MR will be defined and a revision of recent literature on the use of mixed reality in tourism will be carried out. Whereas AR already plays a key role in travel apps and VR has already been used for marketing purposes, MR creates a rich and pleasurable experience, therefore it has the potential to bring about various innovations to the travel and tourism industry. Following this analysis, the study will present the evidence from the experiences and results of the usage of AR, VR and MR, and how these have contributed to the tourism industry in various manners in the context of the COVID-19 pandemic. Finally, the potential impact of a continuous digitalization process will also be discussed and conclusions useful for academia, as well as practitioners, will be outlined.

DIGITALIZATION BEYOND CURRENT REALITY

Todays' travelers have more possibilities than ever to spend their leisure time as well as a universe of information about the activities at their disposal. Just as consumers are more informed nowadays, so are tourists, whose expectations are also higher as most of the consumers expect to immerse themselves in the tourist experience (Neuburger & Egger, 2018: 66), they seek not only information but also entertainment, active participation, and multisensory stimulation in combination with innovative design elements (Pine and Gilmore, in Neuburger & Egger, 2018). In recent times, a debate has appeared regarding the tourists of the future. With the rise of the informed tourist, destinations face the challenge of updating the way they create and deliver value to visitors into the 21st century, especially millennials[1] (born at the turn of the century) and Generation T (born in the 2010s), by asserting their presence in the dual reality of the world surrounding us and the virtual world created by digital technologies, as well as in social media.

AR, VR and MR have their place in the tourism and leisure industries, each with its specific use. VR can inspire travelers before leaving and reconstruct archeological sites, AR can facilitate and enrich the tourist experience at the destination by adding a layer of information to the surrounding environment and MR can help train and support employees in a professional context. In the lines that follow, we will define them and discuss their applications in the tourism and leisure industries, setting the stage for a discussion of how their use has become more relevant than ever against the backdrop of the COVID-19 pandemic.

Virtual Reality

Virtual Reality (VR) creates an alternate reality that is viewed through the lens of a specific pair of glasses, a particular room[2] or an enclosed device[3], although there are other systems that enable virtual reality to occur within a tablet or smartphone, but the sensory experience is not as complete because the user is not fully immersed in the alternate world.

The simulated environment is generated through computer-generated images (CGI) and the user can interact with the environment, meaning the virtual world will change depending on his or her actions. This implies there is a visual or other type of sensory feedback, that enables the user to be immersed in an experience, that would otherwise not be possible in a physical environment. For example, recreating how people used to live in the 19th century can only be achieved through this type of reality, as augmented would not be enough to generate a full experience given that society has significantly changed. The Viking Planet, for example, is a digital museum in Oslo (Norway) that offers several unique experiences based on the Viking heritage[4]. This experiential tool allows individuals to immerse themselves in the Viking era, understanding how life was in the region under Viking ruling.

Immersive exhibitions use this technology to recreate for the public the works of a painter. They attract large crowds, moved for the desire of experiencing art under a new lens, and when effectively managed both from the technical and the cultural point of view, they can result didactical and entertaining for the public. As was the case of the various immersive exhibitions devoted to the Dutch impressionist painter Van Gogh (such as the one entitled *Meet Vincent Van Gogh)* or the VR experience *Klimt's Magic Garden* at the MAK (Museum of Applied Arts)[5], in Vienna (Austria). Through this experience, individuals can "go inside" Klimt's artwork and explore a world inside some of his most famous paintings.

An added advantage for some of these immersive exhibitions is that they can be taken on tour, totally or partially, as was the case with the exhibition on the Belgian surrealist artist René Magritte[6], whose itinerating exhibition Magritte Virtual Reality was installed inside a giant bowler hat in the beach. Unlike other exhibitions, which consist of an immersive room, this one used Head Mounted Devices (HMDs).

In both cases, immersive rooms and HMD, immersive exhibitions are projected and produced by Information Technology (IT) programs that generate an alternative VR designed by a creative team and is enriched with music. The virtual world is digitally generated on a computer, then the images are recorded, animated, and reproduced through projections.

One of the drawbacks of this technology is the need for expensive equipment, which has hindered its generalized adoption. To overcome this obstacle inexpensive versions of HMDs, such as the Google Cardboard, have been designed. One example is the New Museum (New York, USA) and Rhizome Project collaboration, which created a virtual reality art experience for free, using smartphones. By downloading the app ‘First Look: Artists’ VR, anyone can enjoy this immersive experience (New Museum, 2017). In addition to these home-based and low-cost initiatives, new technological developments around smart glasses and wearables will also add a whole new dimension to VR technology.

Augmented Reality

Augmented Reality (AR) involves technologies that incorporate computer-generated elements to the reality viewed by a user. Such elements can be generated through the screen of a mobile phone (in which the superimposed image appears while showing the real background) or by using screens where added features can be seen in the real environment, where the audience is found. Filters on Snapchat, Instagram or Facebook are examples of this type of technology, that can be applied not only to a variety of fields, such as the arts and architecture, but also to diverse industries, including hospitality, travel, and tourism.

AR technology adds to the tourist experience by enabling travelers to plan and preview attractions before they travel, as well as accessing location-based information once they arrive at the destinations in an interactive, and simple manner (Nayyar, Mahapatra, Le, & Suseendran, 2018) without needing advanced digital skills.

As AR enables users to expand the current elements by incorporating computer generated elements to what they are viewing, a richer experience can be generated. For instance, at the Archeological Park of Pompeii (Italy), it is possible to scan ruins and view how the buildings used to look like before the ashes covered the city, while the ArcheoGuide project reconstructs the cultural heritage site of Olympia, in Greece.

Due to their size, tablets have revealed themselves as the most suitable devices for AR in museums and other cultural heritage sites, thus the French company Histovery offers various “augmented visit solutions” supported by its “HistoPad” tablets, both indoors (in monuments and museums) and outdoors (in city tours).

HistoPad rentals have taken the place of audio-guides in various monuments like the Palais des Papes in Avignon, the Castle of Chambord, the Castle of Blois, or the Conciergerie in Paris, historical monuments where visitors can make sense of empty rooms as they follow a tour adapted to their needs (language, time, age). Even though, it is not an immersive experience, it is practical and requires a milder investment than HMDs for VR.

One of the big debates in the use of AR is whether it should be free for users to download in their own mobile devices or provided at a price. In the case of the HistoPad, both the hardware and the software (the application) are rented. The return on the investment (ROI) comes from the fees charged, which makes it possible not only to ensure the quality of the tours by providing "edutainment" (education and entertainment), as well as to offer personalized versions (tours for children, adults, or people with special needs) in several languages, but also to take care of the maintenance and update of the equipment.

The drawback of this business model is that some people are not willing to pay the rental fee and users cannot take the technology with them when they leave, therefore it can only be used in a punctual moment and place. A way to overcome this would be offering the application for free, which is something some tourist offices and visitor attractions already do. This presents the advantage of enabling its use before, during and after the visit, even though certain features will only work with geolocation. The disadvantage is that studies show that some users are reluctant to download new apps and that, without maintenance, these free apps quickly lose their edge. Furthermore, it is still a challenge to develop a stable application for all possible devices on the market.

Although AR applications are free to users, they represent an investment that does not always bring a return. The authors of a study on mobile services offered in several French cities through embedded applications, geolocation applications, QR codes and AR applications, concluded that despite destinations' growing interest and its promising tourist marketing prospects, "the ability of tourist m-services to increase the city's value proposition, perceived by tourists during the pre-travel stage" was limited (Arnaud Rivière & Bourliataux-Lajoinie, 2017).

In between the two options (paying or free), the possibility of including the fee to use a lent device in the price of the entrance ticket would keep some of the advantages (no need to download an app, maintenance is ensured), without the perception of having to pay extra for the service.

Mixed Reality

Mixed Reality, in turn, creates a combination of virtual reality with some augmented features, where new environments are combined with digital objects, co-existing and interacting with each other, which requires immersive technology. It was created in early 1990s and initially used by the US Air Force project "Virtual Fixtures", where it was demonstrated for the first time that human performance could be enhanced and amplified, by combining virtual objects on a person's physical environment. One example of it is Microsoft's HoloLens, an optical see-through display that allows virtual information to appear on the real world.

Figure 1 Simplified representation of a "virtuality continuum"
Source: Milgram and Kishino, 1994.

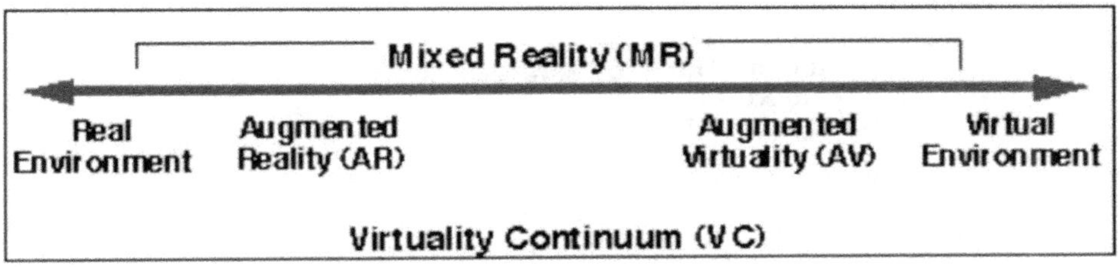

In their 1994 classical article, Milgram and Kishino defined Mixed Reality as "a particular subclass of VR related technologies that involve the merging of real and virtual worlds". Mixed Reality can be represented in a continuum with the real world at one end, VR at the other end and in between different configurations of AR/VR systems, as various levels of integration are possible (Nayyar et al, 2018: 157).

By resorting to MR, tourist attractions and services can see their offer enlarged and enhanced, being able to provide a personalized experience to each consumer. For example, the possibility of enjoying pieces or art in MR from anywhere in the world by accessing large online platforms is an unprecedented feature for cultural institutions. At the Tokyo National Museum, in Japan, 5G, mixed reality and artificial intelligence (AI) are used to allow visitors to enhance their experience of seeing a high-resolution reproduction of the Illustrated biography of Prince Shotoku, a national treasure also reproduced by Google Arts and Culture[7].

In addition, there is greater convenience and comfort, with greater access to information for customers as well as companies. These can learn better who their customers are, how they behave, and which type of experiences can be better suited to their needs, wants, and demands. Furthermore, there can be an immediate update with the latest trends, changes, and even ensuring each user has the best experience thanks to the connectivity and feedback he/she provides. Additionally, marketing and advertising can be tailored to the customers, so they obtain the information they are looking for, as well as promotions and the latest information without having to actively search for it.

To conclude this section, some of the ideas presented here are summarized in a comparative chart that readers can find in Table 1. The following section will discuss the applications that these technologies have found in the travel, tourism and hospitality industries, to set the context for exploring how they have been used to improve the travel experience or make up for it in the context of the COVID-19 pandemic.

Table 1. Comparative table

	AUGMENTED REALITY	VIRTUAL REALITY	MIXED REALITY
DEFINITIONS	Amplifies information about an object in the real world by displaying information in the virtual world	Provides an experience in an alternative reality. It could be the real environment in another location that cannot be visited due to restrictions.	A combination of VR and AR, where new environments can be combined with digital objects and part of the reality users are immersed in
MAINSTREAM APPLICATIONS	Museums, restaurants, hotels, tour operators, destinations	Museums, hotels, tour operators, tourism marketing	Museums
EXAMPLES	Targeting a painting hanging in the wall of a museum, the visitor can get information about the painter, the painting school, the details of the work. Another example could be scanning an image and obtaining a 3D augmented reality vision of how a particular building or archeological site looked in its original form.	Google Maps VR – Users can find their way using the front camera of their mobile devices to integrate the maps with the street view. It can also provide an immersive experience into a particular environment, such as the Museo d'Arte Orientale in Torino (Italy), that allows visitors to explore Beijing's Forbidden City.	By combining the added features of an enhanced reality (AR) with the extension of a virtual exploration (VR) a joint gaming experience can be created. At the Lázaro Galdiano Museum (Madrid, Spain), a mixed digital experience, called Enigma Galdiano, has been created to interact better with visitors, especially children.

Source: Own elaboration

HOW IS THE TOURIST EXPERIENCE ENHANCED BY MIXED REALITY?

Although the recent progress of Information and Communication Technologies (ICTs) and the generalization of the use of smart devices has boosted the use of both AR and VR, the development and adoption of the latter in the tourism industry is slower and more complex. To begin with, VR requires more sophisticated technology than AR, which makes its use more expensive. In any case, the use of AR, VR and MR is still far from being democratic, as it will still take some time for technology to improve, the cost of hardware to go down and for companies to create services that add real value to the customer experience. Nevertheless, destinations would miss a great opportunity if they did not take advantage of new technologies to enhance visitors' experiences.

Mixed realities enhance the travel experience in an effortless way as prospective visitors can preview and book a hotel, navigate around a destination, find dining and entertainment options, or bridge the language barrier by translating written signs and conversations, amongst other. It can all be done in a simple way. No advanced technologies are needed, as all of this can be done with the support of smart device where the proper applications are uploaded. Beyond fads and fashions, VR and AR could bring enormous benefits to people with special needs (people with disabilities and older adults, amongst other) by providing them with unforgettable tourist sensations (O'Sullivan and Kearney, 2018). On the other hand, a truly relevant contribution of these technologies would be the preservation ecosystems, fragile environments, and monuments, which could be closed to the real public but open to a large virtual public.

Figure 2. Customer journey map
Source: Own elaboration

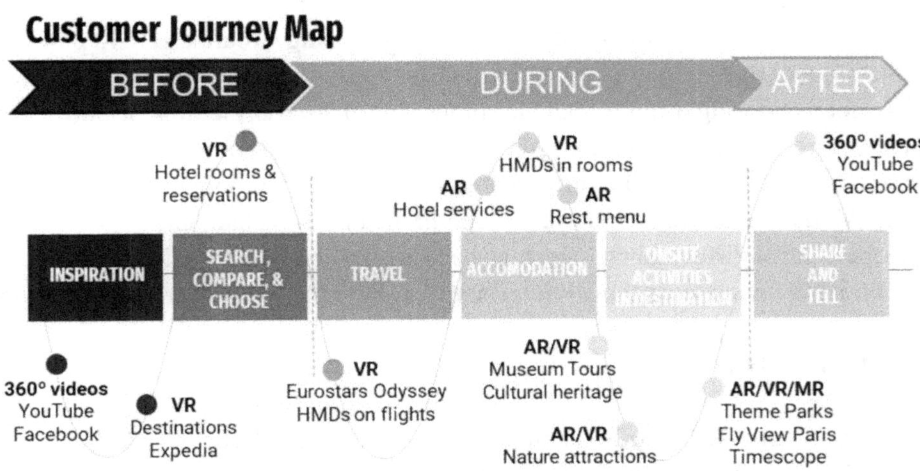

Museums and other artificial attractions can maximize their investment in new technologies by enhancing the visitor experience before, during and after travelling to the destination. Virtual sightseeing of various attractions can have an educational and entertaining function, which helps the public relate to the destination before the trip, but they can also prove helpful in planning the trip itself when they are enriched with valuable information such as the agenda of cultural events and festivals, the location of dining venues near museums or attractions that they are planning to visit or the hotel where they are planning to stay, or useful hints for shopping. During the trip, the travel experience can be enriched with VR applications that enable visitors to always know their whereabouts or communicate with locals in their own language in case they do not speak it.

In the paragraphs that follow, we will examine some examples that show how VR, AR and MR are being used along the customer journey (see Figure 2).

Before the Trip Begins and After it Finishes

As the mental journey of the customer begins before reaching the destination, travel agencies and DMOs are already using 360º videos to inspire prospective visitors in their quest. In this line, the Japan National Tourist Office (JNTO) has created an immersive 360º VR movie[8] entitled "JAPAN Where tradition meets the future VR", which can be enjoyed on the official JNTO website as well as in YouTube, although in order get a fully immersive experience VR goggles, such as Google Cardboard are needed. Professional quality productions like the one in the JNTO 360º VR movie bring the country alive and wake up the desire to visit the destination, they are a powerful promotion tool. On the other hand, 360º videos shared online by travelers already abound in YouTube. Even though the production is not so professional, they still are interactive and succeed in creating vivid impressions.

As for accommodations, Virtual Tours of hotels using VR are used as a powerful marketing tool that enables prospective guests to visit the rooms before setting foot in the hotel, offering a more realistic impression than 2-D pictures or 360º tours. Furthermore, the experience design can include a visit not only to different room types, but also other hotel facilities such as restaurants or outdoor equipment. Hotel chains Marriott and Best Western, and the French resorts chain Club Med already offer this kind of virtual tours. One step ahead are the companies that have integrated their booking processes in the VR interface to capture the customer right on the spot.

During the Trip

Eurostar, the high-speed train connecting France and the UK through the Tunnel Channel, offers its customers the "Eurostar Odyssey"[9], a VR experience of the seabed which can be enjoyed by looking through the windows with HMDs. On the other hand, Air France, British Airways, Garuda Indonesia, and the Paris Airports (Charles de Gaulle and Orly) are amongst the customers of SkyLights[10], a company that offers airlines VR entertainment, both inflight and in the lounge, by providing both the HMDs and the content (early-window blockbusters, series, and documentaries available in Full HD, 2D, 3D and forward-facing 360°, according to the company's website). Even though at the moment they are rented to passengers, in the coming future they could become a "must-have" feature in all long-distance flights and the price included in the ticket, as is the case with onboard entertainment at present.

As for museums and cultural heritage sites, it is the interpretation of their resources what makes visitors appreciate their real value. Without interpretation, the visit becomes an easily forgettable aesthetic experience and the opportunity to educate and communicate, which is part of their mission, is lost. Increasingly, museums are using AR and VR to enrich their visit and augment their experience, making it both educational and entertaining, and contributing to attract visitors, that in turn will fund and support the museums activities, reinforcing in parallel the museums' function.

AR, VR, and MR have also been implemented in leisure attractions, such as theme parks, and nature attractions. Theme parks, like Disneyland amongst other, have adopted virtual environments in rides and simulators, as well as in 3D and 4D theaters. The experience is being exported to Movie Theaters (5D, 6D, 7D) where different special effects go along with images (smoke, light rain, snow, air or water on the face, seat vibrations), projected on the screen or on HMDs. The theme park Futuroscope, located in Poitiers, France, inaugurated in 2018 a 5-D sensorial attraction called "Sebastien Loeb Racing Xperience" which features VR to make visitors feel like seating at the co-pilot's seat driving through Alsace. The attraction, which combines images projected on an HMD (head mounted display) with dynamic seats, is not the only one in the theme park that relies on AR and VR to enhance the visitors' experience[11].

In some destinations, VR experiences can be the attractions by themselves, not the main ones but part of the complementary offer. Not long ago the fashionable way to get a romantic bird's eye view of a city was going up to the top of high-rise building, like the ones in Tokyo or New York, or riding on a Ferris wheel like the London Eye or the Great Berlin Wheel, but now visitors who would like to see Paris from a unique perspective can visit Flight View Paris[12] and opt for one of three VR experiences: The Incredible Flyover, Paris Undercover or Rebuilding Notre-Dame.

Finally, in nature attractions, the use of technology may provide an opportunity to deliver safety and environmental information in a more engaging and targeted way, as well as to enrich the experience on site, point out Potter, Carter, & Coghlan (2016). The authors carried out an experiment in Australia with a HMD VR application, which supported interpretive information about the wildlife that may be found on a nature tour, allowing visitors to identify wildlife when they saw it. The app consisted of five-minute sessions using the Google Cardboard and the Google Cardboard app, with participants able to complete multiple sessions. The authors concluded that, amongst other, it could be used not only to enrich the experience, but also to provide useful information, both static and interactive, to visitors about activities and general safety guidelines.

THE IMPACT OF THE COVID-19 PANDEMIC IN THE TOURISM INDUSTRY

The tourist experience is a continuous journey that starts before the travelling begins and continues after the person returned home. Whereas the enjoyment of the experience is the product of a myriad interactions with the mechanical and human clues at the destination, the memories of the experience echo those episodes that caused a strong impression in the visitor, either because they were highly satisfactory or because they resulted traumatic. Modern technologies can contribute to reduce stress points and amplify the positive aspects of the overall tourist experience, ensuring the satisfaction is at its maximum, and minimizing negative or less positive aspects of it (Yang, et. al, 2021) .

As travel restrictions were eased during the summer of 2020, individuals started considering their alternatives to plan their trips as well as employees within the industry were called to resume their duties, digitalization became a key factor in ensuring the 'new normal' would suit all stakeholders. In this context, many elements begun to change, and the adaptation process was different for each destination.

Conditions to comply with before reaching a destination are amongst the main factors when deciding to visit that particular place, as tourists take into consideration various aspects such as convenience, safety, or cost. In addition, one of the key factors is the local authorities' requirements for visitors, which typically include a visa waiver for tourists or, in its absence, the difficulty or easiness of the process to obtain a tourist visa. During the pandemic, a requirement was added: being considered a non-COVID-19 hazard for the destination. In the Schengen Area[13] most countries begun accepting tourists from within the region without any requirements. Yet, as the cases of individuals who tested positive to the virus increased, new restrictions were imposed.

Some countries, such as Italy, required visitors to conduct a COVID-19 test within the 72 hours prior to entering the country and individuals had to fill out a self-declaration form, holding various personal information, including their whereabouts and if they had been in close contact with anyone diagnosed with the disease. However, there was a complete lack of digitalization in this process, visitors filled out physical forms, personnel from the airport and airlines had to gather all the information, and, in sum, the process was neither fully efficient nor did it ensure the afore-mentioned workers safety.

Other countries, such as Spain, instated a virtual self-declaration where individuals filled out much more detailed information regarding their travelling, including the seat they occupied when arriving by plane, train or boat, and the address at which they would be staying during their visit to the country. When users finished filling up the form, information was automatically sent to a central database and travelers received a QR code, that was scanned at the check point when entering the country. After the visitor arrived and his or her code was scanned, the system automatically sent an SMS with detailed contact numbers and institutions that could be of aid in case they showed any symptoms of COVID-19. In doing so, it became much safer for workers and travelers alike, minimizing the physical contact and therefore risk of contracting the virus.

Outside Europe, other countries also relied on a digitalized entry form, going as far as requiring the COVID-19 test results to be uploaded in the virtual self-declaration form, so that the authorities could confirm the eligibility of people to travel to this country. Such is the case of Argentina, which required visitors from MERCOSUR countries, that wished to enter the nation as tourists, to present a negative COVID-19 test result and had completely digitalized all the process. Hence, when a tourist who had completed the self-declaration and had submitted all documents, had a much easier and simpler process to enter the country.

Complementary to the digitalization and usage of QR codes and similar technologies, organizations in charge of travel infrastructures (such as airports and train stations) and travel services providers (such as airlines or train operators) begun incorporating technology to simplify the tourists' travel experience. This contributed to minimizing the proximity between visitors and employees of these sectors, thus making it safer for all stakeholders. In 2019, using Ocean 3D technology, the first virtual tours of airports were created, achieving an enormous success (Berti, 2019), and in some cases having this pre-existing digitalization contributed to a positive and enhanced travel experience during COVID-19.

However, the feature most organizations have been incorporating within their internal operations is VR and MR. Even though visitors do not get to be in touch with these technologies directly, they have been gaining wider acceptance and during COVID-19 they have proven to be a great element to ensure safer environments and more efficient processes. A few years ago, some airports, such as the one in Helsinki (Finland), had begun digitalizing their processes to incorporate these types of technologies, creating a virtual environment and the results show lowering waste, improving safety, efficiency, and overall employee satisfaction (Ibáñez, 2018; Pushelberg, 2019).

ENHANCING THE TOURIST EXPERIENCE WITH VR, AR, AND MR IN THE CONTEXT OF COVID-19

The travel itself is one of their main concerns when prospective travelers decide to visit a destination, since they wish to minimize the risk of contagion in the times of COVID-19. Other of their major concerns is the accommodation, as they wish to feel safe, in an exceptionally clean and hygienic environment, and make sure that the risk they will be exposed to will also be as low as possible. In this context, organizations which had already developed Virtual Tours of hotels using AR/VR had a great advantage over other firms which had not yet become so digital. While these were used as a powerful marketing tool that enables prospective guests to visit the rooms before setting foot in the hotel, in this context it also provided an insight into the types of measures in place to ensure safety, into how restaurant and breakfast lounges looked like with social distancing and into other relevant aspects.

Along these lines, before COVID-19, Alonso Almeida (2019) had identified that the incorporation of technology should incorporate consumers' viewpoint in designing and building the experience, as well as the different people, processes and operations within the organization that are required to make this new launch a success; whereas Moro et al. (2019), after doing a comprehensive literature review on AR and VR, pointed out that these technologies have proved their capacity to improve and stimulate visitor experiences. Therefore, the design of strategies must be carried out with an overall outlook, and much detail. Those institutions that had already begun this approach were able to better cope with the limitations imposed at the beginning of 2020 and were able to further develop their initiatives, while others who had neither begun, nor considered such a vast usage of new technologies, were forced to adapt at a fast pace.

Some hotels created Mixed Reality tools to allow their guests to navigate the facility and find the right service without getting lost or being wandering around the premises. This allowed to limit the circulation of people, reducing the chances of individuals crossing paths with others that were not travelling in the same group. This is a good example of how these technologies were especially useful in preventing unnecessary physical proximity, protecting travelers and workers alike, while improving the experience for visitors, making them feel much safer and keen to keep travelling.

Incorporating these technologies brought companies that have integrated their booking processes in the VR interface to capture the customer right on the spot and build trust with them before they arrived to the facilities. Further enlarging on safety, AR was used to provide details of the room, digital forms were used to complete the check in, together with QR codes that included in a simple manner much of the information visitors need during their stay such as menu options at restaurants or breakfast buffet, tour alternatives, etc. Most bookings were conducted through digital platforms and payments also were carried out either online, or by credit or debit cards with NFC technology, further digitalizing the accommodation sector.

AR, VR and MR improve the travel experience in an uncomplicated way as prospective visitors can preview and book a hotel, navigate around a destination, find dining and entertainment options, or bridge the language barrier by translating written signs and conversations, amongst other. In the COVID-19 context, this became even more evident and in diverse ways it has created new standards for the industry, further boosting the strategic plans several organizations had already set up and encouraging other institutions to enhance their digitization process. When referring to experiences at the destinations, it had been previously evidenced that museums and other cultural attractions can maximize their investment in new technologies by enhancing the visitor experience before, during and after travelling to the destination.

Travelling Without Leaving your Living-Room

Virtual sightseeing of various attractions can have an educational and entertaining function, which helps the public relate to the destination before the trip, but they can also prove helpful in planning the trip itself when they are enriched with valuable information such as the agenda of cultural events and festivals, the location of dining venues near museums or attractions that they are planning to visit or the hotel where they are planning to stay, or useful hints for shopping. Potter, Carter, & Coghlan (2016) point out that, in nature attractions, the use of technology may provide an opportunity to deliver safety and environmental information in a more engaging and targeted way, as well as to enrich the experience on site.

Shehade & Stylianou-Lambert (2020) conducted thorough interviews to museum professionals during the first months of the COVID-19 pandemic to evaluate the impact of having previously digitalized their cultural offer experiences. Through their study, they confirmed pre-existing usage of the technologies made it much easier for institutions to adapt to the new reality and continue to offer some services even if their facilities had to be closed to the public. Moreover, the experts agree on the need for each organization to incorporate digitalization into their overall strategic business design, as well as understanding the interdisciplinarity of this type of initiatives.

Therefore, choosing the right technologies becomes key for each organization, and understanding that not all tools are necessary nor can provide value to the institution. As an example, 5G, mixed realities and artificial intelligence (AI) used at Tokyo National Museum allowed users to enhance their experience of seeing a high-resolution reproduction of the Illustrated biography of Prince Shotoku, a national treasure also reproduced by Google Arts and Culture. The possibility of viewing these exclusive elements from anywhere in the world by accessing large platforms that are online is an unprecedented feature of the tourism industries.

In Europe, some museums such as The National Gallery in the United Kingdom or the Rijksmuseum in the Netherlands have incorporated virtual tours and visits, including multiple features so tourists can carry out visits from the comfort of their own homes. In this context, the British organization has teamed up with Oculus, a firm specialized in designing VR including specific high-tech goggles. By joining forces with this company, they were able to create a VR experience of one of the gallery's wings which individuals can enjoy from the comfort of their homes (National Gallery, 2020).

Even though the experience is different from an on-site one, the museums that have undertaken these approaches have evidenced an increase in their online traffic and tourists are now keener to visit these institutions when the restrictions allow them to (Network of European Museum Organizations, 2020). While evidence will still be collected during the period in which travel bans are enforced, the overall results from the first months of the pandemic with the stricter restrictions show a

Figure 3. Amazon explore's featured experiences
Source: Amazon Beta (2021)

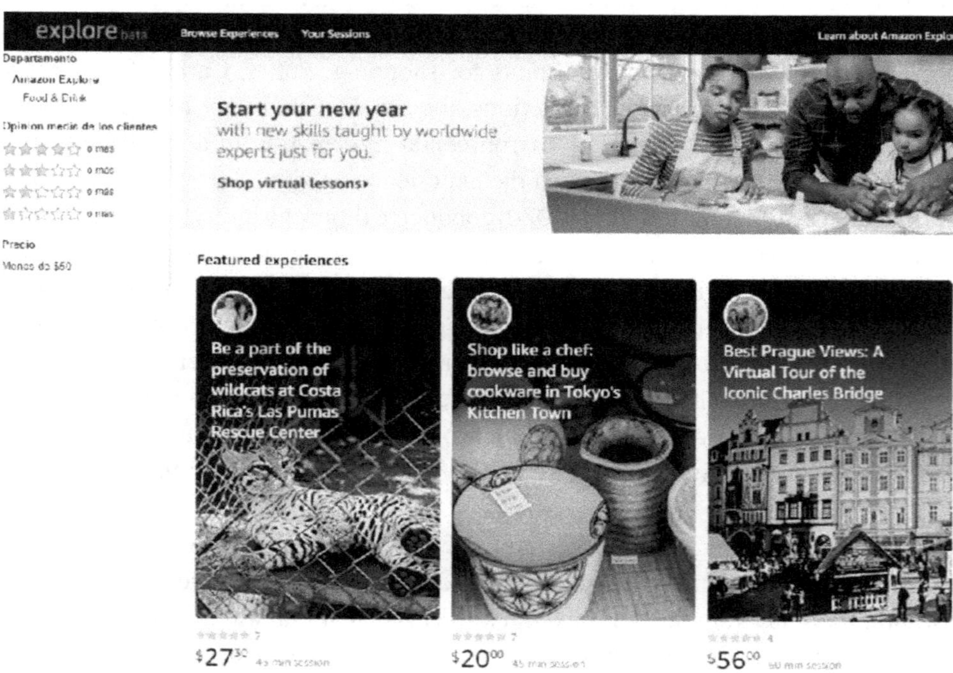

highly positive acceptance from individuals. Therefore, incorporating some digital features can build a relationship between the organization and prospective visitors to encourage them to come to the premises in the future.

In addition to the initiatives that are being set up by cultural institutions, there are other firms that have seen the potential of VR, AR, and MR for this industry. Such is the case of Amazon, that as a result of the pandemic that burst in 2020 evidenced an opportunity to offer a platform for individuals and organizations to promote their virtual experiences and allow consumers to easily find these activities. The initiative is called 'Amazon Explore' - at the end of 2020 it was still in a Beta phase, yet it has been gaining momentum as the travel restrictions continue to be in place. At the moment, an individual can select among a variety of options, as it can be viewed in Image 1, that range from shopping, to sightseeing, to being immersed in a natural environment.

While the current proposal does not include elaborate technological developments such as AR, VR, MR they are working towards converting the service offer into a more immersive experience. This can allow Amazon to enter a new market that had been untapped until this moment and allow visitors to be able to experience a destination in a different manner. It is expected that these types of initiatives will increase, especially if the travel bans are in place, and it could be used as a marketing tool to promote a destinations and specific activities, increasing the travelers' willingness to visit the site in person when this becomes a real possibility.

Considering the abovementioned changes, and that the digitalization process in the tourism industry had started before the 2020 pandemic broke out, yet it has been increasing exponentially since that event, there is a need to update the skillset of workers in the industry (UNWTO, N.D.; WTTC, 2020). This has become a major variable to enable organizations in travel, leisure, hospitality, and other related services, as jobs in the industry are now requiring technical and advanced soft skills. Without this change experts recognize there will be a greater difficulty in effectively achieving a digitalization of the industry.

DISCUSSION AND CONCLUSION

Digitalization of the tourism industry is a process that has been ongoing during the past decades, however Destinations would miss a great opportunity if they did not take advantage of new technologies to enhance their visitors' experiences. However, to take full advantage of technological opportunities, they should act in a proactive way, not only by adopting those new tools, but also by encouraging other actors in the destination to become active users as well. Moreover, these should be introduced with a full understanding of the implications, and with all stakeholders being fully onboard as change aversion could create resistance and limit the positive effects of introducing these new technologies.

COVID-19 has evidenced the need for a greater involvement of all stakeholders in the digitalization process to improve the prospects of the tourism industry, increase trust from visitors and create greater value. Destinations that are able to join forces and develop a strong digitalization process, changing all the institutions, including travel organizations, will be more competitive and most likely chosen by visitors when compared to others. In this sense, VR, AR, and MR can improve efficiency, reduce costs, increase transparency and trust from visitors, as well as the overall experience. Because of this, it is an aspect that should be taken into consideration by all organizations that are currently operating in the industry.

What is more, as it has been evidenced in some cases, it can provide when or where adequate tools to extend the experience at home, adapting the visit to a new environment where the travel restrictions may limit individuals from entering the destination. In doing so, institutions and destinations at large can generate trust and brand awareness so that in the future, when the free flow of tourism returns in the 'new normal,' individuals are more likely to visit sites they have previously had contact with and know can be trusted.

In the specific case of leisure activities, it will be interesting to evaluate during a longer period of time the results of the implementation of AR and VR initiatives that visitors can conduct from the comfort of their homes. Because of the current situation and considering the travel bans and limitations to tourism will continue to be in place at least for another year, it is interesting for those activities that can be reinvented and adapted to an in-home experience to test their implementation. It is expected that, as it was evidenced by the Network of European Museums Organization (2020), the online traffic of visitors will continue to raise, and if the strategic management is able to combine these tools with adequate marketing techniques, they could be a trigger for visitors to come in person whenever it becomes possible.

Moreover, the raise of new actors in the industry, such as Amazon, with disruptive product offerings that are included in a platform that is well known to consumers can be a risk for the remaining stakeholders. Since users have now become much more accustomed to shopping online and platforms like Amazon are now part of many consumers' shopping routine, it will be easier for this company to promote experiences than other agents. In this context, it would be interesting for organizations within the tourism industry to firstly evaluate the digital tools available, secondly select the ones that fit them best, and thirdly evaluate how they can face some of these non-traditional forms of competition.

Therefore, in the current context, there is a greater demand for tourism related sectors to incorporate these novel technologies, to improve workers' skillset to meet these demands, and to continuously improve in providing satisfactory experiences for consumers. Hence, the stakeholders involved will benefit from a constant training strategy that can go hand in hand with their digitalization strategy, to maximize the results of the implementation of these new tools. These changes will affect all sectors within the tourism industry and are a matter that should be tackled. Furthermore, the update of individuals' skillset can be the starting point of the creation of a virtual cycle, constantly updating and improving the digital tools incorporated in the organization.

Taking into consideration VR, AR, MR can be useful for many actors within the tourism industry, and that all of them have been affected in different manners by the halt in tourism flow, exploring new collaborations between those actors could prove to be an advantage. Since many have neither designed an elaborate digital strategy yet, nor are familiar with these tools but show interest and are working towards this goal, working together with other actors in the same situation could be mutually beneficial. However, it is the destination and the organizations that typically manage it (usually public institutions) that can have the networking to gather all the stakeholders involved and generate a holistic proposal.

If a particular destination was able to generate a global strategy and understanding between all the agents included within its tourism industry, it would be able to incorporate VR, AR and MR that generates the ultimate experience for visitors. This is particularly true in the case of a context where a travel ban is enforced, and the main flow of tourists cannot enter the destination due to sanitary restrictions. When diverse agents can work together, from travel agencies to leisure organizations, they are able to combine their expertise and create even better experiences for tourists. If this were to be achieved, those destinations would have a strong basis to consolidate a competitive advantage, visitors would have greater awareness of those locations and they may be at the top of their choices when travel bans are eased in the future.

As it was herein mentioned, the tourist experience is a continuous journey that begins when individuals select the location they wish to visit. Because of this, continuing the interaction with potential tourists even during the pandemic and with all the mobility restrictions in place becomes a key factor. To achieve this objective, AR, VR and MR, together with 3D and other related tools, have so far proven to be highly effective in creating highly interactive experiences. Whether it is on site, or from home, these challenges should be overcome to continue to provide value within the industry and ensure all stakeholders can thrive in the future with the 'new normal.'

ACKNOWLEDGMENT

This chapter is the product of independent research, which has not been funded by any grants.

REFERENCES

Alonso Almeida, M. del M. (2019). Robots, artificial intelligence and virtual reality: An approach to tourism industries. *Cuadernos de Turismo*, (44), 13–26. doi:10.6018/turismo.44.404711

Amazon Explore. (2021). *Amazon Explore's featured experiences*. Retrieved from https://www.amazon.com/-/es/b?ie=UTF8&node=19419898011

Berti, A. (2019). The digital twin: creating virtual airport tours with ocean 3D. *Airport Technology*. Retrieved from https://www.airport-technology.com/features/virtual-reality-at-airports/

Bogicevic, V., Seo, S., Kandampully, J. A., Liu, S. Q., & Rudd, N. A. (2019). Virtual reality presence as a preamble of tourism experience: The role of mental imagery. *Tourism Management*, *74*, 55–64. doi:10.1016/j.tourman.2019.02.009

Correira-Loureiro, S.M.; Guerreiro, J. & Ali, F. (2020). 20 years of research on virtual reality and augmented reality in tourism context: a text-mining approach. *Progress in Tourism Management,* 77. doi:10.1016/j.tourman.2019.104028

Cranmer, E. E., tom Dieck, M. C., & Jung, T. (2018). How can Tourist Attractions Profit from Augmented Reality? In T. Jung & M. C. Dieck (Eds.), *Augmented Reality and Virtual Reality: empowering humans, place and business* (pp. 21–32). Springer International Publishing. doi:10.1007/978-3-319-64027-3_2

Genç, R. (2018). The Impact of Augmented Reality (AR) Technology on Tourist Satisfaction. In T. Jung, & M. C. Dieck, (Eds.), Augmented Reality and Virtual Reality: empowering humans, place and business. Springer International Publishing. doi:10.1007/978-3-319-64027-3_8

Hassan, A., Ekiz, E., Dadwal, S. S., & Lancaster, G. (2018). Augmented Reality Adoption by Tourism Product and Service Consumers: Some Empirical Findings. In T. Jung & M. C. Dieck (Eds.), *Augmented Reality and Virtual Reality: empowering humans, place and business* (pp. 47–64). Springer International Publishing. doi:10.1007/978-3-319-64027-3_4

Ibáñez, A. (2018). Augmented Reality for working in airport control centers. *Ferrovial Blog*. Retrieved from https://blog.ferrovial.com/en/2018/05/augmented-reality-airport-control-centres/

Jung, T., & tom Dieck, M. C. (Eds.). (2018). *Augmented Reality and Virtual Reality: empowering humans, place and business*. Springer International Publishing. doi:10.1007/978-3-319-64027-3

Kim, M. J., Lee, C. K., & Preis, M. W. (2020). The impact of innovation and gratification on authentic experience, subjective well-being, and behavioral intention in tourism virtual reality: The moderating role of technology readiness. *Telematics and Informatics*, *49*, 101349. Advance online publication. doi:10.1016/j.tele.2020.101349

Milgram, P., & Kishino, F. (1994). A taxonomy of mixed reality visual displays. *IEICE Transactions on Information and Systems*, *77*(12), 1321–1329.

Moro, S., Rita, P., Ramos, P., & Esmerado, J. (2019, November 27). Analysing recent augmented and virtual reality developments in tourism. *Journal of Hospitality and Tourism Technology*. doi:10.1108/JHTT-07-2018-0059

National Gallery. (2020). *Sainsbury Wing VR tour*. Retrieved from https://www.nationalgallery.org.uk/visiting/virtual-tours/sainsbury-wing-vr-tour

Nayyar, A., Mahapatra, B., Le, D. N., & Suseendran, G. (2018). Virtual Reality (VR) & Augmented Reality (AR) technologies for tourism and hospitality industry. *IACSIT International Journal of Engineering and Technology*, *7*(2), 156–160. doi:10.14419/ijet.v7i2.21.11858

Network of European Museum Organizations. (2020). *Survey of the impact of the COVID-19 situation on museums in Europe*. Retrieved from https://www.ne-mo.org/fileadmin/Dateien/public/NEMO_documents/NEMO_COVID19_Report_12.05.2020.pdf

Neuburger, L., & Egger, R. (2018). Augmented Reality: providing a different dimension for museum visitors. In T. Jung & M. tom Dieck (Eds.), *Augmented Reality and Virtual Reality. Progress in IS* (pp. 65–77). Springer. doi:10.1007/978-3-319-64027-3_5

New Museum. (2017). First Look: Artists' VR. *New Museum*. Retrieved from https://www.newmuseum.org/calendar/view/1135/first-look-artists-vr

O'Sullivan, M., & Kearney, G. (2018). Virtual Reality (VR) Technology: empowering managers to reduce and eliminate accessibility barriers for people with Autism Spectrum Disorders. In *Transforming our World Through Design, Diversity and Education: Proceedings of Universal Design and Higher Education in Transformation Congress 2018* (*Vol. 256*, p. 253). IOS Press.

Pai, C. K., Liu, Y., Kang, S., & Dai, A. (2020). The role of perceived smart tourism technology experience for tourist satisfaction, happiness and revisit intention. *Sustainability*, *12*(16), 6592. doi:10.3390u12166592

Potter, L. E., Carter, L., & Coghlan, A. (2016). Virtual reality and nature-based tourism: An opportunity for operators and visitors. In *Proceedings of the 28th Australian Computer-Human Interaction Conference, OzCHI 2016* (pp. 652–653). New York: Association for Computing Machinery, Inc. 10.1145/3010915.3011854

Poux, F., Valembois, Q., Mattes, C., Kobbelt, L., & Billen, R. (2020). Initial user-centered design of a virtual reality heritage system: Applications for digital tourism. *Remote Sensing*, *12*(16), 2583. Advance online publication. doi:10.3390/rs12162583

Pushelberg, L. (2019). How airports are using VR to be less wasteful. *Aviation pros*. Retrieved from https://www.aviationpros.com/airports/airport-technology/article/21112224/how-airports-are-using-vr-to-be-less-wasteful

Shehade, M., & Stylianou-Lambert, T. (2020). Virtual Reality in museums: Exploring the experiences of museum professionals. *Applied Sciences (Basel, Switzerland)*, *10*(11), 4031. doi:10.3390/app10114031

United Nations World Tourism Organization. (n.d.). *Digital Transformation*. Retrieved from https://www.unwto.org/digital-transformation

Williams, R. B., & Slak-Valek, N. (2019). Pokémon GO is serious leisure that increases the touristic engagement, physical activity and sense of happiness of players. *Information Technology & Tourism*, *21*(4), 515–533. doi:10.100740558-019-00153-2

WTTC. (2019). *Mobile payments in travel & tourism: unlocking the potential*. Retrieved from www.wttc.org/-/media/files/reports/2019/mobile-payments-in-travel-tourism.pdf

WTTC. (2019). *World transformed: megatrends and their implications for travel and tourism*. Retrieved from www.wttc.org/-/media/files/reports/2019/megatrends-2019---world-transformed.pdf

WTTC. (2020). *To recovery and beyond: the future of travel & tourism in the wake of Covid-19*. Retrieved from https://wttc.org/Research/To-Recovery-Beyond

Wu, H. C., Ai, C. H., & Cheng, C. C. (2019). Virtual reality experiences, attachment and experiential outcomes in tourism. *Tourism Review*, *75*(3), 481–495. doi:10.1108/TR-06-2019-0205

Yang, T., Lai, I., Fan, Z. B., & Mo, Q. M. (2021). The impact of a 360° virtual tour on the reduction of psychological stress caused by COVID-19. *Technology in Society*, *64*, 101514. doi:10.1016/j.techsoc.2020.101514 PMID:33424061

KEY TERMS AND DEFINITIONS

Augmented Reality (AR): An interactive experience in which the objects in the real world surrounding the user, when observed through a visor (for example, a smartphone), are enhanced by computer generated information. While AR recreates the world surrounding us, VR creates a whole new world.

Customer Journey Map: A Design Thinking tool that depicts the positive and negative points of contact during a service encounter.

Destination Branding: Marketing practice which aims at making the destination easily identifiable to increase its value and attractiveness in the public eye.

Destination Management Organization (DMO): Tourist organization in charge of promoting the assets of destination and creating brand awareness in order to attract visitors, who in turn will dynamize the economy with their expenditures.

Head-Mounted Device (HMD): VR device. It is a headset worn on the head (i.e., mounted) or as part of a helmet that has optic displays in front of one or both eyes in which images are projected. Oculus Rift and HTC Vive are HMDs on the high end of the market, whereas Google Cardboard offers a more affordable version, which supports various applications.

HoloLens or Microsoft HoloLens: Mixed reality device. It is a pair of smart glasses developed and manufactured by Microsoft and operating in Windows. The latest version is HoloLens 2.

Mixed Reality (MR): A hybrid reality created by merging the real and the virtual worlds, where physical and digital objects co-exist and interact in real time.

Tourism Diversification: A destination's efforts to create various tourist products that suit different visitor needs and desires, in order to improve tourism-generated income or balance seasonality.

User Experience (UX): The user experience is a concept that appeared in the 2000s to define the result (benefit) and the feeling of the user (experience) in his/her interaction with a good or a service, in person or online (online user experience).

Virtual Environment (VE): A simulated environment is generated through computer-generated images (CGI).

Virtual Reality (VR): A parallel reality created through computer-generated images, which is perceived using a special device (a HMD is the most common one, but could also be a room like the CAVE), and can be interacted with in a realistic way. While AR recreates the world surrounding us, VR creates a whole new world.

ENDNOTES

1 Millennial travelers are said to be looking for authentic and surprising tourism experiences, without ever giving up technology, not in vain they are digital natives. Millennials travel more than other age groups and when they do, they make informed travel decisions, they demand unique experiences, authentic encounters, which they share on social networks.

2 The CAVE (CAVE Automatic Virtual Environment), developed before the appearance of HMDs, despite offering a more immersive experience in VR, it never went mainstream owing to its elevated cost. Nevertheless, it is still in use for industrial purposes, for example, in the automobile industry.

3 A Head Mounted Display (HMD) is the device most commonly used to enable VR. It is a headset worn on the head (i.e., mounted) or as part of a helmet that has optic displays in front of one or both eyes in which images are projected. Oculus Rift and HTC Vive are HMDs on the high end of the market, whereas Google Cardboard offers a more affordable version which supports various applications.

4 https://www.thevikingplanet.com/

5 https://www.mak.at/en/klimtsmagicgarden

6 https://www.bdh.net/immersive/magritte-vr

7 https://artsandculture.google.com/asset/an-illustrated-history-of-prince-shotoku/CwG8nWQNZrHYxQ

8 https://visitjapan-europe.jnto.go.jp/en/experience/vr/

9 https://www.eurostar.com/rw-en/destinations/odyssey

10 https://www.skylights.aero/

11 https://www.futuroscope.com/en/attractions-and-shows/sebastien-loeb-racing-xperience

12 https://www.flyview360.com/en/

13 Made up of 26 European countries (22 of them EU member countries), the area has a common visa policy and works as a single jurisdiction for international travel purposes, therefore there are no border controls when traveling from one of the signatory countries to the other. It is, therefore, the world's largest visa-free area.

Chapter 11
The Influence of Tourist Characteristics on the Perceived Quality of Hostels in Portugal:
An Exploratory Study

Beatriz Fernandes Duarte
Polytechinc Institute of Cávado and Ave, Portugal

Laurentina Cruz Vareiro
https://orcid.org/0000-0001-8945-1593
Polytechnic Institute of Cávado and Ave, Portugal

Bruno Barbosa Sousa
https://orcid.org/0000-0002-8588-2422
Polytechnic Institute of Cávado and Ave, Portugal

Victor Figueira
Polytechinc Institute of Beja, Portugal

ABSTRACT

The hostels sector in Portugal, despite its late appearance compared to other countries in the world, has registered an expressive development. However, the observed growth is not properly accompanied by research that supports and helps in understanding the sector and its future perspective. The chapter intends to be a contribution in this area, having as central focus the analysis of the influence of

DOI: 10.4018/978-1-7998-7239-9.ch011

Copyright © 2021, IGI Global. Copying or distributing in print or electronic forms without written permission of IGI Global is prohibited.

the characteristics of the tourists in relation to the expectations and perceptions regarding the service quality provided in a hostel in Portugal. The methodology used is a descriptive nature of a quantitative nature, based on an online questionnaire survey. The questionnaire essentially applies the SERVQUAL model, not fully following the original structure, presenting reformulated statements that are adequate to the hostel service in dimensions such as tangible aspects, reliability, empathy, promptness, and security.

INTRODUCTION

Hostels represent a segment with increasing importance in the hospitality market, with approximately 20% of all global tourism, being attributed to the youth and student market (Hecht & Martin, 2006). However, although this type of accommodation is part of the tourism system, it has characteristics that differentiate it from this model, representing an alternative means of accommodation to traditional hotels (Thomazi, 2019). As basic characteristics, components related to prices, structures, interaction and sharing of socio-cultural experiences and values between guests are assumed (Mané & Ferreira, 2017).

The issues related to hospitality reveal themselves to be increasingly global, subordinate to the evolution of populations, which change characteristics over time. Thus, the service provided has come to be seen as a need to be met, and not as a luxury item (Kadampully & Suhartanto, 2003). Hospitality is the aspect of the tourism industry charged with improving the quality of services provided during the stay of tourists from a destination (Brotherton, 1999). In this sense, the study of tourist accommodation is of particular importance, given the fact that it plays an elementary role in shaping the perception of the tourist experience and, consequently, in tourist satisfaction (Saraiva, 2013). However, there is a diversification of components in the hostels' public, causing the perceived quality to be influenced by issues such as the characteristics and culture of tourists, making the process of assessing the quality of services of this type of accommodation difficult (Gameiro, 2013). The concept of perceived value has recently gained and popular in the business environment. Perceived value has an effect on consumer behavior and provides strategic implications for the success of companies. According to Permatasari (2020), a perceived value defined as the benefits of customers believe, what are they receive relative to the costs associated with the consumptions. For a service organization, it is important to know how their services are perceived by their customers. Customer Perceived value has been conceptualized in the literature as a trade-off made by a customer between the costs and benefits of a particular product or service (Permatasari, 2020. Hence, perceived value is claimed to be a major tool to help the service company to gain a better competitive position in the market.

With regard to the current market, the quality attributed to this type of accommodation is not based only on the reduced price for the stay, cleanliness, comfort and location. The development of the sector has given rise to the appreciation of other factors such as the attributes of the environment, the staff, privacy, among other elements that require professionalism for their satisfaction (Silva & Köhler, 2015). Marques (2014) states that consumers reveal a growing demand that goes beyond the product, highlighting the demand for differentiation, values, attitudes and experiences. In this sense, taking into account that the customer is increasingly critical in relation to their choices, it is up to the sector to invest in higher levels of quality (Sousa, 2016). The provision of services that meet the demands of consumers provides satisfaction, which may result in their intention to return (Sousa, 2016), representing a competitive advantage, leading to an increase in market share and the success of activity in the tourism sector and industry (Hoffman et al. 1997).

However, despite this trend, the knowledge of the service quality of this type of accommodation is limited, lacking literature that substantially supports it (Musa & Thirumoorthi, 2001). In the Portuguese context, the same is true, with limited access to information and relevant data to support this sector. In this sense, this investigation focuses its analysis on the influence of the characteristics of tourists, with regard to expectations and perceptions of the quality of service provided in a hostel in Portugal.

BACKGROUND

Tourist Accommodation: The Hostel Concept, its Evolution and the Tourist Profile

Competitiveness assumes as a growing phenomenon and quality and innovation are a priority in the business world. Consequently, organizations express the need to create mechanisms capable of providing added value to consumers and occupying a better position in their minds. It is in this perspective that relationship marketing emerges as a strategy of loyalty and creation of strong links between organizations and their consumers (Macedo et al., 2021).

According to UNWTO (1995, p.57), "there is no universal definition for tourist accommodation, however, it can be considered as any facility that regularly (or occasionally) provides overnight accommodation for tourists" (Carrasco, 2017, p.7). In this context, tourist accommodation has a fundamental role, representing the most significant portion of expenses incurred by tourists in the destination, and being a critical element in shaping the perception of the tourist experience, having an impact on tourist satisfaction (Saraiva, 2013).

However, the new class of travelers demands, especially in urban and cultural destinations, that there be more accessible accommodation with a more familiar atmosphere, allowing a personalized stay and different from the global standards of the hotel chains (Pinto, 2018). In order to respond to the needs of the low cost accommodation segment, and as a way to accompany the tourist expansion in Portugal, local accommodations were created, where hostels are inserted in the form of accommodation establishments (Gama, 2018).

Hostel accommodation originated in Germany, shortly before the First World War (Silva et al., 2019). However, in a Portuguese context, the hostel appeared late compared to other countries in the world, starting its activity in 2005 (Silva, 2014). The initial phase of this sector was controversial due to a legal gap in the framework of the first functioning hostels. This gap represented the basis for the creation of the association "Hostels in Portugal" in 2008 (Volante, 2011), a period of growth until today. According to Lima and Vicente (2016), the significant growth in the hostels sector in Portugal, shows to be boosted by two main factors: the growth of low-cost tourism and the international recognition of the quality of the service offered. However, the growth observed is not properly supported by research that helps in understanding the sector and in the perspective of its evolution (R. Lima & Vicente, 2016). In addition, this type of accommodation only registers statistics of its activity in Portugal from the year 2018.

Taking into account the physical characteristics of the hostels, it is considered that there are similarities with the Portuguese Youth Hostels, differing in the fact that the target audience is more comprehensive (Tavares, 2018). In some cities, the opening of hostels contributes to the inversion of the idea of desertification and aging of infrastructure, making urban life more dynamic (Vasconcelos & Sampayo, 2017). According to Law Decree no. 80/2017 of June 30, the hostel can also expand to the level of other activities such as the installation of spaces related to commercial exploitation and provision of services, and can include designations such as "Bed & Breakfast" or "Guest House".

With regard to the profile of the hostel tourist, Pearce (1990) attributed the term backpacker and, together with Loker-Murphy (1995), understood characteristics such as the preference for accommodation at a reduced price, extensive travel time, interaction with different cultures and organization of the trip itself (Thomazi, 2019). In this context, other authors refer that hostel consumers are predominantly young singles, with higher education qualifications. As the main reasons for choosing a hostel, highlight the price and the opportunity to get to know new cultures, traveling with a group of friends. However, even though price is an important factor, it is not the determining factor in choosing accommodation (Mané & Ferreira, 2017; Saraiva, 2013; Tavares & Brea, 2020; Thomazi, 2019). Even though the young audience is the main hostel guest, there is a change in demand in this segment, considering an increase in the age group. Bourget (2012) refers that this type of tourist, designated as flashpacker, concerns old backpackers who, currently do not mind spending a higher monetary value for accommodation, looking for new elements such as quality and privacy, maintaining the atmosphere's appreciation of the social interactions characteristic of hostels (Thomazi, 2019).

Quality in Services: Measuring Quality in Services and the Factors that Affect Perceived Quality

In the tourist business sector, service quality is a key element for customer satisfaction. Therefore, the service provided must be continually innovated, developed and performed based on the needs expressed by customers, which change characteristics over time (Alauddin et al., 2019). Furthermore, Shahin and Dabestani (2010) consider that service quality is an integral and vital element for the success of an organization (Mohamad et al., 2017). For this reason, the notion of the concept of service quality also continues to evolve, and it is essential to create models to assess the quality of service provided by the modern market (Budianto, 2019; Santos et al., 2021).

As one of the main models for assessing quality of service, the SERVQUAL Model stands out, developed by Parasuraman, Zeithaml and Berry (1985). This model represents a very effective tool, used to measure the difference between customers' expectations and their perceptions of the service purchased (Sharma & Srivastava, 2018), in order to obtain "perceived quality" in the hospitality industry (Santos, 2019; Al-Ababneh, 2017; Moura et al., 2021).

Parasuraman et al. (1985) stipulate five dimensions of the service, which include guarantee, empathy, reliability, responsiveness and tangibility. These dimensions are subdivided into 22 categories, being evaluated using the inverted Likert scale based on 7 points, subordinate to the format "Totally disagree" to "Totally agree", regarding the expectations and perceptions of customers regarding the service (Pinto, 2018). The same authors recognize the emergence of gaps or gaps in the quality of services, existing between customer expectations and perceptions, demonstrating that the gap is not only a measure of customer satisfaction, but also a measure of service quality in relation to a specific dimension (Miguel & Salomi, 2004).

The gaps gave rise to the "Model of the 5 Gaps in Service Quality" (Sousa, 2016). This model exposes two distinct segments that influence discrepancies in quality of service: the customer and the supplier. In the context of the supplier, each gap is used as a form of analysis to improve the provision of services and in the case of the customer, it is demonstrated how the customer performs the quality assessment based on the satisfaction attributed to each of the items in the quality dimensions (Miguel & Salomi, 2004). It is understood that the customer has expectations for each of the dimensions of quality, in relation to the service to be purchased. In this way, the service expected by the customer is early defined by three essential elements: personal needs, word of mouth communication and previous experience. It is assumed that these factors guide and determine the expected service, interfering with the idea projected by the client's mind and serving as a quality parameter in the subjective evaluation of what is provided (Dozinete et al., 2017).

In the tourism industry, the quality perceived by customers is also based on impressions about the quality of a destination's substructure, the hospitality service and facilities such as accommodation (Vinh & Nga, 2015). These interpretations of the general quality of the product or service changed the conception that the customer attributes to the price (Asshidin et al., 2016) considering that, when experiencing high quality services in the hospitality sector, you will have a broader notion of value (Chiang, 2014). Pichitphan (2018) enunciates three aspects that make the customer feel that the amount paid is being fairly returned, distinguishing the consistency of the service, the professionalism of the team and the ease of the installations.

Crotti and Moretti (2019) state that the perceived quality of hospitality is directly related to the essence of the consumption experience, in contrast to Verhoef et al. (2009) who defend only the perspective of the perception of the moment. In a complementary way, considering the experience component, Knutson et al. (2009) recognize that quality depends essentially on the participation of the people involved, and not just on tangible aspects. The provision of good service is fundamental for the differentiation and value attributed by the customer, representing the first perception of the service acquired, through the physical environment (Moretti, 2015).

However, Gameiro (2013) points out that the perception of customers is influenced by external factors that hinder the assessment of service quality. As an example of these factors, the author suggests the different needs verified in gender, age, culture, customer preferences and psychological status, which also constitute a significant weight in the formation of service expectations. In this context, Zeithaml (2003) refers that, the non-correspondence to the expectations and needs expressed by the customers, has a significant impact on the satisfaction of the same, being this a component of the quality perceived in services (Yamauchi et al., 2019).

Quality of Service in Hostel: Expectations and Perceptions Regarding the Quality of Service

Based on hostel accommodation, several studies carried out to assess the quality of the service offered to its guests, consider the most valued factors, as can be seen in Table 1:

Table 1. Factors of quality of service in hostel

Autores	Conclusões (elementos mais importantes)
Musa e Thirumoorthi (2001)	Equipamentos; Instalações; Ambiente acolhedor; Segurança; Limpeza; Localização.
Nash et al. (2006)	Conforto; Limpeza; Segurança.
Hecht e Martin (2006)	Limpeza; Localização; Serviço pessoal; Segurança; Instalações.
Chitty et al. (2007)	Imagem da marca; Profissionalismo da equipa.
Brochado e Gameiro (2013)	Profissionalismo da equipa; Localização, Instalações; Facilidades de acesso; Limpeza; Ambiente Social.
Brochado et al. (2015)	Profissionalismo da equipa; Limpeza; Segurança; Instalações; Ambiente social; Localização; Conexão com a cidade.
Brochado e Rita (2018)	Profissionalismo da equipa; Ambiente social; Instalações; Conexão com a cidade.
Brochado et al. (2019)	Profissionalismo da equipa; Instalações; Ambiente social.
Tavares e Brea (2020)	Segurança; Limpeza; Mobiliário; Serviços de apoio; *Design* e ambiente; Localização; Comércio e entretenimento.

Fonte: Elaboração própria a partir de Brochado et al. (2019); Cró e Martins (2017); Tavares e Brea (2020)

According to the investigations described above, the service components most valued by the hostel customer are mainly reflected in the professionalism of the team, facilities, cleanliness and security, being present in most of the conclusions obtained. Santos (2016) mentions that these dimensions of quality are considered the most important for hostel guests worldwide, being influential elements in the prices practiced, justifying the premium prices.

However, another dimension of quality, which proves to be one of the most important factors, is the social environment. Brochado et al. (2015) refer that the social environment is crucial for the creation of the general perception of quality of the hostel's guests, which is a specific dimension of this type of accommodation (Brochado et al., 2019).

Emphasizing the safety factor, Amblee (2015) links the perception of customers about the safety of hostels with cleanliness and location, demonstrating that a sanitized installation implies control and organization, components that are intrinsically linked to safety. Cró and Martins (2017) reveal that guests are willing to pay a premium price for hostels that offer higher levels of security awareness in countries with high crime rates. Hecht and Mantin (2006) refer that this element constitutes a special concern for female guests (Cró & Martins, 2017).

According to Lima and Vicente (2017), guests have positive perceptions about the service provided in hostel accommodation. However, there is a significant difference in the perception of guests with regard to gender. Dimensions such as the competence of the staff, hygiene, comfort, facilities, location, tranquility and privacy are the basis of the satisfaction of the service for female guests. The rest of the extensions, such as price/quality, the environment, services, security and reservations, represent the aspects most selected by male guests. Although, the item with the greatest disparity between genders is location, on the contrary, the environment and services register a greater proximity of opinions (Lima, 2015). Yamauchi et al. (2019) confirm the discrepancy between expectations with regard to the gender of the guests, stating that the female expects a safe place, with a good atmosphere and professionalism on the part of the team, and the male sex is debated on issues such as price/quality and location.

Considering the age group, it appears that younger guests are more satisfied with the service provided by the hostels (Lima, 2015). This group mentions that human resources represent the strongest element for the satisfaction of your stay. In this way, they show to be more dependent on attributes such as the availability of employees to respond to the needs expressed, valuing kindness, friendliness and the ability to solve problems efficiently (R. Lima & Vicente, 2017). Hostels' guests, whose characteristics were based on the acceptance of a basic hosting service at a reduced price, are changing, being less dependent on the price and more influenced by attributes related to the staff (Lima & Vicente, 2016). With regard to older guests, they mostly value cleaning, the price of the service (Lima & Vicente, 2017), good atmosphere and location (Yamauchi et al., 2019).

However, there are other characteristics that influence expectations and perceptions in the hostel service. Pichitphan (2018) highlights the region of origin as an element of this problem, demonstrating that, for American guests, the service provided by employees is the most valued component. In the case of Asian guests, the consistency of the service, the attention paid by the staff, the knowledge of the staff and the ease of the facilities are considered, with regard to European guests, only the ease of the facilities being highlighted (Pichitphan, 2018).

In this context, Marques (2014) refers that the cultural dimensions influence the perceived quality of the service. The discrepancy in the responses of different nationalities, both in terms of service preferences and in the assessment made in relation to them, reveals the generalization of the phenomenon of the influence of cultural values on the perception of perceived quality (Marques, 2014).

RESEARCH METHODOLOGY

The present investigation was based on a case study implemented through an exploratory and descriptive research of a quantitative nature, aiming to verify if the expectation of the quality of the hostel service corresponded to the perception of the tourist after their stay. Data collection was carried out through an online questionnaire survey with objective questions, shared between the months of July and August 2020. For the collection and analysis of the data, only tourists who have already stayed in a hostel unit in Portugal.

Recognizing the lack of literature that underlies the sector of hostels in Portugal, the present study also intended to represent a contribution in this sense, more specifically in the segment of service quality. Therefore, in order to cover all the desired points, the questionnaire survey had four sections.

1st Section: Region of Portugal inherent to the last stay of respondents in a hostel, using as answer options the seven regions of Portugal listed in NUTS II, which include the North, Center, Metropolitan Area of Lisbon, Alentejo, Algarve, Azores Autonomous Region and Madeira Autonomous Region.

2nd Section: Considering as focus the integration of the model of evaluation of the perceived quality of Parasuraman et al. (1985), a second section was designed for this component only. However, despite considering the five dimensions contained in the SERVQUAL model, the structure of the original model was not fully followed. In this sense, the statements were reformulated in a manner appropriate to the characteristics of the hostel service, and respondents were asked to select their degree of disagreement / agreement according to the Likert scale based on five points: strongly disagree (1), disagree partially (2), neither disagree nor agree (3), partially agree (4), totally agree (5).

3rd Section: Definition of some issues considered still relevant. Thus, it was intended to understand how respondents rated their level of satisfaction with hostel accommodation and how they rated hostel service in Portugal. Issues were also considered related to the respondents 'feeling about the hostels' ability to respond to the needs expressed by the customers, the intention to return and the recommendation of this type of accommodation to other tourists. As a last point, respondents were asked about the differentiating aspects they would like to find in a next hostel stay, allowing for freedom of opinion through an open response.

4th Section: The questions destined to the sociodemographic data of the sample were stipulated considering the gender, age, marital status, country of origin and educational qualifications of the respondents, this being the basis for the characteristics to be analyzed.

Regarding the interpretation of results, in addition to the descriptive analysis, the information collected was also introduced in the SPSS statistical software for data processing. In order to meet the intended objectives, several statistical procedures were performed to analyze the variables presented.

CONTEXT AND DISCUSSION

As a starting point for the analysis of the results of the study applied to individuals who stayed in a hostel in Portugal, five variables are inherent to the characteristics of the sample: gender, age, marital status, country of origin and educational qualifications. The application of the questionnaire survey enabled 118 responses to be collected, of which 59.3% were female and 40.7% male. The predominant age group includes respondents between 18 and 24 years old (74.6%), followed by the group between 25 and 34 years old (16.9%) and, with less emphasis, the interval of ages between 45 and 54 years old (5.9%) and between 35 - 44 years old (2.5%). Regarding the domain of nationality, the results obtained reveal that 100% of respondents have Portuguese nationality, and it is not possible to reach any response from foreign tourists.

With regard to the marital status variable, almost all of the respondents were single (87.3%), contrasting with the results obtained by married/unmarried couples (9.3%) and divorced people (3.4%). In terms of educational qualifications, higher education (68.9%) predominates over the total number of responses, revealing a high level of education on the part of respondents. Basic education assumes the second highest percentage value (29.7%) and, lastly, with less academic training, only 1.7% of the total answers are found. In this context, the sample is characterized by being mostly female, of Portuguese nationality and age group concentrated between 18 and 24 years old, with single marital status and higher education qualifications. This analysis comprises a description of the hostel tourist profile, with characteristics similar to those presented by Mané and Ferreira (2017), Saraiva (2013), Tavares and Brea (2020) and Thomazi (2019).

Concerning to the region chosen by the respondents in their last stay in a hostel, the results obtained reveal that the predominant region concerns the North (37.3%), followed by the Metropolitan Area of Lisbon (28.8%) and the Center (17.8%). The Alentejo (9.3%) and Algarve (4.2%) regions are less relevant, with the Autonomous Region of the Azores (1.7%) and the Autonomous Region of Madeira (0.8%) regions with a lower percentage value.

Considering the expectations and perceptions of respondents regarding the dimensions of the SERVQUAL model, there are significant differences in some of the dimensions. In the dimension "Tangible aspects", this difference is present in the issues related to the equipment being modern/easy to use and in the existence of variety of the breakfast offered. In contrast, in the "Reliability" dimension, there are no significantly different values between expectations and perceptions. With regard to the "Empathy" dimension, this has a significant difference between the expectations and perceptions of respondents regarding the issue of individualized attention conceived by the employees of the hostels. On the other hand, in the dimension "Promptness" there are no values that represent significance for the disparity in the perspective of the respondents. Finally, in the "Security" dimension, there is a significant difference between the expectations and perceptions of respondents regarding the fact that the hostel has adequate security measures. In this way, the disparities found reveal that the perception of the services purchased by the respondents is lower than the expected expectation. Furthermore, the averages of expectations are generally higher than the averages of perceptions in all dimensions. Although the differences are not significant, the values obtained may indicate that customer satisfaction and service was below expectations.

In the analysis of gender and age group as an element that influences expectations and perceptions of perceived quality in a hostel, it is important to keep in mind that the small sample size makes it difficult to generalize the conclusions. Still, it appears that women have higher averages for all dimensions, showing a higher level of agreement with the statements. However, the differences are not statistically significant. Both men and women have the highest average for the "Reliability" dimension and the lowest for "Empathy". The analysis of the age group as an influencing element perceptions of perceived quality in a hostel was made considering two age groups: individuals up to 24 years old and individuals aged 25 or over, given the reduced participation of groups of respondents with older age groups. It appears that individuals between 18 and 24 years old have higher averages in the dimensions "Reliability" and "Security" and the older ones in the dimensions "Tangible aspects", "Empathy" and "Promptness". Although, these values do not represent statistically significant differences.

However, the analysis of other issues, proved to be fundamental for understanding other elements of the quality of hostel service in Portugal. In the case of the degree of satisfaction given the experience acquired, it was found that the majority of respondents attributed satisfactory degrees (86.5%) to the service. Regarding the classification of the service, there was a majority in the answers "Good" (44.1%) and "Very good" (28.8%). Although, there is also a significant percentage value attributed to "Reasonable" (22.9%), a figure that should raise some concern for those responsible for the hostels. This should be a reason for reflection on what went less well in the service provided, considering the changes to be adopted to achieve higher ratings. The data obtained also allowed us to understand that almost all respondents recognize that this type of accommodation is prepared to meet the needs expressed by their guests (88.1%), with only 11.9% considering the opposite. In addition, the majority of respondents show an intention to return (88.1%), in contrast to the minority that would not opt for this type of accommodation (11.9%). These data may indicate that the experience obtained was positive enough to provoke the desire to repeat the hostel stay. Regarding the recommendation of this type of accommodation to other tourists, there is an extremely high percentage value, in relation to the positive response (87.3%). However, there was some dissatisfaction with the service provided or with the frustrating expectations that respondents had about the hostel services they attended (12.7%). This percentage should concern those responsible for the hostels, motivating a reflection on what went less well. These values can demonstrate an interconnection understanding that the satisfaction of a guest who considers his needs met, may result in the desire to purchase this type of service again, enhancing the recommendation to other tourists.

Although there is a clear satisfaction regarding most issues, there are elements of hostel service with room for improvement. In this way, there are some different aspects that customers would like to find in their next stay. With regard to the physical facilities of the hostels, aspects such as the need for a more modern environment are mentioned, valuing issues such as the decoration of the space, an improvement in air conditioning and the adoption of recent technology appliances. There is a need for larger dimensions in the facilities, seeking to also have a part of leisure outdoors and with a view to enjoy.

Regarding the comfort of the facilities, the need for privacy is demonstrated, expecting a higher number of single rooms, more comfortable and with private bathrooms, even if this implies the payment of a higher price for the service. Contrary to this perspective, there are guests who just need a place to sleep, not expecting more than clean facilities, looking for a more affordable price for their stay. However, pointing out that there is an evident distinction between low-priced and high-priced hostels, respondents would like to find better conditions for accommodation and service in those initially mentioned. In terms of facilities, aspects related to environmental concern are also considered, showing the need for more sustainable measures and practices by this type of accommodation. Furthermore, there is a concern with the accessibility of places so that barriers to guests who manifest special needs or some type of disability are broken down. In this topic, respondents refer to elements such as improvement of the internet, the need for a car park and a greater number of electrical outlets in the rooms.

The issue of hygiene is a valued aspect, being one of the most mentioned elements, associated with notions such as cleanliness and rigor that are sought by guests during their stay. Given the current situation regarding Covid-19, it is considered that priority should be given to hygiene and safety, specifying and showing guests that they should feel safe with all the hygiene and care within the hostel, improving the preparation for receiving a high number of guests in pandemic times.

With regard to breakfast, some responses reveal the demand for a greater variety or even that it is included, even if this implies a higher price.

Service is another aspect that is strongly highlighted, referring to the need for innovation and enhancing individual and personalized service, investing in staff training. Thus, they mention that they would like the employees of the hostels to be more attentive to the needs of consumers, if they demonstrate more availability and competence to provide a service of superior quality. Regarding the hostel's procedures, there is a need for a free cancellation, also valuing other issues such as flexibility of hours, being able to check out late.

With regard to entertainment, it is suggested that this type of accommodation should have a suggestive card about what can be done/visited in the city where they are staying. Guided tours and group activities were the most supported options for guests' entertainment, especially those traveling alone. There are also proposals such as karaoke nights and live music as activities to liven up guests after dinner, enjoying a moment of conviviality and relaxation. Furthermore, some respondents propose a greater variety of social rooms, which may be related to themes, such as a games room.

CONCLUSION

It is understood that there is no universal definition attributed to the concept of tourist accommodation. However, by providing overnight accommodation to tourists, it plays an important role in creating the perception of the experience by the tourist. Although the tourist accommodation has different types, it does not recognize a place for facilities and/or establishments called "local accommodation". The local accommodation was developed for a new class of travelers, not understanding the global standards of the hotel chains. In this sense, this business model aims at the elimination of illegal accommodation and unfair competition verified in the exploitation of smaller structures, or in those in which the characteristics are not recognized to incorporate the typology of tourism enterprise.

This type of accommodation comprehend four types, of which the so-called "Accommodation establishments" stands out, where hostels are incorporated. The hostel is characterized by providing accommodation at an economical price, configuring a structure with common spaces and shared rooms, creating the opportunity to discover new cultures and share experiences between guests. However, the evolution of this type of accommodation has led to the diversification of the types of hostels, having specific characteristics and adapted to the interests and needs of tourists. The tourist, whose profile is associated with hostel accommodation, is called a backpacker and is composed of young singles, with higher education qualifications, motivated by the opportunity to get to know new cultures and by looking for a reduced price for the service. With the increase in the age group, flashpackers appear, whose characteristics change, assigning a higher value to quality and privacy even though it means a higher monetary value.

When addressing the issue of quality, it is important to recognize that there are instruments that assist in its measurement, as is the case with the SERVQUAL model that demonstrates relevance in the study of customers' expectations and perceptions regarding the service purchased. Focusing on the quality of services at hostels, several authors highlighted the professionalism of the team, the facilities, cleanliness and safety as the factors most valued by guests, having a direct influence on the price. Although, the literature made it possible to verify that expectations and perceptions regarding quality have disparities according to gender, age and region of origin, factors that affect perceived quality and alter the concept of value attributed to service components.

In the Portuguese context, it appears that the hostels sector started its activity late compared to other countries. Despite this, it has been registering positive growth, although it lacks the literature and statistics that support it. Regarding the profile of the hostel tourist, it was possible to prove the characteristics verified in the literature, including elements such as the fact that they are young singles with higher education qualifications. As the region of the last stay, the North region predominates, this being the base region for the assessment of the perceived quality of the hostel service in the case study.

Regarding the study of the respondents' expectations and perceptions, the SERVQUAL model was essential to obtain conclusions. Thus, only three dimensions demonstrate significant differences between the respondents' expectations and perceptions. Therefore, the perception of the services acquired by the respondents reveals to be inferior to the expectations formulated in the dimensions "tangible aspects", "empathy" and "security". In general, the average expectations are also higher than the average perceptions of the dimensions under study. Although the differences are not significant, the values obtained may indicate that customer satisfaction and service was below expectations.

When analyzing the influence of gender and age group, there were no statistically significant differences to verify disparities in the perspective of the respondents, despite showing higher averages in females and over 25 years of age. This is due to the fact that the sample is small and makes it difficult to conclude the results.

However, in general, respondents are satisfied with the experience acquired in the hostel, attributing positive ratings to the service and feeling that there is a correspondence to the expressed needs. Consequently, these elements result in the respondents' intention of returning and recommending this type of accommodation to other tourists. Regarding the intention to return, it is recognized the need for concern on the part of those responsible for the hostels to understand what went less well in the experience of the number of tourists who do not show an intention to return. This number represents an alert for some dissatisfaction with the service provided or with the frustrating expectations that respondents had about the hostel services they attended.

However, considering that there is scope to improve the service, the study allowed us to understand the differentiating aspects that would be important in future stays. Therefore, it is suggested an improvement in the facilities, comfort, equipment, hygiene, cleanliness and schedules of the hostels. As differentiating aspects, components related to service also stand out, demonstrating a lack of professionals focused on individual and personalized service. In addition, new elements are suggested, such as environmental concern, the search for accessibility for people with special needs or disabilities and the preparation of activities for the entertainment of guests.

In a conclusive way, it is recognized that the present investigation allowed to explore the quality issue, understanding only the levels in a national context. This is due to the small sample size, which made it difficult to extrapolate the results. In this sense, it was not possible to obtain answers by region of origin (national), or by foreign tourists, not allowing the analysis of expectations and perceptions, taking into account the influence of the culture / region of origin. Likewise, although it was possible to analyze it in terms of categories such as gender and age group, there was little diversity in its components, resulting in values that do not represent significant differences in the respondents' perspectives.

FUTURE RESEARCH DIRECTIONS

This study is an exploratory research with a quantitative approach and its conclusions are based on the analysis of data resulting from the Portuguese case study. In this context, taking into account possible future lines of research, it is expected to analyze issues related to the influence of the culture/country of origin (national or international) and the influence of educational qualifications on the perceived quality of the hostel service in Portugal. The development of research in this sector of the tourism industry is extremely important, considering the lack of literature that justifies the evolution, characteristics and components of hostels in Portugal.

REFERENCES

Al-Ababneh, M. M. (2017). Service quality in the hospitality industry. *Journal of Tourism & Hospitality (Los Angeles, Calif.)*, *6*(1), 3633089. doi:10.4172/2167-0269.1000e133

Alauddin, M., Ahsan, S. H., Mowla, D. M. M., Islam, D. M., & Hossain, M. M. (2019). Investigating the relationship between service. *Global Journal of Management and Business Research. Administrative Management*, *19*(1), 6. doi:10.24411/1994-2796-2019-10911

Amblee, N. (2015). The impact of cleanliness on customer perceptions of security in hostels: A WOM-based approach. *International Journal of Hospitality Management*, *49*, 37–39. doi:10.1016/j.ijhm.2015.04.011

Asshidin, N. H. N., Abidin, N., & Borhan, H. B. (2016). Perceived quality and emotional value that influence consumer's purchase intention towards american and local products. *Procedia Economics and Finance*, *35*(3), 639–643. doi:10.1016/S2212-5671(16)00078-2

Brochado, A., Rita, P., & Moro, S. (2019). Discovering patterns in online reviews of beijing and lisbon hostels. *Journal of China Tourism Research*, *15*(2), 172–191. doi:10.1080/19388160.2018.1543065

Budianto, A. (2019). The graduate program of Universitas Galuh Master of Management Studies Program Customer Loyalty: Quality of service. *Journal of Management Review*, *3*(1), 299–305. doi:10.25157/jmr.v3i1.1808

Carrasco, J. (2017). *Plano de negócio: Algés Hostel* [Instituto Universitátio de Lisboa]. http://hdl.handle.net/10071/14997

Cró, S., & Martins, A. M. (2017). The importance of security for hostel price premiums: European empirical evidence. *Tourism Management*, *60*, 159–165. doi:10.1016/j.tourman.2016.11.021

Crotti, M. S. R., & Moretti, S. L. do A. (2019). A hospitalidade e a influência do atendimento na percepção de satisfação dos hóspedes em hotéis de luxo no Brasil. *Turismo: Visão e Ação, 21*(3).

Dozinete, K. C., Medeiros, A. S., dos Reis, A. L. G., Milan, W. W., & Ferracini, M. J. F. (2017). Análise dos cinco gaps pelo método da pesquisa SERVQUAL. *Organizações & Sociedade*, *6*(6), 55. doi:10.29031/ros.v6i6.311

Feng, C. (2018). *Factors affecting purchase intention of young chinese travelers in booking hostel online* [President University]. http://repository.president.ac.id/xmlui/handle/123456789/784

Gama, R. N. (2018). *A influência da paixão empreendedora e suas origens no contexto do segmento turístico low cost: hostels. Um estudo de caso dos melhores hostels do mundo* [Universidade da Beira Interior]. http://hdl.handle.net/10400.6/9699

Gameiro, C. (2013). *Service quality in hostels* [Instituto Universitário de Lisboa]. http://hdl.handle.net/10071/6668

Knutson, B. J., Beck, J. A., Kim, S., & Cha, J. (2009). Identifying the dimensions of the guest's hotel experience. *Cornell Hospitality Quarterly, 50*(1), 44–55. doi:10.1177/1938965508326305

Lima, R., & Vicente, P. (2016). A qualidade do serviço como determinante da satisfação dos turistas nos hostels de Lisboa. *Revista de Gestão Dos Países de Língua Portuguesa, 15*(3), 4. doi:10.12660/rgplp.v15n3.2016.78398

Lima, R., & Vicente, P. (2017). Exploring the relationship between hostel service quality and guests' satisfaction: A comparison between younger and older guests. *Review of Tourism Research, 14*(5), 214–240.

Lima, R. D. (2015). A qualidade do serviço nos hostels de Lisboa: Impacto na satistafação e lealdade dos hóspedes. *Revista de Gestão dos Países de Língua Portuguesa*. http://hdl.handle.net/10071/11133

Macedo, J., Gonçalves, M., & Sousa, B. B. (2021). Relationship Marketing and Communication as a Driver of Fundraising Loyalty and Success in Non-Profit Organizations: A Theoretical Perspective. In M. D. Olvera-Lobo, J. Gutiérrez-Artacho, I. Rivera-Trigueros, & M. Díaz-Millón (Eds.), *Innovative Perspectives on Corporate Communication in the Global World* (pp. 62–77). IGI Global. doi:10.4018/978-1-7998-6799-9.ch004

Mané, A., & Ferreira, L. (2017). El perfil del consumidor de hostel en Brazil y sus motivaciones. *Estudos e Perspectivas Em Turismo, 26*(4), 995–943. https://dialnet.unirioja.es/servlet/articulo?codigo=6329748

Marques, J. de O. (2014). *A influência das diferenças culturais na perceção da qualidade do serviço pelo consumidor de um hostel: estudo de caso comparativo entre Coreanos e Franceses.* Instituto Português de Administração de Marketing. http://hdl.handle.net/10400.26/10210

Miguel, P. A. C., & Salomi, G. E. (2004). Uma revisão dos modelos para medição da qualidade em serviços. *Revista Produção, 14*(1), 12–30. doi:10.15900103-65132004000100003

Mohamad, H. A. D., Yazid, M. S. A., Khatibi, A., & Azam, S. M. F. (2017). Service quality, customer satisfaction and customer loyalty ot the hotel industry in United Arab Emirates (UAE): A measurement model. *European Journal of Management and Marketing Studies, 2*(4), 1–26. doi:10.5281/zenodo.1066572

Moura, M., Sousa, B., Malheiro, A., & Veloso, C. (2021). The Quality of Services in Tourism: A Study Applied to Transport Companies. *Quality - Access to Success, 22*(183), 124–129.

Permatasari, A. (2020). The Influence of Perceived Value Towards Customer Satisfaction in Hostel Business: A Case of Young Adult Tourist in Indonesia. *International Journal of Tourism and Hospitality Management in the Digital Age, 4*(2), 11–22. doi:10.4018/IJTHMDA.2020070102

Pichitphan, P. (2018). *A study of service quality dimensions in hostel in Bangkok and its impact on backpackers' satisfaction.* Thammasat University.

Pinto, I. C. A. (2018). *A qualidade e o alojamento local no Porto: caracterização e perspetivas futuras.* Instituto Politécnico do Porto. http://hdl.handle.net/10400.22/12657

Santos, G. E. de O. (2016). Worldwide hedonic prices of subjective characteristics of hostels. *Tourism Management, 52*, 451–454. doi:10.1016/j.tourman.2015.07.001

Santos, V., Ramos, P., Sousa, B., & Valeri, M. (2021). Towards a Framework for the Global Wine Tourism System. *Journal of Organizational Change Management, ahead-of-print*(ahead-of-print). Advance online publication. doi:10.1108/JOCM-11-2020-0362

Saraiva, A. (2013). *Hostels independentes: o caso de Lisboa.* Escola Superior de Hotelaria e Turismo do Estoril. http://comum.rcaap.pt/handle/123456789/4425

Sharma, S., & Srivastava, S. (2018). Exploring the relationship between service quality and customer satisfaction in croatian hotel industry. *Tourism and Hospitality Management, 2*(1), 2598–9839.

Silva, M. B. (2014). *O segmento low cost na indústria hoteleira em Coimbra: o caso dos hostels.* https://estudogeral.sib.uc.pt/bitstream/10316/27600/3/RelatórioMestrado2014-2.pdf

Silva, S. C., Morano, R. P., Santiago, Z. M. P., & Viilarouco, V. (2019). *Hostel: uma forma de hospitalidade aberta*. doi:10.5151/ergodesign2019-2.37

Silva, T., & Köhler, A. F. (2015). O mercado de albergues/ hostels do município de São Paulo: Caracterização e avaliação de estabelecimentos e empreendedores. *Revista Iberoamericana de Turismo*, *5*(1), 54–78.

Silva, T. F. F. (2019). *O valor da qualidade de serviço e a satisfação em hóteis de cinco estrelas em Portugal: um estudo exploratório*. Instituto Politécnico do Cávado e do Ave.

Sousa, B. L. (2016). *Avaliação de serviços com base no Servqual: uma aplicação prática*. Universidade dos Açores. http://hdl.handle.net/10400.3/4096

Tavares, F. A. de O. (2018). *Hostels e Hóteis Boutique: análise de fatores de seleção na cidade de Oporto*. Universidade de Vigo. http://hdl.handle.net/11093/1124

Tavares, F. O., & Brea, J. A. F. (2020). Análise confirmatória à preferência nos hostels na cidade do Porto, em Portugal. *Revista Investigaciones Turísticas, 19*, 104–220. doi:10.14198/INTURI2020.19.05

Thomazi, M. R. (2019). *Hostel: Território de hospedagem marcado pela trama turístico-comunicacional*. Universidade de Caxias do Sul. https://repositorio.ucs.br/11338/4706

Tinoco, M. A. C., & Ribeiro, J. L. D. (2007). Uma nova abordagem para a modelagem das relações entre os determinantes da satisfação dos clientes de serviços. *Produção*, *17*(3), 454–470. doi:10.15900103-65132007000300005

Vasconcelos, M., & de Sampayo, M. T. (2017). Hostels e a nova tendência do turismo urbano. O caso de estudo do centro histórico de Lisboa de 2005 a 2014. In Actas do 4.o Congresso Internacional Casa Nobre - Um património para o futuro (pp. 1195–1205). Academic Press.

Verhoef, P. C., Lemon, K. N., Parasuraman, A., Roggeveen, A., Tsiros, M., & Schlesinger, L. A. (2009). Customer experience creation: Determinants, dynamics and management strategies. *Journal of Retailing*, *85*(1), 31–41. doi:10.1016/j.jretai.2008.11.001

Vinh, T. T., & Nga, V. T. Q. (2015). The relationship between components of customer-based brand equity for destination: Conceptual framework and preliminary testing for scales. *South East Asia Journal of Contemporary Business, Economics and Law*, *7*(2), 47–53.

Volante, P. J. T. (2011). *O segmento low-cost da indústria hoteleira em Portugal: o caso dos hostels*. https://repositorio.iscte-iul.pt/handle/10071/4068

Yamauchi, D. A. M., Reis, B. C. M., & Rodrigues, M. D. (2019). Análise da qualidade dos serviços de hostels em Ouro Preto - Minas Gerais. *Revista Eletrónica de Gestão e Serviços*, *10*(1), 2599–2623. doi:10.15603/2177-7284/regs.v10n1p2599-2626

KEY TERMS AND DEFINITIONS

Communication: Communication is a process that involves the exchange of information between two or more interlocutors.

Loyalty: Consideration of the precepts that concern honor, decency, and honesty.

Relationship Marketing: Relationship marketing corresponds to marketing work focused on the relationship with customers and suppliers.

SERVQUAL Model: SERVQUAL is a multi-dimensional research instrument, designed to capture consumer expectations and perceptions of a service along the five dimensions that are believed to represent service quality. SERVQUAL is built on the expectancy-disconfirmation paradigm, which in simple terms means that service quality is understood as the extent to which consumers' pre-consumption expectations of quality are confirmed or disconfirmed by their actual perceptions of the service experience.

Compilation of References

Abdul Kadir, M. A. B., & Sarif, S. M. (2016). Social entrepreneurship, social entrepreneur and social enterprise: A review of concepts, definitions and development in Malaysia. *Journal of Emerging Economies and Islamic Research*, *4*(2), 1–16. doi:10.24191/jeeir.v4i2.9086

Abu-Saifan, S. (2012). Social entrepreneurship: Definition and boundaries. *Technology Innovation Management Review*, *2*(2), 22–27. doi:10.22215/timreview/523

Ahmed, M. Z., Ahmed, O., Aibao, Z., Hanbin, S., Siyu, L., & Ahmad, A. (2020). Epidemic of COVID-19 in China and associated Psychological Problems. *Asian Journal of Psychiatry*, *51*, 102092. doi:10.1016/j.ajp.2020.102092 PMID:32315963

Aji, R. R. (2020). Tourism social entrepreneurship in community-based tourism: A case study of Pentingsari tourism village. *Earth and Environmental Science*, *447*, 1–9. doi:10.1088/1755-1315/447/1/012009

Akama, J. S. (2002). The Role of Government in the Development of Tourism in Kenya. *International Journal of Tourism Research*, *4*, 1-13.

Aktan, C. C., & Börü, D. (2007). Kurumsal sosyal sorumluluk. In C. C. Aktan (Ed.), *Kurumsal Sosyal Sorumluluk*. İktisadi Girişim ve İş Ahlakı Derneği Yayını.

Al-Ababneh, M. M. (2017). Service quality in the hospitality industry. *Journal of Tourism & Hospitality (Los Angeles, Calif.)*, *6*(1), 3633089. doi:10.4172/2167-0269.1000e133

Alauddin, M., Ahsan, S. H., Mowla, D. M. M., Islam, D. M., & Hossain, M. M. (2019). Investigating the relationship between service. *Global Journal of Management and Business Research. Administrative Management*, *19*(1), 6. doi:10.24411/1994-2796-2019-10911

Alegre, I., & Berbegal-Mirabent, J. (2016). Social innovation success factors: Hospitality and tourism social enterprises. *International Journal of Contemporary Hospitality Management*, *28*(6), 1155–1176. doi:10.1108/IJCHM-05-2014-0231

Alexander, R., Arago, O. R., Bookwala, J., Cherbuin, N., Gatt, J. M., Kahrilas, I. J., Kästner, N., Lawrence, A., Lowe, L., Morrison, R. G., Mueller, S. V., Nusslock, R., Papadelis, C., Polnaszek, K. L., Richter, S. H., Silton, R. L., & Styliadis, C. (2021). The neuroscience of positive emotions and affect: Implications for cultivating happiness and wellbeing. *Neuroscience and Biobehavioral Reviews*, *121*, 220–249. doi:10.1016/j.neubiorev.2020.12.002 PMID:33307046

Alexis-Thomas, C. (2020). An examination of issues related to tourism and health and well-being as a sustainable development goal by tourism providers in Tobago. *Worldwide Hospitality And Tourism Themes, 12*(3), 293–303. doi:10.1108/WHATT-02-2020-0006

Algoe, S. B., & Zhaoyang, R. (2016). Positive psychology in context: Effects of expressing gratitude in ongoing relationships depend on perceptions of enactor responsiveness. *The Journal of Positive Psychology*, *11*(4), 399–415. doi:10.1080/17439760.2015.1117131 PMID:27800009

Ali, F., & Amin, M. (2014). The influence of physical environment on emotions, customer satisfaction and behavioural intentions in Chinese resort hotel industry. *Journal for Global Business Advancement*, *7*(3), 249–266. doi:10.1504/JGBA.2014.064109

Ali, F., Kim, W. G., Li, J., & Jeon, H. M. (2018). Make it delightful: Customers' experience, satisfaction and loyalty in Malaysian theme parks. *Journal of Destination Marketing & Management*, *7*, 1–11. doi:10.1016/j.jdmm.2016.05.003

Alighieri, D. (1896). Vita Nova. Houghton, Mifflin and Company. (original work published 1294)

Almond, J. (2006). *A shared biography of nursing. The second half of the twentieth century.* Coimbra: Formasau.

Alonso Almeida, M. del M. (2019). Robots, artificial intelligence and virtual reality: An approach to tourism industries. *Cuadernos de Turismo*, (44), 13–26. doi:10.6018/turismo.44.404711

Al-Seheel, A. Y., & Noor, N. M. (2016). Effects of an Islamic based gratitude strategy on Muslim students' level of happiness. *Mental Health, Religion & Culture*, *19*(7), 686–703. doi:10.1080/13674676.2016.1229287

Amaro, S., Antunes, A., & Henriques, C. (2018). A closer look at Santiago de Compostela's pilgrims through the lens of motivations. *Tourism Management*, *64*, 271–280. doi:10.1016/j.tourman.2017.09.007

Amazon Explore. (2021). *Amazon Explore's featured experiences*. Retrieved from https://www.amazon.com/-/es/b?ie=UTF8&node=19419898011

Amblee, N. (2015). The impact of cleanliness on customer perceptions of security in hostels: A WOM-based approach. *International Journal of Hospitality Management*, *49*, 37–39. doi:10.1016/j.ijhm.2015.04.011

Amin, Z., & Akbar, K. P. (2013). Analysis of psychological well-being and turnover intentions of hotel employees: An empirical study. *International Journal of Innovation and Applied Studies*, *3*(3), 662–671.

Ansarah, M. G. d R. (2002). Formação e capacitação do profissional em turismo e hotelaria: reflexões e cadastro das instituições educacionais no Brasil. São Paulo: Aleph.

Anuar, M. A. H., Salleh, N. H. M., & Jeevan, J. (2020). Developing dive site risk assessment model (DSRAM) to enhance tourism safety and sustainability in Perhentian Island. *Journal of Sustainability Science and Management*, *15*(1), 125–135.

An, W., & Alarcón, S. (2020). How Can Rural Tourism Be Sustainable? A Systematic Review. *Sustainability*, *12*(18), 7758. doi:10.3390u12187758

Aquino, R. S., Lück, M., & Schänzel, H. A. (2018). A conceptual framework of tourism social entrepreneurship for sustainable community development. *Journal of Hospitality and Tourism Management*, *37*, 23–32. doi:10.1016/j.jhtm.2018.09.001

Asshidin, N. H. N., Abidin, N., & Borhan, H. B. (2016). Perceived quality and emotional value that influence consumer's purchase intention towards american and local products. *Procedia Economics and Finance*, *35*(3), 639–643. doi:10.1016/S2212-5671(16)00078-2

Austin, J., Wei-Skillern, J., & Stevenson, H. (2006). Social and commercial entrepreneurship: Same, different, or both? *Entrepreneurship Theory and Practice*, *30*(1), 1–22. doi:10.1111/j.1540-6520.2006.00107.x

Ayazlar, G., & Ayazlar, R. A. (2015). Rural tourism: A conceptual approach. In Tourism, Environment and Sustainability. St. Kliment Ohridski University Press.

Aydin, D., & Omuris, E. (2020). The mediating role of meaning in life in the relationship between memorable tourism experiences and subjective well-being. *Advances in Hospitality and Tourism Research-AHTR*, *8*(2), 314–337. doi:10.30519/ahtr.656469

Bâc, D. P. (2012). The Impacts Of Tourism On Society. *The Annals of the University of Oradea*, *21*(1), 500-506. Retrieved from: https://url2.cl/jpZNE

Baker, R., & Slemp, G. (2018, December). *An introduction to micro job crafting: How job crafting for 12 minutes or less a day contributes to sustainable positive behaviour change*. Paper presented at the CIPD Applied Research Conference, Nottingham, UK.

Bakker, A. B., & Demerouti, E. (2007). The Job Demands-Resources model: State of the art. *Journal of Managerial Psychology*, *22*(3), 309–328. doi:10.1108/02683940710733115

Ballina, F. J. (2020). Is there rural smart tourism? A Spanish experience. *Management Theory and Studies for Rural Business and Infrastructure Development*, *42*(3), 369–380. doi:10.15544/mts.2020.37

Barber, R. (1993). *Pilgrimages*. The Boydell Press.

Barki, E., Comini, G., Cunliffe, A., Hart, S., & Rai, S. (2015). Social entrepreneurship and social business: Retrospective and prospective research. *Revista de Administração de Empresas*, *55*(4), 380–384. doi:10.1590/S0034-759020150402

Barnett, M. D., Van Vleet, S. C., & Cantu, C. (2021). Gratitude mediates perceptions of previous generations' prosocial behaviors and prosocial attitudes toward future generations. *The Journal of Positive Psychology*, *16*(1), 54–59. doi:10.1080/17439760.2019.1676459

Barreira, I. (2012). *Cidades narradas: memória, representações e práticas de turismo*. Pontes Editores.

Barreto, M. (2004). Manual de iniciação ao estudo do Turismo (13th ed.). Campinas/SP. Publisher: Papirus.

Barros, J. L. G. (2008). *A satisfação com a qualidade de serviços na fidelização dos destinos turísticos* (Master's Dissertation). Faculdade de Economia da Universidade do Porto.

Basterretxea-Iribar, I., Sotés, I., & Maruri, M. (2019). Managing bathers' capacity at overcrowded beaches: A case on the Spanish North Atlantic coast. *Tourism Management*, *71*, 453–465. doi:10.1016/j.tourman.2018.10.016

Baum, T. (2013). *International perspectives on women and work in hotels, catering and tourism*. International Labour Office.

BBC. (2020). *Londoners 'experiencing racism over coronavirus'*. BBC News. Retrieved from: https://www.bbc.com/news/av/uk-england-london-51361930

Bear, M. F., Connors, B. W., & Paradiso, M. A. (2002). *Neurociências: desvendando o sistema nervoso*. Artmed Editora.

Beedie, C. J., Terry, P. C., & Lane, A. M. (2005). Distinctions between emotion and mood. *Cognition and Emotion*, *19*(6), 847–878. doi:10.1080/02699930541000057

Beni, M. C. (2003). Structural analysis of tourism (9ª ed). São Paulo: Senac.

Berchicci, L. (2009). *Innovating for sustainability: green entrepreneurship in personal mobility*. Routledge.

Bergson, A. (1954). On the concept of social welfare. *The Quarterly Journal of Economics*, *68*(2), 233–252. doi:10.2307/1884448

Berti, A. (2019). The digital twin: creating virtual airport tours with ocean 3D. *Airport Technology*. Retrieved from https://www.airport-technology.com/features/virtual-reality-at-airports/

Besler, S. (2010). Sosyal girişimcilik. In S. Besler (Ed.), *Sosyal Girişimcilik* (pp. 3–30). Beta Yayıncılık.

Biddulph, R. (2017). Social enterprise and inclusive tourism. Five cases in Siem Reap, Cambodia. *Tourism Geographies*, 1–20.

Bimonte, S., & Faralla, V. (2016). Does residents' perceived life satisfaction vary with tourist season? A two-step survey in a Mediterranean destination. *Tourism Management*, *55*, 199–208. doi:10.1016/j.tourman.2016.02.011

Blackwell, R. (2014). Motivation for pilgrimage: Using theory to explore motivations. *Scripta Instituti Donneriani Aboensis*, *22*, 24–37. doi:10.30674cripta.67360

Blancas, F. J., Lozano-Oyola, M., González, M., Guerrero, F. M., & Caballero, R. (2011). How to use sustainability indicators for tourism planning: The case of rural tourism in Andalusia (Spain). *The Science of the Total Environment*, *412*, 28–45. doi:10.1016/j.scitotenv.2011.09.066 PMID:22033358

Blanco, J. (2015). *Libro blanco de los destinos turísticos inteligentes*. Estrategias y soluciones para fomentarla innovación en el turismo digital. Madrid, Spain. LID Editorial Empresarial.

Blom, T., Nilsson, M., & Santos, X. (2016). The way to Santiago beyond Santiago. Fisterra and the pilgrimage's post-secular meaning. *European Journal of Tourism Research*, *12*, 133–146.

Boes, K., Buhalis, D., & Inversini, A. (2015). Conceptualizing smart tourism destination dimensions. In *Information and communication technologies in tourism 2015* (pp. 391–403). Springer. doi:10.1007/978-3-319-14343-9_29

Bogicevic, V., Seo, S., Kandampully, J. A., Liu, S. Q., & Rudd, N. A. (2019). Virtual reality presence as a preamble of tourism experience: The role of mental imagery. *Tourism Management*, *74*, 55–64. doi:10.1016/j.tourman.2019.02.009

Borges, V. P. C. (2014). Políticas Públicas e o Desenvolvimento Segregador no Turismo: Uma discussão conceitual e reflexiva referente ao turismo de Parnaíba-Piauí. *Turismo: Estudos e Práticas*, *3*, 23–40.

Borges, V. P. C. (2017). *Política Pública e Planejamento Estratégico do Turismo no Território Polo Costa do Delta: Um Modelo de Desenvolvimento Turístico*. Tese de Doutoramento. Universidade de Coimbra.

Bosnjak, M., Brown, C. A., Lee, D.-J., Yu, G. B., & Sirgy, M. J. (2016). Self-Expressiveness in Sport Tourism: Determinants and Consequences. *Journal of Travel Research*, *55*(1), 125–134. doi:10.1177/0047287514535845

Boukas, N. (2019). Rural tourism and residents' wellbeing in Cyprus: Towards a conceptualized framework of the appreciation of rural tourism for islands' sustainable development and competitiveness. *International Journal of Tourism Anthropology*, *7*(1), 60–86. doi:10.1504/IJTA.2019.098105

Boukas, N., & Chourides, P. (2016). Niche tourism in Cyprus: Conceptualising the importance of social entrepreneurship for the sustainable development of islands. *International Journal of Leisure and Tourism Marketing*, *5*(1), 26–43. doi:10.1504/IJLTM.2016.077128

Boyatzis, R. E. (1998). Transforming Qualitative Information: Thematic Analysis and Code Development. *Sage (Atlanta, Ga.)*.

Bramwell, B. (1994). Rural tourism and sustainable rural tourism. *Journal of Sustainable Tourism*, *2*(1-2), 1–6. doi:10.1080/09669589409510679

Brazililian Ministry of Tourism. (2021). *Ministério do Turismo seleciona roteiros para o Projeto Experiências do Brasil Rural*. Retrieved from: https://www.gov.br/turismo/pt-br/assuntos/noticias/ministerio-do-turismo-seleciona-roteiros-para-o-projeto-experiencias-do-brasil-rural

Brochado, A., Rita, P., & Moro, S. (2019). Discovering patterns in online reviews of beijing and lisbon hostels. *Journal of China Tourism Research*, *15*(2), 172–191. doi:10.1080/19388160.2018.1543065

Brouwer, W. B., Culyer, A. J., van Exel, N. J. A., & Rutten, F. F. (2008). Welfarism vs. extra-welfarism. *Journal of Health Economics*, *27*(2), 325–338. doi:10.1016/j.jhealeco.2007.07.003 PMID:18179835

Budianto, A. (2019). The graduate program of Universitas Galuh Master of Management Studies Program Customer Loyalty: Quality of service. *Journal of Management Review*, *3*(1), 299–305. doi:10.25157/jmr.v3i1.1808

Budisusila, A., Sarungu, J. J., Wiyono, V. H., & Soesilo A.M. (2019). Supply chain analysis of tourism social enterprises in Bejiharjo tourism village. *International Journal of Recent Technology and Engineering, 8*(2S9), 305-313.

Buhalis, D. (2000). Tourism and information technologies: Past, present and future. *Tourism Recreation Research*, *25*(1), 41–58. doi:10.1080/02508281.2000.11014899

Buhalis, D., & Amaranggana, A. (2013). Smart tourism destinations. In *Information and communication technologies in tourism 2014*. Springer. doi:10.1007/978-3-319-03973-2_40

Buhalis, D., Harwood, T., Bogicevic, V., Viglia, G., Beldona, S., & Hofacker, C. (2019). Technological disruptions in services: Lessons from tourism and hospitality. *Journal of Service Management*, *30*(4), 484–506. doi:10.1108/JOSM-12-2018-0398

Buhalis, D., & Sinarta, Y. (2019). Real-time co-creation and nowness service: Lessons from tourism and hospitality. *Journal of Travel & Tourism Marketing*, *36*(5), 563–582. doi:10.1080/10548408.2019.1592059

Bull, A. (2007). Oceans and coastal tourism: Integrating tourism into bioeconomic models. *Tourism in Marine Environments*, *4*(1), 29–39. doi:10.3727/154427307784835688

Bulsara, H. P., Gandhi, J., & Chandwani, S. (2015). Social entrepreneurship in India: An exploratory study. *International Journal of Innovation*, *3*(1), 7–16. doi:10.5585/iji.v3i1.20

Bundick, M. J., Remington, K., Morton, E., & Colby, A. (2021). The contours of purpose beyond the self in midlife and later life. *Applied Developmental Science*, *25*(1), 62–68. doi:10.1080/10888691.2018.1531718

Bursztyn, I., Bartholo, R., & Delamaro, M. (2009). Turismo pra quem? On development paths and alternatives for tourism in Brazil. In R. Bartholo (Ed.), *Turismo de base comunitaria: diversidade de olhares* (Vol. 144-162, pp. 76–107). Letra e Imagem.

Bushueva, I. V. (2020). Trends of Wellness Tourism Development in Russia. *Turismo-Estudos E Praticas, 4*.

Cacho, A., Mendes-Filho, L., Estaregue, D., Moura, B., Cacho, N., Lopes, F., & Alves, C. (2016). Mobile tourist guide supporting a smart city initiative: a Brazilian case study. *International Journal of Tourism Cities*. doi:10.1108/IJTC-12-2015-0030

Caffyn, A. (2012). Advocating and implementing slow tourism. *Tourism Recreation Research, 37*(1), 77–80. doi:10.1080/02508281.2012.11081690

Cai, Y., Ma, J., & Lee, Y.-S. (2020). How do Chinese travelers experience the Arctic? Insights from a hedonic and eudaimonic perspective. *Scandinavian Journal Of Hospitality And Tourism, 20*(2), 144–165. doi:10.1080/15022250.2020.1744185

Capocchi, A., Vallone, C., Pierotti, M., & Amaduzzi, A. (2019). Overtourism: A literature review to assess implications and future perspectives. *Sustainability (Switzerland), 11*(12), 3303. Advance online publication. doi:10.3390u11123303

Carrasco, J. (2017). *Plano de negócio: Algés Hostel* [Instituto Universitátio de Lisboa]. http://hdl.handle.net/10071/14997

Carvache-Franco, M., Carvache-Franco, W., Carvache-Franco, O., & Hernández-Lara, A. B. (2020). Analysis of Push and Pull Motivations and the Intentions to Return and Recommend a Coastal or Marine Destination. *Journal of Coastal Research, 36*(6), 1313–1322. doi:10.2112/JCOASTRES-D-20-00006.1

Carvache-Franco, M., Carvache-Franco, W., Carvache-Franco, O., Hernández-Lara, A. B., & Buele, C. V. (2020). Segmentation, motivation, and sociodemographic aspects of tourist demand in a coastal marine destination: A case study in Manta (Ecuador). *Current Issues in Tourism, 23*(10), 1234–1247. doi:10.1080/13683500.2019.1600476

Carvache-Franco, W., Carvache-Franco, M., Carvache-Franco, O., & Hernández-Lara, A. B. (2019). Segmentation of foreign tourist demand in a coastal marine destination: The case of Montañita, Ecuador. *Ocean and Coastal Management, 167*(1), 236–244. doi:10.1016/j.ocecoaman.2018.10.035

Carvache-Franco, W., Carvache-Franco, M., & Hernández-Lara, A. B. (2020). From motivation to segmentation in coastal and marine destinations: A study from the Galapagos Islands, Ecuador. *Current Issues in Tourism, 0*(0), 1–17. doi:10.1080/13683500.2020.1811651

Carvalho, M. B. (2010). *A Felicidade na agenda da administração e suas relações com conceitos organizacionais* (Dissertação de mestrado). Minas Gerais: FUMEC College.

Castellani, P., Bonfanti, A., Canestrino, R., & Magliocca, P. (2020). Dimensions and triggers of memorable tourism experiences: Evidence from Italian social enterprises. *The TQM Journal, 32*(6), 1115–1138. doi:10.1108/TQM-11-2019-0259

Cater, C. I. (2008). The life aquatic: Scuba diving and the experiential imperative. *Tourism in Marine Environments, 5*(4), 233–244. doi:10.3727/154427308788714777

Catford, J. (1998). Social entrepreneurs are vital for health promotion but they need supportive environments too. *Health Promotion International*, *13*(2), 95–97. doi:10.1093/heapro/13.2.95

Cavalheiro, M. B., Luz, A. B. T., & Mayer, V. F. (2021). *Turismo Náutico em Niterói -Levantamento Documental*. http://turismonautico.uff.br/2021/05/07/relatorio-do-levantamento-documental-sobre-esportes-nauticos-em-niteroi-e-sobre-iniciativas-de-turismo-nautico-no-brasil/

Çelebi, D., Pirnar, I., & Eriş, E. D. (2020). Bibliometric analysis of social entrepreneurship in gastronomy tourism. *Tourism: An International Interdisciplinary Journal*, *68*(1), 58–67. doi:10.37741/t.68.1.5

Chan, D. W. (2013). Counting blessings versus misfortunes: Positive interventions and subjective well-being of Chinese school teachers in Hong Kong. *Educational Psychology*, *33*(4), 504–519. doi:10.1080/01443410.2013.785046

Chandra, Y., & Leenders, M. A. A. M. (2012). User innovation and entrepreneurship, in the virtual world: A study of second life residents. *Technovation*, *32*(7-8), 464–476. doi:10.1016/j.technovation.2012.02.002

Chapapría, V. E. (2000). El planteamiento de infraestructuras para el turismo náutico. *Cuadernos de Turismo*, *6*, 29–44.

Chen, C. Y., Yen, C. H., & Tsai, F. C. (2014). Job crafting and job engagement: The mediating role of person-job fit. *International Journal of Hospitality Management*, *37*(2), 21–28. doi:10.1016/j.ijhm.2013.10.006

Cheng, J. C., & O-Yang, Y. (2018). Hotel employee job crafting, burnout, and satisfaction: The moderating role of perceived organizational support. *International Journal of Hospitality Management*, *72*, 78–85. doi:10.1016/j.ijhm.2018.01.005

Cheng, S.-T., Tsui, P. K., & Lam, J. H. M. (2015). Improving mental health in health care practitioners: Randomized controlled trial of a gratitude intervention. *Journal of Consulting and Clinical Psychology*, *83*(1), 177–186. doi:10.1037/a0037895 PMID:25222798

Chen, L. H., & Wu, C. H. (2014). Gratitude enhances change in athletes' self-esteem: The moderating role of trust in coach. *Journal of Applied Sport Psychology*, *26*(3), 349–362. doi:10.1080/10413200.2014.889255

Chen, Y., Fu, X., & Lehto, X. Y. (2016). Chinese Tourist Vacation Satisfaction and Subjective Well-being. *Applied Research in Quality of Life*, *11*(1), 49–64. doi:10.100711482-014-9354-y

Chen, Y., Lehto, X. Y., & Cai, L. (2013). Vacation and well-being: A study of chinese tourists. *Annals of Tourism Research*, *42*, 284–310. doi:10.1016/j.annals.2013.02.003

Chia, K. W., & Liao, Y. M. (2020). An Exploratory Study of Factors Influencing Chinese Outbound Medical Tourism. *Journal of China Tourism Research*, 1–19. Advance online publication. doi:10.1080/19388160.2020.1780177

Chia, Y. M., & Chu, M. J. T. (2016). Moderating effects of presenteeism on the stress-happiness relationship of hotel employees: A note. *International Journal of Hospitality Management*, *55*, 52–56. doi:10.1016/j.ijhm.2016.02.005

Chi, C. G., Chi, O. H., & Ouyang, Z. (2020). Wellness hotel: Conceptualization, scale development, and validation. *International Journal of Hospitality Management*, *89*, 102404. Advance online publication. doi:10.1016/j.ijhm.2019.102404

Chinazzi, M., Davis, J. T., Ajelli, M., Gioannini, C., Litvinova, M., Merler, S., Pastore y Piontti, A., Mu, K., Rossi, L., Sun, K., Viboud, C., Xiong, X., Yu, H., Halloran, M. E., Longini, I. M. Jr, & Vespignani, A. (2020). The effect of travel restrictions on the spread of the 2019 novel coronavirus (COVID-19) outbreak. *Science*, *368*(6489), 395–400. doi:10.1126cience.aba9757 PMID:32144116

Christoph, G., Laeis, M., & Lemke, S. (2016). Social entrepreneurship in tourism: Applying sustainable livelihoods approaches. *International Journal of Contemporary Hospitality Management*, *28*(6), 1–32.

Clancy, M. (2015). What's at stake in the slow movement? Consumption, responsibility and citizenship. *The Association of American Geographers Annual Meeting*.

CLIA. (2020). *State Of The Cruise Industry Outlook 2021*. https://cruising.org/en/news-and-research/research/2020/december/state-of-the-cruise-industry-outlook-2021

Coelho, M. F., & Mayer, V. F. (2020). Gestão de Serviços Pós-Covid: O que se pode aprender com o setor de turismo e viagens? *Revista Eletrônica Gestão & Sociedade*, *14*(39), 3698-3706. Retrieved from: https://ges.emnuvens.com.br/gestaoesociedade/article/view/3306

Coelho, M. F., de Sevilha Gosling, M., & de Almeida, A. S. A. (2018, December). Tourism experiences: Core processes of memorable trips. *Journal of Hospitality and Tourism Management*, *37*, 11–22. doi:10.1016/j.jhtm.2018.08.004

Coelho, S. D., Mota, K. C. N., & Vasconcelos, F. P. (2015). The vision of the community in the implementation of urban development projects for tourism and quality of life: Public policies in the Grande Pirambu neighborhood, Fortaleza-CE. Revista Turismo - Visão e Ação. *Balneário Camboriú*, *17*(1), 210–240.

Collins-Kreiner, N. (2010a). Researching pilgrimage: Continuity and transformations. *Annals of Tourism Research*, *37*(2), 440–456. doi:10.1016/j.annals.2009.10.016

Collins-Kreiner, N. (2010b). The geography of pilgrimage and tourism: Transformations and implications for applied geography. *Applied Geography (Sevenoaks, England)*, *30*(1), 153–164. doi:10.1016/j.apgeog.2009.02.001

Collins-Kreiner, N. (2016). Life cycle of concepts: The case of pilgrimage tourism. *Tourism Geographies*, *18*(3), 322–334. doi:10.1080/14616688.2016.1155077

Collins-Kreiner, N. (2020a). Pilgrimage tourism-past, present and future rejuvenation: A perspective article. *Tourism Review*, *75*(1), 145–148. doi:10.1108/TR-04-2019-0130

Collins-Kreiner, N. (2020b). A review of research into religion and tourism Launching the Annals of Tourism Research Curated Collection on religion and tourism. *Annals of Tourism Research*, •••, 82.

Cooke, G. B., Burns, J. K., Mann, S. L., Vardy, K. W. J., & Cass, B. L. (2016). The case of Ireland's County Donegal: Stimulating rural labour markets via training, tourism and nurturing social enterprise and entrepreneurship. *E-Journal of International and Comparative Labour Studies*, *5*(1), 1–37.

Cooper, M., & Buckley, R. (2021). Tourist Mental Health Drives Destination Choice, Marketing, and Matching. *Journal of Travel Research*. Advance online publication. doi:10.1177/00472875211011548

Correia, A. B. da C., & Sebastião, S. P. (2017). Os desportos náuticos e o turismo em Portugal. O potencial dos centros de treino de remo. *Revista Investigaciones Turísticas*, *13*(13), 114–139. doi:10.14198/INTURI2017.13.06

Correia, M. I. G., Lopez, L., Gonzalez, R. C. L., & Santos, X. M. (2017). The Challenges of the First European Cultural Itinerary: The Way to St. James. *AlmaTourism - Journal of Tourism. Culture and Territorial Development*, *8*(6), 1–19.

Correira-Loureiro, S.M.; Guerreiro, J. & Ali, F. (2020). 20 years of research on virtual reality and augmented reality in tourism context: a text-mining approach. *Progress in Tourism Management,* 77. doi:10.1016/j.tourman.2019.104028

Courtney, M. (2015). A review of the literature exploring the occupation of contemporary pilgrimage. *Journal of Occupational Science*, *22*(2), 170–182. doi:10.1080/14427591.2013.764816

Cranmer, E. E., tom Dieck, M. C., & Jung, T. (2018). How can Tourist Attractions Profit from Augmented Reality? In T. Jung & M. C. Dieck (Eds.), *Augmented Reality and Virtual Reality: empowering humans, place and business* (pp. 21–32). Springer International Publishing. doi:10.1007/978-3-319-64027-3_2

Cró, S., & Martins, A. M. (2017). The importance of security for hostel price premiums: European empirical evidence. *Tourism Management*, *60*, 159–165. doi:10.1016/j.tourman.2016.11.021

Crotti, M. S. R., & Moretti, S. L. do A. (2019). A hospitalidade e a influência do atendimento na percepção de satisfação dos hóspedes em hotéis de luxo no Brasil. *Turismo: Visão e Ação, 21*(3).

Csikszentmihalyi, M. (1988). The flow experience and its significance for human psychology. In M. Csikszentmihalyi & I. S. Csikszentmihalyi (Eds.), *Optimal experience: Psychological studies of flow in consciousness* (pp. 15–35). Cambridge University Press. doi:10.1017/CBO9780511621956.002

Csikszentmihalyi, M. (1990). *Flow: The psychology of optimal experience*. Harper & Row.

Csikszentmihalyi, M. (2013). *Flow: The psychology of happiness*. Random House.

Csirmaz, É., & Pető, K. (2015). International Trends in Recreational and Wellness Tourism. *Procedia Economics and Finance*, *32*(15), 755–762. doi:10.1016/S2212-5671(15)01458-6

Cunha, L. (2006). Turismo de Saúde – conceitos e mercados. *Revista Lusófona de Humanidades e Tecnologias: estudos e ensaios*, *10*, 79-84. Retrieved from: https://recil.grupolusofona.pt/handle/10437/2577

Cunha, S. K. & Cunha, J. C. (2005). Competitiveness and sustainability of a tourism cluster: a proposed systemic model for measuring the impact of tourism on local development. *Journal of Contemporary Administration*, *9*(2).

Dahles, H., Khieng, S., Verver, M., & Manders, I. (2019). Social entrepreneurship and tourism in Cambodia: Advancing community engagement. *Journal of Sustainable Tourism*, 1–18.

Damari, C., & Mansfeld, Y. (2016). Reflections on pilgrims' identity, role and interplay with the pilgrimage environment. *Current Issues in Tourism*, *19*(3), 199–222. doi:10.1080/13683500.2014.957660

daSilva, A. W., & Heatherton, T. F. (2018). Social neuroscience of subjective well-being and life satisfaction. In J. E. Maddux (Ed.), Frontiers of social psychology. Subjective well-being and life satisfaction (p. 32–49). Routledge/Taylor & Francis Group.

D'Atti, M., & Cinti, F. (2014). *Guida alla Via Francigena: Oltre 900 Chilometri a Piedi sulle Strade del Pellegrinaggio verso Roma*. Terre di mezzo.

Davis, D. E., Choe, E., Meyers, J., Wade, N., Varjas, K., Gifford, A. Z., Quinn, A., Hook, J. N., Tongeren, D. R., Griffin, B. J., & Worthington, E. (2016). Thankful for the little things: A meta-analysis of gratitude interventions. *Journal of Counseling Psychology*, *63*(1), 20–31. doi:10.1037/cou0000107 PMID:26575348

Davis, D., & Tisdell, C. (1995). Recreational scuba-diving and carrying capacity in marine protected areas. *Ocean and Coastal Management*, *26*(1), 19–40. doi:10.1016/0964-5691(95)00004-L

De Bloom, J., Geurts, S. A., & Kompier, M. A. (2013). Vacation (after-) effects on employee health and well-being, and the role of vacation activities, experiences and sleep. *Journal of Happiness Studies*, *14*(2), 613–633. doi:10.100710902-012-9345-3

de Devotto, R. P., & Wechsler, S. M. (2019). Job crafting interventions: Systematic review. *Trends in Psychology*, *27*(2), 371–383. doi:10.9788/TP2019.2-06

De Knop, P. (1990). Sport for All and Active Tourism. *World Leisure & Recreation*, *32*(3), 30–36. doi:10.1080/10261133.1990.10559120

De Lange, D., & Dodds, R. (2017). Increasing sustainable tourism through social entrepreneurship. *International Journal of Contemporary Hospitality Management*, *29*(7), 1–57. doi:10.1108/IJCHM-02-2016-0096

Dees, J. G. (2001). *The meaning of "social entrepreneurship"*. Draft Paper. http://www.redalmarza.cl/ing/pdf/ TheMeaningofsocialEntrepreneurship.pdf

Demerouti, E., Xanthopoulou, D., Petrou, P., & Karagkounis, C. (2017). Does job crafting assist dealing with organizational changes due to austerity measures? Two studies among Greek employees. *European Journal of Work and Organizational Psychology*, *26*(4), 574–589. doi:10.1080/1359432X.2017.1325875

Deng, Y., Xiang, R., Zhu, Y., Li, Y., Yu, S., & Liu, X. (2019). Counting blessings and sharing gratitude in a Chinese prisoner sample: Effects of gratitude-based interventions on subjective well-being and aggression. *The Journal of Positive Psychology*, *14*(3), 303–311. doi:10.1080/17439760.2018.1460687

Desbiens, Y. & Ferreira de Lima, J. (2004). Cadrage du Devel6ppment Regional. *Revista Interfaces*, *4*, 179-192.

DeSteno, D. (2009). Social emotions and intertemporal choice: 'Hot' mechanisms for building social and economic capital. *Current Directions in Psychological Science*, *18*(5), 280–284. doi:10.1111/j.1467-8721.2009.01652.x

DeSteno, D., Li, Y., Dickens, L., & Lerner, J. S. (2014). Gratitude: A tool for reducing economic impatience. *Psychological Science*, *25*(6), 1262–1267. doi:10.1177/0956797614529979 PMID:24760144

Devereux, C., & Carnegie, E. (2006). Pilgrimage: Journeying beyond self. *Tourism Recreation Research*, *31*(1), 47–56. doi:10.1080/02508281.2006.11081246

DeWall, C. N., Lambert, N. M., Pond, R. S. Jr, Kashdan, T. B., & Fincham, F. D. (2012). A grateful heart is a nonviolent heart: Cross-sectional, experience sampling, longitudinal, and experimental evidence. *Social Psychological & Personality Science*, *3*(2), 232–240. doi:10.1177/1948550611416675

Dias, E. C., & Mendes, R. (2002). *Contribution to the expansion and operationalization of the concepts of occupational health promotion and healthy workplaces for informal sector workers*. Mimeographed.

Dickens, L. R. (2017). Using gratitude to promote positive change: A series of meta-analyses investigating the effectiveness of gratitude interventions. *Basic and Applied Social Psychology*, *39*(4), 193–208. doi:10.1080/01973533.2017.1323638

Dickinson, J. E., Robbins, D., & Lumsdon, L. M. (2010). Holiday travel discourses and climate change. *Journal of Transport Geography*, *18*(3), 482–489. doi:10.1016/j.jtrangeo.2010.01.006

Diener, E. (1984). Subjective Well-being. *Psychological Bulletin*, *75*(3), 542–575. doi:10.1037/0033-2909.95.3.542 PMID:6399758

Diener, E. (2000). Subjective well-being: The science of happiness and a proposal for a national index. *The American Psychologist*, *55*(1), 34–43. doi:10.1037/0003-066X.55.1.34 PMID:11392863

Diener, E., Oishi, S., & Lucas, R. E. (2003). Personality, culture and subjective well-being: Emotional and cognitive evaluations of life. *Annual Review of Psychology*, *54*(1), 403–425. doi:10.1146/annurev.psych.54.101601.145056 PMID:12172000

Diener, E., & Suh, E. (1997). Measuring quality of life: Economic, social, and subjective indicators. *Social Indicators Research*, *40*(1/2), 189–216. doi:10.1023/A:1006859511756

Diener, E., Suh, E. M., Lucas, R. E., & Smith, H. E. (1999). Subjective well-being: Three decades of progress. *Psychological Bulletin*, *125*(2), 276–302. doi:10.1037/0033-2909.125.2.276

Digance, J. (2003). Pilgrimage at contested sites, tourism management. *Annals of Tourism Research*, *30*(1), 143–159. doi:10.1016/S0160-7383(02)00028-2

Digdon, N., & Koble, A. (2011). Effects of constructive worry, imagery distraction, and gratitude interventions on sleep quality: A pilot trial. *Applied Psychology. Health and Well-Being*, *3*(2), 193–206. doi:10.1111/j.1758-0854.2011.01049.x

Dik, B. J., Duffy, R. D., Allan, B. A., O'Donnell, M. B., Shim, Y., & Steger, M. F. (2015). Purpose and meaning in career development applications. *The Counseling Psychologist*, *43*(4), 558–585. doi:10.1177/0011000014546872

Dillette, A. K., Douglas, A. C., & Andrzejewski, C. (2020). Dimensions of holistic wellness as a result of international wellness tourism experiences. *Current Issues in Tourism*, *24*(6), 794–810. doi:10.1080/13683500.2020.1746247

do Nascimento, C. A. X., Meneghatti, M. R., Hsu, P. L., & da Silva, D. (2019). Revisiting Co-creation - Bibliometric Study. *Marketing & Tourism Review*, *4*(1), 1–25.

Dodds, R., & Butler, R. (2019). The phenomena of overtourism: A review. *International Journal of Tourism Cities*, *5*(4), 519–528. doi:10.1108/IJTC-06-2019-0090

Dolnicar, S., Yanamandram, V., & Cliff, K. (2012). The contribution of vacations to quality of life. *Annals of Tourism Research*, *39*(1), 59–83. doi:10.1016/j.annals.2011.04.015

Dossey, L. (2018). The helper's high. *Explore (New York, N.Y.)*, *14*(6), 393–399. doi:10.1016/j.explore.2018.10.003 PMID:30424992

Dozinete, K. C., Medeiros, A. S., dos Reis, A. L. G., Milan, W. W., & Ferracini, M. J. F. (2017). Análise dos cinco gaps pelo método da pesquisa SERVQUAL. *Organizações & Sociedade*, *6*(6), 55. doi:10.29031/ros.v6i6.311

Dunn, E. W., Aknin, L. B., & Norton, M. I. (2008). Spending money on others promotes happiness. *Science*, *319*(5870), 1687–1688. doi:10.1126cience.1150952 PMID:18356530

Dunn, E. W., Aknin, L. B., & Norton, M. I. (2014). Spending money on others: Using money to benefit others pays off. *Current Directions in Psychological Science*, *23*(1), 41–47. doi:10.1177/0963721413512503

Dunn, E., & Norton, M. (2014). *Happy money: The science of smarter spending*. Simon & Schuster Paperbacks.

Duro, J. A., & Farré, F. X. (2015). Tourist Seasonality in Spanish Provinces: Measurement and Analysis. *Cuadernos de Turismo*, *36*, 157–174. doi:10.6018/turismo.36.230921

Dye, T. R. (2005). Understanding Public Policy (11th ed.). Pearson Education Inc.

Egoshi, K. (2009). The Fallacy of Sustainable Development. In A. J. H. Guevara (Ed.), *Consciousness and Sustainable Development in Organizations* (pp. 115–144). Elsevier.

Ehrlich, C. (2018). The development of an extended goal-striving reasons framework: Evidence for its relevance in the workplace, for its theoretical difference to self-concordance and for its buffering effect on work intensity. *Journal of Positive Psychology and Wellbeing*, *2*(2), 1–23.

Ehrlich, C. (2019). The relevance of identified and intrinsic goal motivation for work engagement. *European Journal of Applied Positive Psychology*, *3*(5), 1–5.

Eid, M., & Diener, E. (2004). Global judgement of subjective well-being: Situational variability and long-term stability. *Social Indicators Research*, *65*(3), 245–277. doi:10.1023/B:SOCI.0000003801.89195.bc

Eisenberg, N. (1986). *Altruistic emotion, cognition and behavior*. Erlbaum.

Ekinci, M. B. (2014). The Cittaslow philosophy in the context of sustainable tourism development: The case of Turkey. *Tourism Management*, *41*, 178–189. doi:10.1016/j.tourman.2013.08.013

Elías, S., Fernandez, M. R., & Castro, M. (2012). Proposal for a system of indicators to evaluate the impact of tourism on poverty in the town of Monte Hermoso. *Tourism and Society Yearbook*, *8*, 41–53.

Elliot, J. (1997). *Tourism: politics and public sector management*. Routledge.

Emmons, R. A., & McCullough, M. E. (2003). Counting blessings versus burdens: An experimental investigation of gratitude and subjective well-being in daily life. *Journal of Personality and Social Psychology*, *84*(2), 377–389. doi:10.1037/0022-3514.84.2.377 PMID:12585811

Emmons, R. A., & Stern, R. (2013). Gratitude as a Psychotherapeutic Intervention. *Journal of Clinical Psychology*, *69*(8), 846–855. doi:10.1002/jclp.22020 PMID:23775470

Erbaş, E. (2019). Social business model and sharing economy for community based tourism development: A case study of Lisinia Doğa. *Journal of Tourism and Gastronomy Studies*, *7*(4), 2399–2417. doi:10.21325/jotags.2019.478

Ergül, M., Johnson, C., & Kleinrichert, D. (2012). Social entrepreneurship: A hospitality and tourism perspective. *European Journal of Tourism. Hospitality and Recreation*, *3*(3), 115–126.

European Central Bank. (2021). *Prices for travel during the COVID-19 pandemic: is there commonality across countries and items?* Retrieved from: https://www.ecb.europa.eu/pub/economic-bulletin/focus/2021/html/ecb.ebbox202101_06~bcb28cb255.en.html

European Commission. (2017). Commission Staff Working Document on Nautical tourism. In EPSC. doi:10.2495/RAV110541

FAO, Food and Agriculture Organization of Nations. (2021). *Launch of the Digital Villages initiative in Latin America and the Caribbean.* Retrieved from: http://www.fao.org/americas/eventos/ver/en/c/1395198/

Farber, M. E., & Hall, T. E. (2007). Emotion and environment: Visitors' extraordinary experiences along the Dalton Highway in Alaska. *Journal of Leisure Research, 39*(2), 248–270. doi:10.1080/00222216.2007.11950107

Faulkner, J., O'Brien, W. J., McGrane, B., Wadsworth, D., Batten, J., Askew, C. D., Badenhorst, C., Byrd, E., Coulter, M., Draper, N., Elliot, C., Fryer, S., Hamlin, M. J., Jakeman, J., Mackintosh, K. A., McNarry, M. A., Mitchelmore, A., Murphy, J., Ryan-Stewart, H., . . . Lambrick, D. (2020). Physical activity, mental health and well-being of adults during initial COVID-19 containment strategies: A multi-country cross-sectional analysis. *Journal of Science and Medicine in Sport.*

Feng, C. (2018). *Factors affecting purchase intention of young chinese travelers in booking hostel online* [President University]. http://repository.president.ac.id/xmlui/handle/123456789/784

Fereday, J., & Muir-Cochrane, E. (2006). Demonstrating rigor using thematic analysis: A hybrid approach of inductive and deductive coding and theme development. *International Journal of Qualitative Methods, 5*(1), 80–92. doi:10.1177/160940690600500107

Fernandes, L. M. M. (2014). O Ceará turístico: Regionalization policy and governance in inducing destinations. *Journal GeoUECE, 3*(5), 213–214.

Fernandez, M. J., Gullette, G. S., Luque, A. C., Mendoza-Otiveros, M., & Monterrubio, J. C. (2012). Social impacts of tourism as perceived by state-planned tourism destination residents: The case of Huatulco, Mexico. *International Journal of Tourism Anthropology, 2*(1), 34–52. doi:10.1504/IJTA.2012.046094

Figueiredo, V., & Franco, M. (2018). Wine cooperatives as a form of social entrepreneurship: Empirical evidence about their impact on society. *Land Use Policy, 79*, 812–821. doi:10.1016/j.landusepol.2018.09.022

Filep, S. (2014). Moving Beyond Subjective Well-Being A Tourism Critique. *Journal of Hospitality & Tourism Research (Washington, D.C.), 38*(2), 266–274. doi:10.1177/1096348012436609

Filep, S., Laing, J., & Csikszentmihalyi, M. (2017). What Is Positive Tourism? Why Do We Need It? In S. Filep, J. Laing, & M. Csikszentmihalyi (Eds.), *Positive Tourism* (pp. 3–15). Routledge.

Filimonau, V., Lemmer, C., Marshall, D., & Bejjani, G. (2017). 'Nudging'as an architect of more responsible consumer choice in food service provision: The role of restaurant menu design. *Journal of Cleaner Production, 144*, 161–170. doi:10.1016/j.jclepro.2017.01.010

Fistola, R., & La Rocca, R. A. (2018). Slow Mobility and Cultural Tourism. Walking on Historical Paths. In R. Papa, R. Fistola, & C. Gargiulo (Eds.), *Smart Planning: Sustainability and Mobility in the Age of Change* (pp. 301–322). Springer. doi:10.1007/978-3-319-77682-8_18

Fragoso, R. M. D. S. (2015). Habitat Features and Strategies for the Sustainable Development in the Alentejo Region. *Revista de Economia e Sociologia Rural*, *53*(suppl 1), 103–114. doi:10.1590/1234-56781806-94790053s01008

Fredrickson, B. L. (2004). The broaden-and-build theory of positive emotions. *Philosophical Transactions of the Royal Society of London. Series B, Biological Sciences*, *359*(1449), 1367–1378. doi:10.1098/rstb.2004.1512 PMID:15347528

Fredrickson, B. L., Cohn, M. A., Coffey, K. A., Pek, J., & Finkel, S. M. (2008). Open hearts build lives: Positive emotions, induced through loving-kindness meditation, build consequential personal resources. *Journal of Personality and Social Psychology*, *95*(5), 1045–1062. doi:10.1037/a0013262 PMID:18954193

Freire, K. M. F., Tubino, R. A., Monteiro-Neto, C., Andrade-Tubino, M. F., Belruss, C. G., Tomás, A. R. G., Tutui, S. L. S., Castro, P. M. G., Maruyama, L. S., Catella, A. C., Crepaldi, D. V., Daniel, C. R. A., Machado, M. L., Mendonça, J. T., Moro, P. S., Motta, F. S., Ramires, M., Silva, M. H. C., & Vieira, J. P. (2016). Brazilian recreational fisheries: Current status, challenges and future direction. *Fisheries Management and Ecology*, *23*(3–4), 276–290. doi:10.1111/fme.12171

Frey, B. & Stutzer, A. (2017). *Public Choice and Happiness.* CREMA Workpaper, nº2017-03. Center for Research Economics, Management and the Arts.

Frey, K. (2000). Public Policies; A Conceptual Debate and Reflections concerning the Practice of Public Policy Analysis in Brazil. *Planejamento e Políticas Públicas*, *21*.

Fullagar, S., Markwell, K., & Wilson, E. (2012). Starting slow: Thinking through slow mobilities and experiences. In S. Fullagar, K. Markwell, & E. Wilson (Eds.), *Slow tourism: Experiences and mobilities* (pp. 1–8). Channel View Publications. doi:10.21832/9781845412821-003

Galinha, I., & Ribeiro, J. L. P. (2005). History and Evolution of the Concept of Subjective Well-Being. *Psychology. Health & Illness*, *6*(2), 203–214.

Gama, R. N. (2018). *A influência da paixão empreendedora e suas origens no contexto do segmento turístico low cost: hostels. Um estudo de caso dos melhores hostels do mundo* [Universidade da Beira Interior]. http://hdl.handle.net/10400.6/9699

Gameiro, C. (2013). *Service quality in hostels* [Instituto Universitário de Lisboa]. http://hdl.handle.net/10071/6668

Gazzola, P., Grechi, D., Romanelli, M., & Pollice, F. (2020). Il turismo religioso: La via Francigena, una opportunità per l'Italia. *Economia Aziendale Online*, *11*(3), 341–358.

Geldenhuys, L.-L., Van der Merwe, P., & Slabbert, E. (2014). Who is the scuba diver that visits sodwana bay and why? *S.A. Journal for Research in Sport Physical Education and Recreation*, *36*(2), 91–104. https://www.researchgate.net/publication/274376878_Who_is_the_scuba_diver_who_visits_Sodwana_Bay_and_why

Genç, R. (2018). The Impact of Augmented Reality (AR) Technology on Tourist Satisfaction. In T. Jung, & M. C. Dieck, (Eds.), Augmented Reality and Virtual Reality: empowering humans, place and business. Springer International Publishing. doi:10.1007/978-3-319-64027-3_8

Germann Molz, J. (2009). Representing pace in tourism mobilities: Staycations, slow travel and the amazing race. *Journal of Tourism and Cultural Change*, *7*(4), 270–286. doi:10.1080/14766820903464242

Gesler, W. (1996). Lourdes: Healing in a place of pilgrimage. *Health & Place*, *2*(2), 95–105. doi:10.1016/1353-8292(96)00004-4

Ghandeharioun, A., Azaria, A., Taylor, S., & Picard, R. W. (2016). "Kind and Grateful": A context-sensitive smartphone app utilizing inspirational content to promote gratitude. *Psychology of Well-Being*, *6*(9), 9. Advance online publication. doi:10.118613612-016-0046-2 PMID:27441170

Gibson, D., Pratt, S., & Movono, A. (2012). Tribe tourism: A case study of the Tribewanted project in Vorovoro, Fiji. In S. Fullagar, K. Markwell, & E. Wilson (Eds.), *Slow Tourism: Experiences and mobilities* (pp. 185–200). Channel View Publications. doi:10.21832/9781845412821-016

Gibson, H. J. (1998). Sport Tourism: A Critical Analysis of Research. *Sport Management Review*, *1*(1), 45–76. doi:10.1016/S1441-3523(98)70099-3

Gilbert, D., & Abdullah, J. (2004). Holidaytaking and the sense of well-being. *Annals of Tourism Research*, *31*(1), 103–121. doi:10.1016/j.annals.2003.06.001

Giudici, E., Dettori, A., & Caboni, F. (2017). Neurotourism: Futuristic perspective or today's reality? 20th Excellence. In *Services International Conference Verona*. University of Verona.

Gladwell, M. (2008). *Outliers: The story of success*. Little, Brown and Company.

GNH Centre Bhutan. (n.d.). *The 9 domains of GNH*. Accessed on 21/01/2021. Available at http://www.gnhcentrebhutan.org/?page_id=833

Goeldner, C. (1989). 39th Congress AIEST: English Workshop Summary. *Revue de Tourisme, 44*(4), 6-7.

Goldstein, E., Allen, R., Allaby, M., Davoll, J., & Lawrence, S. (Eds.). (1972). *A blueprint for survival. The Ecologist, 2(1)* .

Gomes, R.S.E., Pinto, H.E.R.S.C., & Almeida, C.M.B.R. (2017). Residential tourism in the Algarve: the perception of public sector managers. *Revista Bras. Pesq. Tur., 11*(2), 197-217.

Gomes, M., & Ferreira, L. J. (2018). Public policies and the goals of sustainable development. *Law And Development*, *9*(2), 155–178.

González, L., Yen, E., González, L., Carmelo, J., & Ledesma, D. L. (2015). Highlights of consumption and satisfaction in nautical tourism. A comparative study of visitors to the Canary Islands and Morocco. *Gestion y Ambiente*, *18*(1), 129–145.

Goodricj, J. N., & Goodrich, G. E. (1987, September). Health-care Tourism. *Tourism Management, 8*(3), 217–222. doi:10.1016/0261-5177(87)90053-7

Goodwin, H. (2017). *The Challenge of Overtourism. Responsible Tourism Partnership.* Working Paper 4. Retrieved from http://www.millennium-destinations.com/uploads/4/1/9/7/ 41979675/ rtpwp4overtourism012017.pdf

Gordon, S., Tang, C. H., Day, J., & Adler, H. (2019). Supervisor support and turnover in hotels: Does subjective well-being mediate the relationship? *International Journal of Contemporary Hospitality Management, 31*(1), 496–512. doi:10.1108/IJCHM-10-2016-0565

Gössling, S., Scott, D., & Hall, C. M. (2021). Pandemics, tourism and global change: A rapid assessment of COVID-19. *Journal of Sustainable Tourism, 29*(1), 1–20. doi:10.1080/09669582.2020.1758708

Gowdy, J. (2005). Toward a new welfare economics for sustainability. *Ecological Economics, 53*(2), 211–222. doi:10.1016/j.ecolecon.2004.08.007

Gowreesunkar, G. B. V., Van der Sterren, J., & Séraphin, H. (2015). Social entrepreneurship as a tool for promoting global citizenship in island tourism destination management. Academy for Touris, 7-23.

Greenhaus, J. H., & Kossek, E. E. (2014). The contemporary career: A work–home perspective. *Annual Review of Organizational Psychology and Organizational Behavior, 1*(1), 361–388. doi:10.1146/annurev-orgpsych-031413-091324

Greffe, X. (1994). Is rural tourism a lever for economic and social development? *Journal of Sustainable Tourism, 2*(1-2), 22–40. doi:10.1080/09669589409510681

Gretzel, U., Sigala, M., Xiang, Z., & Koo, C. (2015). Smart tourism: Foundations and developments. *Electronic Markets, 25*(3), 179–188. doi:10.100712525-015-0196-8

Guedes, A. S. (2014). *O Domínio Cultural no Turismo Organizado - O Caso Portugês: Da polarização territorial à valorização endógena e local* (Doctoral dissertation). Programa de Doctorado: New Resources and Sustainability in Tourism. Universidad de Salamanca.

Guiver, J., & McGrath, P. (2017). Slow Tourism: Exploração de discursos. *Dos Algarves: A Multidisciplinary e-Journal*, (27), 11-34.

Günlü, E. (2015). The conceptual analysis of social entrepreneurship and social entrepreneurship in tourism. *Journal of Social Sciences and Humanities Researches, 35*, 23–42.

Hadavi, S. (2017). Direct and indirect effects of the physical aspects of the environment on mental well-being. *Environment and Behavior, 49*(10), 1071–1104. doi:10.1177/0013916516679876

Hall, C. M. (2011). Health and medical tourism: a kill or cure for global public health? *Tourism Review, 66*(1-2), 4-15.

Han, H., Kiatkawsin, K., Koo, B., & Kim, W. (2020). Thai wellness tourism and quality: Comparison between Chinese and American visitors' behaviors. *Asia Pacific Journal of Tourism Research, 25*(4), 424–440. doi:10.1080/10941665.2020.1737551

Hassan, A., Ekiz, E., Dadwal, S. S., & Lancaster, G. (2018). Augmented Reality Adoption by Tourism Product and Service Consumers: Some Empirical Findings. In T. Jung & M. C. Dieck (Eds.), *Augmented Reality and Virtual Reality: empowering humans, place and business* (pp. 47–64). Springer International Publishing. doi:10.1007/978-3-319-64027-3_4

Hawe, P., Webster, C., & Shiell, A. (2004). A glossary of terms for navigating the field of social network analysis. *Journal of Epidemiology and Community Health, 58*(12), 971–975. doi:10.1136/jech.2003.014530 PMID:15547054

Heberger, E. A., Christie, A. C., & Alkin, C. M. (2010). A bibliometric analysis of the academic influences of and on evaluation theorists' published works. *The American Journal of Evaluation, 31*(1), 24–44. doi:10.1177/1098214009354120

Heidmann, F. & Salm, J. F. (2010). *Políticas públicas e desenvolvimento: bases epistemológicas e modelos de análise*. Brasília: UnB.

Heitmann, S., Robinson, P., & Povey, G. (2011). Slow food, slow cities and slow tourism. In P. Robinson, S. Heitmann, & P. Dieke (Eds.), *Research Themes for Tourism* (pp. 114–127). CABI., doi:10.1079/9781845936846.0114

He, L., Zhou, W., He, M., Nie, X., & He, J. (2021). Openness and COVID-19 induced xenophobia: The roles of trade and migration in sustainable development. *PLoS One, 16*(4), e0249579. doi:10.1371/journal.pone.0249579 PMID:33831012

Hendriks, T., Schotanus-Dijkstra, M., Hassankhan, A., de Jong, J., & Bohlmeijer, E. (2020). The efficacy of multi-component positive psychology interventions: A systematic review and meta-analysis of randomized controlled trials. *Journal of Happiness Studies, 21*(1), 357–390. doi:10.100710902-019-00082-1

Hernández, C., León, C., & Moreno-González, A. A. (2020). Revalorización del producto turístico a través del turismo de salud. Turismo pos-COVID-19. Reflexiones, retos y oportunidades.

Honoré, C. (2005). *In Praise of slow: How a worldwide movement is challenging the cult of speed.* Orion Books Ltd.

Honorio, J.B. & Mattos, P.L.C.L. (2010). Organizational roles: What the pragmatics of language leads us to think. *Alcance Journal - Electronic, 17*(1), 22-33.

Hosany, S., Prayag, G., Deesilatham, S., Caušević, S., & Odeh, K. (2015). Measuring tourists' emotional experiences: Further validation of the destination emotion scale. *Journal of Travel Research, 54*(4), 482–495. doi:10.1177/0047287514522878

Houge Mackenzie, S., & Hodge, K. (2020). Adventure recreation and subjective well-being: a conceptual framework. *Leisure Studies, 39*(1), 26–40. doi:10.1080/02614367.2019.1577478

Howard, C. (2012). Speeding up and slowing down: Pilgrimage and slow travel through time. In S. Fullagar, K. Markwell, & E. Wilson (Eds.), *Slow Tourism: Experiences and Mobilities* (pp. 11–24). Channel View Publications.

Huang, S., van der Veen, R., & Song, Z. (2018). The impact of coping strategies on occupational stress and turnover intentions among hotel employees. *Journal of Hospitality Marketing & Management*, *27*(8), 926–945. doi:10.1080/19368623.2018.1471434

Hu, B., Liu, J., & Qu, H. (2019). The employee-focused outcomes of CSR participation: The mediating role of psychological needs satisfaction. *Journal of Hospitality and Tourism Management*, *41*, 129–137. doi:10.1016/j.jhtm.2019.10.012

Hu, D., & Yang, Y. (2020). The Development of Marine Sports Tourism Industry Based on Low-carbon Economy. *Journal of Coastal Research*, *112*(sp1), 97–99. doi:10.2112/JCR-SI112-028.1

Hudson, J. (2013). Welfare. In An introduction to social policy. Sage Publications. doi:10.4135/9781526435835.n1

Hume, T. M., Mulcahy, N., & Mead, S. T. (2019). An Overview of Changing Usage and Management Issues in New Zealand's Surf Zone Environment. *Journal of Coastal Research*, *87*(sp1), 1–12. doi:10.2112/SI87-001.1

Huta, V., & Waterman, A. (2014). Eudaimonia and Its Distinction from Hedonia: Developing a Classification and Terminology for Understanding Conceptual and Operational Definitions. *Journal of Happiness Studies, Springer*, *15*(6), 1425–1456. doi:10.100710902-013-9485-0

Hwang, J., Lee, J. J., Park, S., Chang, H., & Kim, S. S. (2014). The impact of occupational stress on employee's turnover intentions in the luxury hotel segment. *International Journal of Hospitality & Tourism Administration*, *15*(1), 60–77. doi:10.1080/15256480.2014.872898

Ibáñez, A. (2018). Augmented Reality for working in airport control centers. *Ferrovial Blog*. Retrieved from https://blog.ferrovial.com/en/2018/05/augmented-reality-airport-control-centres/

Iorgulescua, M., & Ravara, A. S. (2015). The contribution of social enterprises to the development of tourism. The case of Romania. *Procedia Economics and Finance*, *32*, 672–679. doi:10.1016/S2212-5671(15)01448-3

Isik, S., & Erguner-Tekinalp, B. (2017). The effects of gratitude journaling on Turkish first year college students' college adjustment, life satisfaction and positive affect. *International Journal for the Advancement of Counseling*, *39*(2), 164–175. doi:10.100710447-017-9289-8

Jackowski, A., & Smith, V. L. (1992). Polish pilgrim-tourists. *Annals of Tourism Research*, *19*(1), 92–106. doi:10.1016/0160-7383(92)90109-3

Jepson, A., Stadler, R., & Spencer, N. (2019). Making positive family memories together and improving quality-of-life through thick sociality and bonding at local community festivals and events. *Tourism Management*, *75*, 34–50. doi:10.1016/j.tourman.2019.05.001

Jiménez-García, M., Ruiz-Chico, J., Peña-Sánchez, A. R., & López-Sánchez, J. A. (2020). A bibliometric analysis of sports tourism and sustainability (2002-2019). *Sustainability (Switzerland), 12*(7), 1–18. doi:10.3390u12072840

Johnson, P., & Thomas, B. (1992). Tourism research and policy. In P. Johnson & B. Thomas (Eds.), *Perspectives on tourism policy*. Mansell Publishing Limited.

Jones, A., & Navarro, C. (2018). Events and the blue economy: Sailing events as alternative pathways for tourism futures – the case of Malta Andrew. *International Journal of Event and Festival Management, 9*(2), 204–222. doi:10.1108/IJEFM-09-2017-0055

Judge, T. A., Thoreson, C. J., Bono, J. E., & Patton, G. K. (2001). The job satisfaction-job performance relationship: A qualitative and quantitative review. *Psychological Bulletin, 127*(3), 376–407. doi:10.1037/0033-2909.127.3.376 PMID:11393302

Jung, T., & tom Dieck, M. C. (Eds.). (2018). *Augmented Reality and Virtual Reality: empowering humans, place and business*. Springer International Publishing. doi:10.1007/978-3-319-64027-3

Kabil, M., Priatmoko, S., Magda, R., & Dávid, L. D. (2021). Blue Economy and Coastal Tourism: A Comprehensive Visualization Bibliometric Analysis. *Sustainability, 13*(7), 3650. doi:10.3390u13073650

Kaczmarek, L. D., Kashdan, T. B., Drazkowski, D., Enko, J., Kosakowski, M., Szaefer, A., & Bujacz, A. (2015). Why do people prefer gratitude journaling over gratitude letters? The influence of individual differences in motivation and personality on web-based interventions. *Personality and Individual Differences, 75*, 1–6. doi:10.1016/j.paid.2014.11.004

Kahneman, D. (1999). Objective happiness. In D. Kahneman, E. Diener, & N. Schwarz (Eds.), *Well-being: The foundations of hedonic psychology* (pp. 3–25). Russell Sage Foundation.

Kandel, E. R. (2020). *Mentes diferentes: O que cerébros incomuns revelam sobre nós*. Manole.

Kandel, E. R., Schwartz, J. H., Jessell, T. M., Siegelbaum, S. A., & Hudspeth, A. J. (2014). *Princípios de Neurociências*. 5o ed. Porto Alegre. Mc Graw Hill Education. *The Art of Medication*.

Kaneko, M., & Nakamura, K. (1979). The Nash social welfare function. *Econometrica, 47*(2), 423–435. doi:10.2307/1914191

Karatepe, O. M., & Ehsani, E. (2012). Work-related depression in frontline service jobs in the hospitality industry: Evidence from Iran. *Journal of Human Resources in Hospitality & Tourism, 11*(1), 16–35. doi:10.1080/15332845.2012.621052

Kashdan, T. B., Mishra, A., Breen, W. E., & Froh, J. J. (2009). Gender differences in gratitude: Examining appraisals, narratives, the willingness to express emotions, and changes in psychological needs. *Journal of Personality, 77*(3), 691–730. doi:10.1111/j.1467-6494.2009.00562.x PMID:20078735

Kato, K., & Progano, R. N. (2017). Spiritual (walking) tourism as a foundation for sustainable destination development: Kumano-kodo pilgrimage, Wakayama, Japan. *Tourism Management Perspectives*, *24*, 243–251. doi:10.1016/j.tmp.2017.07.017

Kelly, C. M., Strauss, K., Arnold, J., & Stride, C. (2020). The relationship between leisure activities and psychological resources that support a sustainable career: The role of leisure seriousness and work-leisure similarity. *Journal of Vocational Behavior*, *117*, 1–15. doi:10.1016/j.jvb.2019.103340

Kelly, C., & Smith, M. K. (2008). Holistic tourism: Integrating body, mind and spirit. In J. Jafari, R. Bushell, & P. Sheldon (Eds.), *Wellness and Tourism: Mind, Body, Spirit, Place*. Innovation and Tourism Book Series.

Kerlin, J. (2006). Social enterprise in the United States and Europe: Understanding and learning from the differences. *Voluntas*, *17*(3), 247–263. doi:10.100711266-006-9016-2

Khangarot, G. (2019). An overview of social entrepreneurship in tourism. *Think India Journal,* 110-117.

Kidd, C. V. (1992). The evolution of sustainability. *Journal of Agricultural & Environmental Ethics*, *5*(1), 1–26. doi:10.1007/BF01965413

Kim, B., & Kim, S. (2019). Hierarchical value map of religious tourists visiting the Vatican City/ Rome. *Tourism Geographies*, *21*(3), 529–550. doi:10.1080/14616688.2018.1449237

Kim, B., Kim, S. S., & King, B. (2016). The sacred and the profane: Identifying pilgrim traveler value orientations using means-end theory. *Tourism Management*, *56*, 142–155. doi:10.1016/j.tourman.2016.04.003

Kim, B., Kim, S., & King, B. (2019). Religious tourism studies: Evolution, progress, and future prospects. *Tourism Recreation Research*, *45*(2), 185–203. doi:10.1080/02508281.2019.1664084

Kim, H., Lee, S., Uysal, M., Kim, J., & Ahn, K. (2015). Nature-based tourism: Motivation and subjective well-being. *Journal of Travel & Tourism Marketing*, *32*(sup1), S76–S96. doi:10.1080/10548408.2014.997958

Kim, J., Lee, J., Jhang, J., Park, J., & Lee, J. C. (2021). The impact of the COVID-19 threat on the preference for high versus low quality/price options. *Journal of Hospitality Marketing & Management*, 1–17. doi:10.1080/19368623.2021.1884163

Kim, M. J., & Hall, C. M. (2019). A hedonic motivation model in virtual reality tourism: Comparing visitors and non-visitors. *International Journal of Information Management*, *46*, 236–249. doi:10.1016/j.ijinfomgt.2018.11.016

Kim, M. J., Lee, C. K., & Preis, M. W. (2020). The impact of innovation and gratification on authentic experience, subjective well-being, and behavioral intention in tourism virtual reality: The moderating role of technology readiness. *Telematics and Informatics*, *49*, 101349. Advance online publication. doi:10.1016/j.tele.2020.101349

King, L. A., Richards, J. H., & Stemmerich, E. (1998). Daily goals, life goals, and worst fears: Means, ends, and subjective well-being. *Journal of Personality*, *66*(5), 713–744. doi:10.1111/1467-6494.00030 PMID:9802231

Knutson, B. J., Beck, J. A., Kim, S., & Cha, J. (2009). Identifying the dimensions of the guest's hotel experience. *Cornell Hospitality Quarterly*, *50*(1), 44–55. doi:10.1177/1938965508326305

Ko, K., Margolis, S., Revord, J., & Lyubomirsky, S. (2019). Comparing the effects of performing and recalling acts of kindness. *The Journal of Positive Psychology*. doi:10.1080/17439760.2019.1663252

Kock, F., Nørfelt, A., Josiassen, A., Assaf, A. G., & Tsionas, M. G. (2020). Understanding the COVID-19 tourist psyche: The evolutionary tourism paradigm. *Annals of Tourism Research*, *85*, 103053. doi:10.1016/j.annals.2020.103053 PMID:32921847

Koncul, N. (2012). Wellness: A New Mode of tourism. *Economic Research-Ekonomska Istraživanja*, *25*(2), 525–534. doi:10.1080/1331677X.2012.11517521

Kooij, D. T. A. M., van Woerkom, M., Wilkenloh, J., Dorenbosch, L., & Denissen, J. J. A. (2017). Job crafting towards strengths and interests: The effects of a job crafting intervention on person-job fit and the role of age. *The Journal of Applied Psychology*, *102*(6), 971–981. doi:10.1037/apl0000194 PMID:28277726

Korstanje, M. E. (2020). El COVID-19 y El turIsmo rural: una perspectiva antropológica. *Dimensiones Turísticas, 4*, 179-196. doi:10.47557/CKDK5549

Koydemir, S., Sökmez, A. B., & Schütz, A. (2020). *A Meta-Analysis of the Effectiveness of Randomized Controlled Positive Psychological Interventions on Subjective and Psychological Well-Being*. Applied Research in Quality of Life. doi:10.100711482-019-09788-z

Krippendorf, J. (2001). *Sociology of Tourism - for a new understanding of leisure and travel*. Aleph.

Kubacka, K. E., Finkenauer, C., Rusbult, C. E., & Keijsers, L. (2011). Maintaining Close Relationships: Gratitude as a Motivator and a Detector of Maintenance Behavior. *Personality and Social Psychology Bulletin*, *37*(10), 1362–1375. doi:10.1177/0146167211412196 PMID:21670220

Kubickova, M., & Campbell, J. M. (2020). The role of government in agro-tourism development: A top-down bottom-up approach. *Current Issues in Tourism*, *23*(5), 587–604. doi:10.1080/13683500.2018.1551338

Kummitha, H. R. (2020). Stakeholders involvement towards sustaining ecotourism destinations: The case of social entrepreneurship at Mangalajodi ecotourism trust in India. *Geo Journal of Tourism and Geosites*, *29*(2), 636–648. doi:10.30892/gtg.29220-495

Lam González, Y. E., De León Ledesma, J., & León González, C. J. (2015). European nautical tourists: Exploring destination image perceptions. *Tourism and Hospitality Management*, *21*(1), 33–49. doi:10.20867/thm.21.1.3

Lancia, F. (2012). *The Logic of the T- LAB Tools Explained*. www.tlab.it/en/toolsexplained.php

Latusi, S., & Fissore, M. (2021). The Seven Sisters of Piedmont: a case study of potentially worthwhile cooperation. In M. Della Lucia & E. Giudici (Eds.), *Humanistic Management and Sustainable Tourism: Human, Social and Environmental Challenges*. Routledge. doi:10.4324/9781003108955-3

Law, R., Sun, S., & Chan, I. C. C. (2019). Hotel technology: A perspective article. *Tourism Review, 75*(1), 286–289. Advance online publication. doi:10.1108/TR-05-2019-0150

Lee, L. Y.-S., Lam, K. Y.-C., & Lam, M. Y. C. (2020). Urban wellness: the space-out moment. *Journal Of Tourism Futures, 6*(3), 247–250. doi:10.1108/JTF-10-2019-0111

Lee, K. J., & Hwang, S. (2018). Serious leisure qualities and subjective well-being. *The Journal of Positive Psychology, 13*(1), 48–56. doi:10.1080/17439760.2017.1374437

Lemmi, E. (2020). Heritage and new communication technologies: development perspectives on the basis of the Via Francigena experience. In *Conference Preceedings of Heritage, tourism and Hospitality international Conference HTHIC* (pp. 43–63). USI-Università della Svizzera Italiana.

Lemos, C. C. (2013). Tourism planning at the federal level: an analysis of the instruments used and investments in the sector. *Revista de Administração Pública, 47*(6), 1.401-1.427.

Leslie, D. (2012). *Tourism Enterprises and the Sustainability Agenda across Europe. Surray.* Ashgate Publishing.

Li, C.-L. (2020). Quality of life: The perspective of urban park recreation in three Asian cities. *Journal of Outdoor Recreation and Tourism, 29*, 100260. doi:10.1016/j.jort.2019.100260

Lickorish, L. J., & Jenkins, C. L. (2000). Introduction to Tourism. Rio de Janeiro: Campus.

Lickorish, L. J., & Jenkins, C. L. (2002). *Introduction to tourism.* Campus.

Li, H., Wang, G., & Niu, Y. (2020). The Development Strategy of Marine Sports Tourism Based on Cost Theory under the Belt and Road Strategy. *Journal of Coastal Research, 112*(sp1), 112–115. doi:10.2112/JCR-SI112-033.1

Lillebø, A. I., Teixeira, H., Morgado, M., Martínez-López, J., Marhubi, A., Delacámara, G., Strosser, P., & Nogueira, A. J. A. (2019). Ecosystem-based management planning across aquatic realms at the Ria de Aveiro Natura 2000 territory. *The Science of the Total Environment, 650*, 1898–1912. doi:10.1016/j.scitotenv.2018.09.317 PMID:30286356

Lima, R. D. (2015). A qualidade do serviço nos hostels de Lisboa: Impacto na satistafação e lealdade dos hóspedes. *Revista de Gestão dos Países de Língua Portuguesa*. http://hdl.handle.net/10071/11133

Lima, R., & Vicente, P. (2016). A qualidade do serviço como determinante da satisfação dos turistas nos hostels de Lisboa. *Revista de Gestão Dos Países de Língua Portuguesa, 15*(3), 4. doi:10.12660/rgplp.v15n3.2016.78398

Lima, R., & Vicente, P. (2017). Exploring the relationship between hostel service quality and guests' satisfaction: A comparison between younger and older guests. *Review of Tourism Research*, *14*(5), 214–240.

Lipovetsky, G., & Serroy, J. (2011). *A cultura-mundo, respostas a uma sociedade desorientada*. Companhia das Letras.

Liutikas, D. (2015). In search of miracles: Pilgrimage to the miraculous places. *Tourism Review*, *70*(3), 197–213. doi:10.1108/TR-08-2013-0046

Liu, Z. (2003). Sustainable Tourism Development: A Critique. *Journal of Sustainable Tourism*, *11*(6), 459–475. doi:10.1080/09669580308667216

Liu, Z., Li, H., & Cao, Q. (2019). The Construction of a Marine Tourism Image Based on Grounded Theory and the Construction of Sports Thinking. *Journal of Coastal Research*, *98*(sp1), 267–270. doi:10.2112/SI98-066.1

Liu, Z., Li, H., & Cao, Q. (2020). Grey Prediction and Development Path of the Fusion of Marine Sports and Tourism. *Journal of Coastal Research*, *107*(sp1), 153–156. doi:10.2112/JCR-SI107-039.1

Li, Z., & Niu, Y. (2020). Development Strategy of Marine Sports Tourism: A Perspective of Cost Theory. *Journal of Coastal Research*, *107*(sp1), 17–20. doi:10.2112/JCR-SI107-005.1

Lohmann, M., & Bloom, J. (2015). *Happiness in a Tourism Context*. Paper presented to the International Conference on Gross National Happiness, Paro, Bhutan.

Loke, Z. (2020). Investigation Of Medical- And Wellness Tourists Of A Hungarian Spa To Explore Relationships Between Service Quality, Customer Satisfaction And Loyalty. *Deturope-The Central European Journal Of Regional Development And Tourism*, *12*(1), 102–118.

Loosemore, M. (2015). Barriers to social enterprise in the UK construction industry. In A. B. Raidén, & E. Aboagye-Nimo (Eds.), *Procs 31st Annual ARCOM Conference*. Association of Researchers in Construction Management.

Lopez, L. (2013). How Long Does the Pilgrimage Tourism Experience to Santiago de Compostela Last? *International Journal of Religious Tourism and Pilgrimage*, *1*, 1–14.

Lopez, L., González, R. C. L., & Fernández, B. M. C. (2017). Spiritual tourism on the way of Saint James the current situation. *Tourism Management Perspectives*, *24*, 225–234. doi:10.1016/j.tmp.2017.07.015

Lövheim, H. (2012). A new three-dimensional model for emotions and monoamine neurotransmitters. *Medical Hypotheses*, *78*(2), 341–348. doi:10.1016/j.mehy.2011.11.016 PMID:22153577

Lowi, T. (1972). Four Systems of Policy, Politics and Choice. *Public Administration Review*, *22*.

Lubowiecki-Vikuk, A., Dąbrowska, A., & Machnik, A. (2021). Responsible consumer and lifestyle: Sustainability insights. *Sustainable Production and Consumption, 25*, 91–101.

Lucarno, G. (2016). The Camino de Santiago de Compostela (Spain) and the Via Francigena (Italy): A comparison between two important historic pilgrimage routes in Europe. *International Journal of Religious Tourism and Pilgrimage*, *4*(7), 48–58.

Luck, M. (2007, Feb.). Nautical tourism development: opportunities and threats. *Nautical Tourism: Concepts and Issues,* 3–13.

Luković, T. (2012). Nautical Tourism and Its Function in the Economic Development of Europe. In M. Kasimoglu (Ed.), *Visions for Global Tourism Industry - Creating and Sustaining Competitive Strategies*. InTech. http://www.intechopen.com/books/visions-for-global-tourism-industry-creating-and-sustaining-competitive-strategies/nautical-tourism-in-the-function-of-the-economic-development-of-europe

Lumsdon, L. M., & McGrath, P. (2011). Developing a conceptual framework for slow travel: A grounded theory approach. *Journal of Sustainable Tourism*, *19*(3), 265–279. doi:10.1080/09669582.2010.519438

Lustosa, A. E., & Melo, L. F. (2010). Gross Internal Happiness (GNH) - Index of sustainable development. Brasília: Conjuntura econômica goiana.

Lv, Q., & Xie, X. (2017). Community involvement and place identity: The role of perceived values, perceived fairness, and subjective well-being. *Asia Pacific Journal of Tourism Research*, *22*(9), 951–964. doi:10.1080/10941665.2017.1345772

Lyubomirski, S., Sheldon, K. M., & Schkade, D. (2005). Pursuing happiness: The architecture of sustainable change. *Review of General Psychology*, *9*(2), 111–131. doi:10.1037/1089-2680.9.2.111

Lyubomirsky, S. (2010). *The How of Happiness*. Piatkus Books.

Lyubomirsky, S., King, L. A., & Diener, E. (2005). The benefits of frequent positive affect: Does happiness lead to success? *Psychological Bulletin*, *131*(6), 803–855. doi:10.1037/0033-2909.131.6.803 PMID:16351326

Lyu, J., Huang, H., & Mao, Z. (2021). Middle-aged and older adults' preferences for long-stay tourism in rural China. *Journal of Destination Marketing & Management*, *19*, 100552. doi:10.1016/j.jdmm.2020.100552

Macedo, J., Gonçalves, M., & Sousa, B. B. (2021). Relationship Marketing and Communication as a Driver of Fundraising Loyalty and Success in Non-Profit Organizations: A Theoretical Perspective. In M. D. Olvera-Lobo, J. Gutiérrez-Artacho, I. Rivera-Trigueros, & M. Díaz-Millón (Eds.), *Innovative Perspectives on Corporate Communication in the Global World* (pp. 62–77). IGI Global. doi:10.4018/978-1-7998-6799-9.ch004

Machado, M. B. T. (2012). Social fear and tourism in Rio de Janeiro. *Tourism & Management Studies, Faro*, (8), 48–54.

Malisch, J. L., Harris, B. N., Sherrer, S. M., Lewis, K. A., Shepherd, S. L., McCarthy, P. C., Spott, J. L., Karam, E. P., Moustaid-Moussa, N., Calarco, J. M. C., Ramalingam, L., Talley, A. E., Cañas-Carrell, J. E., Ardon-Dryer, K., Weiser, D. A., Bernal, X. E., & Deitloff, J. (2020). Opinion: In the wake of COVID-19, academia needs new solutions to ensure gender equity. *Proceedings of the National Academy of Sciences of the United States of America*, *117*(27), 15378–15381. doi:10.1073/pnas.2010636117 PMID:32554503

Malkina-Pykh, I. G., & Pykh, Y. A. (2014). Subjective well-being and personality: Implications for wellness tourism. *WIT Transactions on Ecology and the Environment*, *187*, 51–64. doi:10.2495/ST140051

Mané, A., & Ferreira, L. (2017). El perfil del consumidor de hostel en Brazil y sus motivaciones. *Estudos e Perspectivas Em Turismo, 26*(4), 995–943. https://dialnet.unirioja.es/servlet/articulo?codigo=6329748

Mannheim, K. (1976). *Ideologia e Utopia*. Zahar Editores.

Ma, Q. G., Hu, L. F., Pei, G. X., Ren, P. Y., & Ge, P. (2014). Applying Neuroscience to Tourism Management: A Primary Exploration of Neurotourism. *Applied Mechanics and Materials*, *670–671*, 1637–1640. . doi:10.4028/www.scientific.net/AMM.670-671.1637

Maracajá, K. F. B.; & Pinheiro, I. F. S. (2020). Environmental perception in tourism planning: proposals for the development of Seridó, RN. *Revista Observatório de Inovação do Turismo, 14*(1), 41-59.

Marcon, E. M. (2007). Tourism as an Agent of Social Development and the Guarani Community in the Jesuit Ruins of São Miguel das Missões. *Pasos Revista de Turismo y Patrimonio Cultural*, *5*, 343–352. doi:10.25145/j.pasos.2007.05.025

Marques, J. de O. (2014). *A influência das diferenças culturais na perceção da qualidade do serviço pelo consumidor de um hostel: estudo de caso comparativo entre Coreanos e Franceses.* Instituto Português de Administração de Marketing. http://hdl.handle.net/10400.26/10210

Marques, R. (2000). *O livro das virtudes de sempre*. Ed. Asa.

Martinez-Marti, M. L., Avia, M. D., & Hernández-Lloreda, M. J. (2010). The effects of counting blessings on subjective well-being: A gratitude intervention in a Spanish sample. *The Spanish Journal of Psychology*, *13*(2), 886–896. doi:10.1017/S1138741600002535 PMID:20977036

Martin, R. L., & Osberg, S. (2007). Social entrepreneurship: The case for definition. *Stanford Social Innovation Review*, 1–17.

Martins, S. R. O. (2002, September). Local development: Conceptual and methodological issues. *International Journal of Local Development, Campo Grande*, *3*(5), 51–59.

Marušić, E., Šoda, J., & Krčum, M. (2020). The Three-Parameter Classification Model of Seasonal Fluctuations in the Croatian Nautical Port System. *Sustainability*, *12*(12), 5079. doi:10.3390u12125079

Max-Neef, M. (1993). *Desarollo a Escala Humana: Conceptos, Aplicaciones y Algunas Reflexiones*. ICARIA.

Max-Neef, M. (2012). *Development at the Human Scale: ConceptionApplication-Further Reflections*. EDFURB.

Mayer, H., & Knox, P. L. (2006). Slow cities: Sustainable places in a fast world. *Journal of Urban Affairs*, *28*(4), 321–334. doi:10.1111/j.1467-9906.2006.00298.x

Mayer, V. F., & Coelho, M. D. F. (2021). Interrupted dreams: Memories and emotions of travel experiences during Covid-19 breakthrough. *Revista Brasileira de Pesquisa em Turismo*, *15*(1). Advance online publication. doi:10.7784/rbtur.v15i1.2192

Mayer, V. F., & Coelho, M. de F. (2021). Interrupted dreams: Memories and emotions of travel experiences during Covid-19 breakthrough. *RBTUR*, *15*(1), 1–20.

Mayer, V. F., Machado, J. D. S., Marques, O., & Nunes, J. M. G. (2020). Mixed feelings?: Fluctuations in wellbeing during tourist travels. *Service Industries Journal*, *40*(1-2), 158–180. doi:10.1080/02642069.2019.1600671

McCabe, S., & Johnson, S. (2013). The happiness factor in tourism: Subjective well-being and social tourism. *Annals of Tourism Research*, *41*, 42–65. doi:10.1016/j.annals.2012.12.001

McCarthy, B. (2012). From fishing and factories to cultural tourism: The role of social entrepreneurs in the construction of a new institutional field. *Entrepreneurship & Regional Development: An International Journal*, *24*(3-4), 259–282. doi:10.1080/08985626.2012.670916

McCullough, M. E., Eammons, R. A., & Tsang, J. A. (2002). The grateful disposition: A conceptual and empirical topography. *Journal of Personality and Social Psychology*, *82*(1), 112–127. doi:10.1037/0022-3514.82.1.112 PMID:11811629

McCullough, M. E., Kilpatrick, S. D., Emmons, R. A., & Larson, D. B. (2001). Is gratitude a moral affect? *Psychological Bulletin*, *127*(2), 249–266. doi:10.1037/0033-2909.127.2.249 PMID:11316013

McCullough, M. E., Kimeldorf, M. B., & Cohen, A. D. (2008). An adaptation for altruism? The social causes, social effects, and social evolution of gratitude. *Current Directions in Psychological Science*, *17*(4), 281–285. doi:10.1111/j.1467-8721.2008.00590.x

McCullough, M. E., Tsang, J.-A., & Emmons, R. A.McCullough. (2004). Gratitude in intermediate affective terrain: Links of grateful moods to individual differences and daily emotional experience. *Journal of Personality and Social Psychology*, *86*(2), 295–309. doi:10.1037/0022-3514.86.2.295 PMID:14769085

McGrath, P., & Sharpley, R. (2016). Slow travel and slow tourism: New concept or new label? In M. Clancy (Ed.), *Slow tourism, food and cities: Pace and the search for the 'good life'*. Routledge.

McMahan, E. A., & Estes, D. (2011). Hedonic versus eudaimonic conceptions of wellbeing: Evidence of differential associations with experienced well-being. *Social Indicators Research*. Advance online publication. https://psycnet.apa.org/doi/10.1007/s11205-010-9698-0

Mdee, A., & Emmott, R. (2008). Social enterprise with international impact: The case for fair trade certification of volunteer tourism. *Education, Knowledge & Economy*, *2*(3), 191–201. doi:10.1080/17496890802426188

Melief, A. E. F., Schulp, J. A., & de Boer, M. R. (2020). Happiness at work: A phenomenological study of the experiences of hospitality industry employees. *Research in Hospitality Management*, *10*(1), 63–66. doi:10.1080/22243534.2020.1790212

Miguel, P. A. C., & Salomi, G. E. (2004). Uma revisão dos modelos para medição da qualidade em serviços. *Revista Produção*, *14*(1), 12–30. doi:10.15900103-65132004000100003

Milano, C., Novelli, M., & Cheer, J. M. (2019). Overtourism and degrowth: A social movements perspective. *Journal of Sustainable Tourism*, *27*(12), 1–19. doi:10.1080/09669582.2019.1650054

Mileti, A., Guido, G., & Prete, I. M. (2016). Nanomarketing: A new frontier of neuromarketing. *Psychology and Marketing*, *33*(8), 664–674. doi:10.1002/mar.20907

Milgram, P., & Kishino, F. (1994). A taxonomy of mixed reality visual displays. *IEICE Transactions on Information and Systems*, *77*(12), 1321–1329.

Moein, R. (2014). Investigation and prioritization of social entrepreneurship components based on tourism industry development approach using TOPSIS technique. *International Journal of Scientific Management and Development*, *2*(11), 602–613.

Mohamad, H. A. D., Yazid, M. S. A., Khatibi, A., & Azam, S. M. F. (2017). Service quality, customer satisfaction and customer loyalty ot the hotel industry in United Arab Emirates (UAE): A measurement model. *European Journal of Management and Marketing Studies*, *2*(4), 1–26. doi:10.5281/zenodo.1066572

Moore, K. (2012). On the periphery of pleasure: Hedonics, eudaimonics and slow travel. In S. Fullagar, K. Markwell, & E. Wilson (Eds.), *Slow Tourism: Experiences and Mobilities* (pp. 25–35). Channel View Publications. doi:10.21832/9781845412821-005

Moreira, C. C. O. (2013). *Turismo, Território e Desenvolvimento: Competitividade e Gestão Estratégica de Destinos* (Doctoral dissertation), Coimbra, Universidade de Coimbra.

Moreno, M. J., & Otamendi, F. J. (2017). Fostering Nautical Tourism in the Balearic Islands. *Sustainability*, *9*(12), 1–20. doi:10.3390u9122215

Moro, S., Rita, P., Ramos, P., & Esmerado, J. (2019, November 27). Analysing recent augmented and virtual reality developments in tourism. *Journal of Hospitality and Tourism Technology*. doi:10.1108/JHTT-07-2018-0059

Morris, M. H., Avila, R. A., & Allen, J. (1993). Individualism and the modern corporation: Implications for innovation and entrepreneurship. *Journal of Management*, *19*(3), 595–612. doi:10.1177/014920639301900305

Mort, G., Weerawardena, J., & Carnegie, K. (2002). Social entrepreneurship: Towards conceptualisation. *International Journal of Nonprofit and Voluntary Sector Marketing*, *8*(1), 76–88. doi:10.1002/nvsm.202

Mottiar, Z. (2016). Exploring the motivations of tourism social entrepreneurs: The role of a national tourism policy as a motivator for social entrepreneurial activity in Ireland. *International Journal of Contemporary Hospitality Management*, *28*(6), 1137–1154. doi:10.1108/IJCHM-06-2014-0278

Moura, M., Sousa, B., Malheiro, A., & Veloso, C. (2021). The Quality of Services in Tourism: A Study Applied to Transport Companies. *Quality - Access to Success*, *22*(183), 124–129.

Moyle, B. D., Moyle, C., Bec, A., & Scott, N. (2019). 'The next frontier in tourism emotion research'. *Current Issues in Tourism*, *22*(12), 1393–1399. doi:10.1080/13683500.2017.1388770

Moyle, B., Moyle, C. L., Ruhanen, L., Weaver, D., & Hadinejad, A. (2020). Are we really progressing sustainable tourism research? A bibliometric analysis. *Journal of Sustainable Tourism*, *29*(1), 106–122. doi:10.1080/09669582.2020.1817048

Muarifuddin, & Kriswanto, H. D. (2020). Social entrepreneurship learning based on community empowerment of tourism village. *PEOPLE: International Journal of Social Sciences*, *6*(1), 81-92.

Müller, C., & Friederich, D. B. (2020). A Dinâmica do Federalismo Brasileiro no Tema das Políticas Públicas, Controle Social e COVID-19. *Revista de Direitos Fundamentais & Democracia*, *25*(3), 49–77.

Mullins, L. J. (2013). *Management & Organisational Behaviour* (10th ed.). Pearson.

Mundt, L. (2011). *Tourism and Sustainable Development: Reconsidering a Concept of Vague Policies*. Berlin: ESV.

Naderi, A., Vosta, N. L., Ebrahimi, A., & Jalilvand, M. R. (2019). The contributions of social entrepreneurship and transformational leadership to performance insights from rural tourism in Iran. *The International Journal of Sociology and Social Policy*, *39*(9/10), 719–737. doi:10.1108/IJSSP-06-2019-0124

Nagaraju, L. G., & Chandrashekara, B. (2014). Rural tourism and rural development in India. *International Journal of Interdisciplinary and Multidisciplinary Studies*, *1*(6), 42–48.

Naghiu, A., Vázquez, J. L., & Georgiev, I. (2005). Rural development strategies through rural tourism activities in romania: Chance for an internal demand? *International Review on Public and Nonprofit Marketing*, *2*(1), 85–95. doi:10.1007/BF02893253

Nash, J. F. Jr. (1950). The Bargaining Problem. *Econometrica*, *18*(2), 155–162. doi:10.2307/1907266

National Gallery. (2020). *Sainsbury Wing VR tour*. Retrieved from https://www.nationalgallery.org.uk/visiting/virtual-tours/sainsbury-wing-vr-tour

National Wellness Institute. (2020). *The six dimensions of wellness*. Retrieved from https://cdn.ymaws.com/members.nationalwellness.org/resource/resmgr/pdfs/sixdimensionsfactsheet.pdf

Navío-Marco, J., Ruiz-Gómez, L. M., & Sevilla-Sevilla, C. (2018, December). Progress in information technology and tourism management: 30 years on and 20 years after the internet-Revisiting Buhalis & Law's landmark study about eTourism. *Tourism Management*, *69*, 460–470. doi:10.1016/j.tourman.2018.06.002

Nawijn, J. (2011). Determinants of Daily Happiness on Vacation. *Journal of Travel Research*, *50*(5), 559–566. doi:10.1177/0047287510379164

Nawijn, J., Marchand, M. A., Veenhoven, R., & Vingerhoets, A. J. (2010a). Vacationers Happier, but Most not Happier After a Holiday. *Applied Research in Quality of Life*, *5*(1), 35–47. doi:10.100711482-009-9091-9 PMID:20234864

Nawijn, J., & Mitas, O. (2012). Resident Attitudes to Tourism and Their Effect on Subjective Well-Being: The Case of Palma de Mallorca. *Journal of Travel Research*, *51*(5), 531–541. doi:10.1177/0047287511426482

Nawijn, J., & Peeters, P. M. (2010b). Travelling 'green': Is tourists' happiness at stake? *Current Issues in Tourism*, *13*(4), 381–392. doi:10.1080/13683500903215016

Nayyar, A., Mahapatra, B., Le, D. N., & Suseendran, G. (2018). Virtual Reality (VR) & Augmented Reality (AR) technologies for tourism and hospitality industry. *IACSIT International Journal of Engineering and Technology*, *7*(2), 156–160. doi:10.14419/ijet.v7i2.21.11858

Network of European Museum Organizations. (2020). *Survey of the impact of the COVID-19 situation on museums in Europe*. Retrieved from https://www.ne-mo.org/fileadmin/Dateien/public/NEMO_documents/NEMO_COVID19_Report_12.05.2020.pdf

Neuburger, L., & Egger, R. (2018). Augmented Reality: providing a different dimension for museum visitors. In T. Jung & M. tom Dieck (Eds.), *Augmented Reality and Virtual Reality. Progress in IS* (pp. 65–77). Springer. doi:10.1007/978-3-319-64027-3_5

New Museum. (2017). First Look: Artists' VR. *New Museum*. Retrieved from https://www.newmuseum.org/calendar/view/1135/first-look-artists-vr

Newman, D. B., Tay, L., & Diener, E. (2014). Leisure and subjective well-being: A model of psychological mechanisms as mediating factors. *Journal of Happiness Studies*, *15*(3), 555–578. doi:10.100710902-013-9435-x

Ng, T. W. H. (2016). Embedding employees early on: The importance of workplace respect. *Personnel Psychology*, *69*(3), 599–633. doi:10.1111/peps.12117

Nilsson, M. (2018). Wanderers in the shadow of the sacred myth: Pilgrims in the 21st century. *Social & Cultural Geography*, *19*(1), 21–38. doi:10.1080/14649365.2016.1249398

Nilsson, M., & Tesfahuney, M. (2016). Performing the "post-secular" in Santiago de Compostela. *Annals of Tourism Research*, *57*, 18–30. doi:10.1016/j.annals.2015.11.001

Noy, C. (2004). This trip really changed me: Backpackers' narratives of self-change. *Annals of Tourism Research*, *31*(1), 78–102. doi:10.1016/j.annals.2003.08.004

O'Leary, K., & Dockray, S. (2015). The Effects of Two Novel Gratitude and Mindfulness Interventions on Well-Being. *Journal of Alternative and Complementary Medicine (New York, N.Y.)*, *21*(4), 243–245. doi:10.1089/acm.2014.0119 PMID:25826108

O'Neill, J. W., & Davis, K. (2011). Work stress and well-being in the hotel industry. *International Journal of Hospitality Management*, *30*(2), 385–390. doi:10.1016/j.ijhm.2010.07.007 PMID:23794780

O'Sullivan, M., & Kearney, G. (2018). Virtual Reality (VR) Technology: empowering managers to reduce and eliminate accessibility barriers for people with Autism Spectrum Disorders. In *Transforming our World Through Design, Diversity and Education: Proceedings of Universal Design and Higher Education in Transformation Congress 2018* (*Vol. 256*, p. 253). IOS Press.

OECD - Organization for Economic Co-operation and Development. (2020). *The territorial impact of COVID-19: managing the crisis across levels of government*. Retrieved from: https://www.oecd.org/coronavirus/policy-responses/theterritorial-impact-of-covid-19-managing-the-crisis-across-levels-ofgovernment-d3e314e1/

OECD. (2020). *OECD Tourism Trends and Policies 2020*. OECD Publishing., doi:10.1787/6b47b985-

OECD. (2020). Rebuilding tourism for the future: COVID-19 policy responses and recovery. In *OECD Economic Outlook* (Vol. 2020, Issue 2). https://www.oecd.org/coronavirus/policy-responses/rebuilding-tourism-for-the-future-covid-19-policy-responses-and-recovery-bced9859/

Oficina del Peregrino. (2019). *Informe estadístico Año 2019*. Publisher.

Oh, H., Assaf, A. G., & Baloglu, S. (2016). Motivations and goals of slow tourism. *Journal of Travel Research*, *55*(2), 205–219. doi:10.1177/0047287514546228

Okech, R., Haghiri, M., & George, B. (2012). Rural tourism as a sustainable development alternative: an analysis with special reference to Luanda, Kenya. *CULTUR, 6*(3).

Oliveira, C. T. F., Zouain, D. M. (2011). Turismo rural e agricultura familiar: desafios e perspectivas para o campo. *Revista Acadêmica Observatório de Inovação do Turismo, 6*(2).

Olsen, D. H., & Wilkinson, G. (2016). Are fast pilgrims true pilgrims? The Shikoku pilgrimage. *Annals of Tourism Research*, *61*, 228–230. doi:10.1016/j.annals.2016.08.004

ONU - Organização das Nações Unidas. (2002). *Publicação Johannesbourg Summit. World Summit on Sustainable Development*. United Nations Department of Public Information.

Orri, M., Pingault, J. B., Rouquette, A., Lalanne, C., Falissard, B., Herba, C., Côté, S. M., & Berthoz, S. (2017). Identifying affective personality profiles: A latent profile analysis of the Affective Neuroscience Personality Scales. *Scientific Reports*, *7*(1), 4548. doi:10.103841598-017-04738-x PMID:28674393

Oviedo, L., de Courcier, S., & Farias, M. (2014). Rise of pilgrims on the Camino to Santiago: Sign of change or religious revival? *Review of Religious Research*, *56*(3), 433–442. doi:10.100713644-013-0131-4

Ozdemir, G., & Celebi, D. (2018). Exploring dimensions of slow tourism motivation. *Anatolia*, *29*(4), 540–552. doi:10.1080/13032917.2018.1460854

P. S., S., & Abraham, S. (2017). Wellness Tourism - an Analysis of Tourist Preferences and Perceptions in Puducherry. *Journal of Hospitality Application & Research, 12*(2), 61–79. https://search.ebscohost.com/login.aspx?direct=true&db=buh&AN=128135305&site=ehost-live&scope=site

Pai, C. K., Liu, Y., Kang, S., & Dai, A. (2020). The role of perceived smart tourism technology experience for tourist satisfaction, happiness and revisit intention. *Sustainability*, *12*(16), 6592. doi:10.3390u12166592

Paixao, D. L. D. (2007). Thermae et Ludus: O início do turismo de saúde no Brasil e no mundo. *Turismo em Análise*, *18*(2), 133–147. doi:10.11606/issn.1984-4867.v18i2p133-147

Pan, T., Shu, F., Kitterlin-Lynch, M., & Beckman, E. (2021). Perceptions of cruise travel during the COVID-19 pandemic: Market recovery strategies for cruise businesses in North America. *Tourism Management*, *85*, 104275. doi:10.1016/j.tourman.2020.104275

Panyik, E., & Gonçalves, J. (2017). Neurotourism: A paradigm shift in travel on the intersection of e-tourism and neuroscience. *Revista Turismo & Desenvolvimento*, (27/28), 321-323.

Park, J. & Min, H. (2020). Turnover intention in the hospitality industry: A meta-analysis. *International Journal of Hospitality Management, 90*, 1-11. doi:10.1016/j.ijhm.2020.102599

Park, H., Lee, M., & Back, K.-J. (2020). Exploring the roles of hotel wellness attributes in customer satisfaction and dissatisfaction: Application of Kano model through mixed methods. *International Journal of Contemporary Hospitality Management*, *33*(1), 263–285. doi:10.1108/IJCHM-05-2020-0442

Paula, A. T. (2013). *State role, poverty alleviation and tourism: theoretical relations with the National Tourism Plan 2007/2010* (Master's dissertation). Programa de Pós- Graduação em Turismo, Universidade de Brasília.

Pawson, M. G., Glenn, H., & Padda, G. (2008). The definition of marine recreational fishing in Europe. *Marine Policy*, *32*(3), 339–350. doi:10.1016/j.marpol.2007.07.001

Pearce, P. L. (2020). Tourists' perception of time: Directions for design. *Annals of Tourism Research*, *83*, 102932. doi:10.1016/j.annals.2020.102932

Peng, K., & Lin, P. M. C. (2016). Social entrepreneurs: Innovating rural tourism through the activism of service science. *International Journal of Contemporary Hospitality Management, 28*(6), 1225–1244. doi:10.1108/IJCHM-12-2014-0611

Pennington-Gray, L., Schroeder, A., Wu, B., Donohoe, H., & Cahyanto, I. (2014). Travelers' perceptions of crisis preparedness certification in the United States. *Journal of Travel Research, 53*(3), 353–365. doi:10.1177/0047287513496470

Perdue, R., Tyrrel, T., & Uysal, M. (2010). Understanding the value of tourism: a conceptual divergence. In D. Pearce & R. Butler (Eds.), *Tourism Research: A 20:20 Vision* (pp. 123–134). Goodfellow Publishers.

Peredo, B., & Wurzelmann, S. (2015). Indigenous tourism and social entrepreneurship in the Bolivian Amazon: Lessons from San Miguel del Bala. *International Indigenous Policy Journal, 6*(4), 1–26. doi:10.18584/iipj.2015.6.4.5

Pereira, E. C. S., Mascarenhas, M. V. M., Flores, A. J. G., & Pires, G. M. V. S. (2015). Nautical small-scale sports events portfolio: A strategic leveraging approach. *European Sport Management Quarterly, 15*(1), 27–47. doi:10.1080/16184742.2015.1007883

Permatasari, A. (2020). The Influence of Perceived Value Towards Customer Satisfaction in Hostel Business: A Case of Young Adult Tourist in Indonesia. *International Journal of Tourism and Hospitality Management in the Digital Age, 4*(2), 11–22. doi:10.4018/IJTHMDA.2020070102

Peterson, C., & Seligman, M. E. P. (2004). *Character strengths and virtues: A handbook and classification*. Oxford University Press.

Petrou, P., & Bakker, A. B. (2016). Crafting one's leisure time in response to high job strain. *Human Relations, 69*(2), 507–529. doi:10.1177/0018726715590453

Petrou, P., Bakker, A. B., & van den Heuvel, M. (2017). Weekly job crafting and leisure crafting: Implications for meaning-making and work engagement. *Journal of Occupational and Organizational Psychology, 90*(2), 129–152. doi:10.1111/joop.12160

Philippe, M. (2020). Social and Associative Sports Tourism in France: The Glénans and the National Union of Outdoor Sports Centres (UCPA). *The International Journal of the History of Sport, 37*(5–6), 433–450. doi:10.1080/09523367.2020.1754197

Pichitphan, P. (2018). *A study of service quality dimensions in hostel in Bangkok and its impact on backpackers' satisfaction*. Thammasat University.

Pinchot, G. (1985). *Intrapreneuring*. Harper and Row.

Pine, B. J. II, & Gilmore, J. H. (1998, July). Welcome to the Experience Economy. *Harvard Business Review*.

Pinto, I. C. A. (2018). *A qualidade e o alojamento local no Porto: caracterização e perspetivas futuras*. Instituto Politécnico do Porto. http://hdl.handle.net/10400.22/12657

Plaice, K. T. (2004). *State Public Agencies and the Development of Tourism in Brazil* (Doctoral dissertation). Universidade de São Paulo. São Paulo.

Polukhina, A., Sheresheva, M., Efremova, M., Suranova, O., Agalakova, O., & Antonov-Ovseenko, A. (2021). The Concept of Sustainable Rural Tourism Development in the Face of COVID-19 Crisis: Evidence from Russia. *Journal of Risk and Financial Management*, *14*(1), 38. doi:10.3390/jrfm14010038

Ponting, J., & O'Brien, D. (2014). Liberalizing Nirvana: An analysis of the consequences of common pool resource deregulation for the sustainability of Fiji's surf tourism industry. *Journal of Sustainable Tourism*, *22*(3), 384–402. doi:10.1080/09669582.2013.819879

Popoviciu, I., & Popoviciu, S. (2011). Social entrepreneurship, social enterprise and the principles of a community of practice. *Revista de Cercetare si Interventie Sociala*, *33*, 44–55.

Postman, A., & Schmuecker, D. (2017). Understanding and Overcoming Negative Impacts of Tourism in City Destinations: Conceptual Model and Strategic Framework. *Journal of Tourism Futures*, *3*(2), 144–156. doi:10.1108/JTF-04-2017-0022

Potter, L. E., Carter, L., & Coghlan, A. (2016). Virtual reality and nature-based tourism: An opportunity for operators and visitors. In *Proceedings of the 28th Australian Computer-Human Interaction Conference, OzCHI 2016* (pp. 652–653). New York: Association for Computing Machinery, Inc. 10.1145/3010915.3011854

Poux, F., Valembois, Q., Mattes, C., Kobbelt, L., & Billen, R. (2020). Initial user-centered design of a virtual reality heritage system: Applications for digital tourism. *Remote Sensing*, *12*(16), 2583. Advance online publication. doi:10.3390/rs12162583

Pushelberg, L. (2019). How airports are using VR to be less wasteful. *Aviation pros*. Retrieved from https://www.aviationpros.com/airports/airport-technology/article/21112224/how-airports-are-using-vr-to-be-less-wasteful

Queiroz, A. (2017). Planning and metropolization of maritime leisure in Fortaleza-Ceará, Northeast Brazil. Eure. *Santiago de Chile*, *43*(128), 153–173.

Quintela, J. A., Costa, C., & Correia, A. (2016). Health, wellness and medical tourism – a conceptual approach. *Enlightening Tourism: A Pathmaking Journal*, 1-18. Retrieved from: https://dialnet.unirioja.es/servlet/articulo?codigo=5559522

Quintero Santos, J. L. (2004). Los impactos económicos, socioculturales y medioambientales del turismo y sus vínculos con el turismo sostenible. *Anales del Museo de América*, (12), 263–274.

Radic, A., Law, R., Lück, M., Kang, H., Ariza-Montes, A., Arjona-Fuentes, J. M., & Han, H. (2020). Apocalypse now or overreaction to coronavirus: The global cruise tourism industry crisis. *Sustainability (Switzerland)*, *12*(17), 1–19. doi:10.3390u12176968

Radić, M. M., Jukić, I., & Roje, A. (2020). Increasing tourism through social entrepreneurship – the case of Croatia. *Journal of Corporate Governance, Insurance, and Risk Management*, *7*(1), 25–38. doi:10.51410/jcgirm.7.1.3

Rahman, N. S. F. A., Bazaruddin, N. H., Shah, S. F. S. S. J., Said, M. H., & Othman, M. K. (2020). The sustainability of nautical sports tourism in terengganu using a cost benefit analysis. *Journal of Sustainability Science and Management*, *15*(1), 148–161.

Raj, R., Griffin, K., & Blackwell, R. (2015). Motivations for religious tourism, pilgrimage, festivals and events. In R. Raj & K. Griffin (Eds.), *Religious Tourism and Pilgrimage Management: An International Perspective* (pp. 103–117). CABI. doi:10.1079/9781780645230.0103

Reader, I. (2007). Pilgrimage growth in the modern world: Meanings and implications. *Religion*, *37*(3), 210–229. doi:10.1016/j.religion.2007.06.009

Reeves, M., Vogt, F., & Marks, E. (2021). Dispositional mindfulness, gratitude and self-compassion: Factors affecting tinnitus distress. *Mindfulness*, *12*(4), 1002–1008. Advance online publication. doi:10.100712671-020-01569-2

Rendle, E. J., & Rodwell, L. D. (2014). Artificial surf reefs: A preliminary assessment of the potential to enhance a coastal economy. *Marine Policy*, *45*, 349–358. doi:10.1016/j.marpol.2013.09.004

ReportLinker. (2020). *The Global Boat Rental Market is expected to grow from USD 17,620.57 Million in 2019 to USD 23,160.11 Million by the end of 2025 at a Compound Annual Growth Rate (CAGR) of 4.66% Market*. https://www.globenewswire.com/news-release/2020/07/03/2057347/0/en/The-Global-Boat-Rental-Market-is-expected-to-grow-from-USD-17-620-57-Million-in-2019-to-USD-23-160-11-Million-by-the-end-of-2025-at-a-Compound-Annual-Growth-Rate-CAGR-of-4-66.html

Rinschede, G. (1992). Forms of religious tourism. *Annals of Tourism Research*, *19*(1), 51–67. doi:10.1016/0160-7383(92)90106-Y

Rivera, M., Croes, R., & Lee, S. H. (2016). Tourism development and happiness: A residents' perspective. *Journal of Destination Marketing & Management*, *5*(1), 5–15. doi:10.1016/j.jdmm.2015.04.002

Roberto, K. J., Johnson, A. F., & Rauhaus, B. M. (2020). Stigmatization and prejudice during the COVID-19 pandemic. *Administrative Theory & Praxis*, *42*(3), 364–378. doi:10.1080/10841806.2020.1782128

Robson, K., Plangger, K., Kietzmann, J., McCarthy, I., & Pitt, L. (2015). Is it all a game? Understanding the principles of gamification. *Business Horizons*, *58*(4), 411–420. doi:10.1016/j.bushor.2015.03.006

Rogerson, C., & Rogerson, J. (2021). COVID-19 and Changing Tourism Demand: Research Review and Policy Implications for South Africa. *African Journal of Hospitality, Tourism and Leisure*, *10* (1), 1–21. doi:10.46222/ajhtl.19770720-83

Rooney, A. (2018). *A História da Neurociência*. M.Books.

Roper, J., & Cheney, G. (2005). The meanings of social entrepreneurship today. *Corporate Governance International Journal of Business in Society*, *5*(3), 95–104. doi:10.1108/14720700510604733

Rosa, M. E. T., & Kanikadan, A. Y. S. (2021). Tourism in Cape Verde: development for whom? Turismo: Estudos & Práticas. *Mossoró*, *10*(1), 1–19.

Ruhanen, L., Weiler, B., Moyle, B. D., & McLennan, C. L. J. (2015). Trends and patterns in sustainable tourism research: A 25-year bibliometric analysis. *Journal of Sustainable Tourism*, *23*(4), 517–535. doi:10.1080/09669582.2014.978790

Sagawa, S., & Segal, E. (2000). Common interest, common good: Creating value through business and social sector partnership. *California Management Review*, *42*(2), 105–122. doi:10.1177/000812560004200201

Sakuraya, A., Shimazu, A., Imamura, K., Namba, K., & Kawakami, N. (2016). Effects of a job crafting intervention program on work engagement among Japanese employees: A pretest-posttest study. *BMC Psychology*, *4*(1), 49. doi:10.118640359-016-0157-9 PMID:27776553

Salces-Cubero, I. M., Ramírez-Fernández, E., & Ortega-Martínez, A. R. (2018). Strengths in older adults: Differential effect of savoring, gratitude and optimism on well-being. *Aging & Mental Health*, *23*(8), 1017–1024. doi:10.1080/13607863.2018.1471585 PMID:29781723

Salviati, M. E. (2017). *Manual do Aplicativo Iramuteq*. Planaltina. http://iramuteq.org/documentation/fichiers/manual-do-aplicativo-iramuteq-par-maria-elisabeth-salviati

Sam, I., John, O. O., & Bernadett, K. (2016). Social Entrepreneurship in Hospitality Industry. The Case of Operators at the Coastal Tourism Circuit in Kenya. *International Journal of Humanities Social Sciences and Education*, *3*(2), 115–121.

Sampaio, C. A. C. (2019). Homage to a Barefoot Economist: Manfred Max-Neef. *Development and the Environment*, *51*, 370–372.

Santos, F. M. (2012). A positive theory of social entrepreneurship. *Journal of Business Ethics*, *111*(3), 335–351. doi:10.100710551-012-1413-4

Santos, G. E. de O. (2016). Worldwide hedonic prices of subjective characteristics of hostels. *Tourism Management*, *52*, 451–454. doi:10.1016/j.tourman.2015.07.001

Santos, V., Ramos, P., Sousa, B., & Valeri, M. (2021). Towards a Framework for the Global Wine Tourism System. *Journal of Organizational Change Management*, *ahead-of-print*(ahead-of-print). Advance online publication. doi:10.1108/JOCM-11-2020-0362

Santos, X. M. (2002). Pilgrimage and Tourism at Santiago de Compostela. *Tourism Recreation Research*, *27*(2), 41–50. doi:10.1080/02508281.2002.11081219

Saraiva, A. (2013). *Hostels independentes: o caso de Lisboa*. Escola Superior de Hotelaria e Turismo do Estoril. http://comum.rcaap.pt/handle/123456789/4425

Sarıbaş, Ö., Kömürcü, S., & Akbaba, A. (2020). The concept of social entrepreneurship in tourism: A bibliometric analysis for tourism literature. *Euroasia Journal of Social Science and Humanities*, *7*(4), 51–58.

Scheske, C., Arroyo Rodriguez, M., Buttazzoni, J. E., Strong-Cvetich, N., Gelcich, S., Monteferri, B., Rodríguez, L. F., & Ruiz, M. (2019). Surfing and marine conservation: Exploring surf-break protection as IUCN protected area categories and other effective area-based conservation measures. *Aquatic Conservation, 29*(S2), 195–211. doi:10.1002/aqc.3054

Schlüter, R. G. (2003). Metodologia da pesquisa em turismo e hotelaria. *Aleph (Jerusalem).*

Schuhbauer, A., & Koch, V. (2013). Assessment of recreational fishery in the Galapagos Marine Reserve: Failures and opportunities. *Fisheries Research, 144*, 103–110. doi:10.1016/j.fishres.2013.01.012

Scliar, M. (2007). História do Conceito de Saude. *Physis (Rio de Janeiro, Brazil), 17*(1), 29–41. doi:10.1590/S0103-73312007000100003

Scott, N. (2020). Cognitive psychology and tourism – surfing the "cognitive wave": A perspective article. *Tourism Review, 75*(1), 49–51. doi:10.1108/TR-06-2019-0217

Seaman, A. N., & Depper, G. L. (2019). Visiting scuttled ships: An examination of the important elements of the wreck diving experience. *Tourism in Marine Environments, 14*(1), 31–44. doi:10.3727/154427319X15567670161919

Seligman, M. E. P., Steen, T. A., Park, N., & Peterson, C. (2005). Positive psychology progress: Empirical validation of interventions. *American Psychologist, 60*(5), 410-421. doi:10.1037/0003-066X.60.5.410

Seligman, M. E. P. (2011). *Flourish: A visionary new understanding of happiness and well-being*. Free Press.

Seligman, M. E. P., & Csikszentmihalyi, M. (2000). Positive Psychology: An introduction. *The American Psychologist, 55*(1), 5–14. doi:10.1037/0003-066X.55.1.5 PMID:11392865

Sen, A. (2000). Development as Freedom. *Revista Gaceta Ecológica*, 55.

Senf, K., & Liau, A. K. (2013). The effects of positive interventions on happiness and depressive symptoms, with an examination of personality as a moderator. *Journal of Happiness Studies: An Interdisciplinary Forum on Subjective Well-Being, 14*(2), 591–612. doi:10.100710902-012-9344-4

Senge, P. (2018). *The Fifth Discipline: The Art and Practice of the Learning Organization*. Best Seller Publisher.

Serdane, Z. (2020). Slow philosophy in tourism development in Latvia: The supply side perspective. *Tourism Planning & Development, 17*(3), 295–312. doi:10.1080/21568316.2019.1650103

Shah, K., Kamrai, D., Mekala, H., Mann, B., Desai, K., & Patel, R. S. (2020). Focus on Mental Health During the Coronavirus (COVID-19) Pandemic: Applying Learnings from the Past Outbreaks. *Cureus, 12*(3), e7405. doi:10.7759/cureus.7405 PMID:32337131

Shamim, S., Cang, S., Yu, H., & Li, Y. (2017). Examining the feasibilities of Industry 4.0 for the hospitality sector with the lens of management practice. *Energies, 10*(4), 499. Advance online publication. doi:10.3390/en10040499

Shang, W., Qiao, G., & Chen, N. (2020). Tourist experience of slow tourism: From authenticity to place attachment–a mixed-method study based on the case of slow city in China. *Asia Pacific Journal of Tourism Research, 25*(2), 170–188. doi:10.1080/10941665.2019.1683047

Shao, Y., & Sun, Y. (2020). SWOT Analysis of Coastal Sports Tourism. *Journal of Coastal Research, 112*(sp1), 103–105. doi:10.2112/JCR-SI112-030.1

Sharma, G. D., Thomas, A., & Paul, J. (2021). Reviving tourism industry post-COVID-19: A resilience-based framework. *Tourism Management Perspectives, 37*, 1–11. doi:10.1016/j.tmp.2020.100786 PMID:33391988

Sharma, S., & Srivastava, S. (2018). Exploring the relationship between service quality and customer satisfaction in croatian hotel industry. *Tourism and Hospitality Management, 2*(1), 2598–9839.

Sharpley, R., & Jepson, D. (2011). Rural tourism: A spiritual experience? *Annals of Tourism Research, 38*(1), 52–71. doi:10.1016/j.annals.2010.05.002

Shehade, M., & Stylianou-Lambert, T. (2020). Virtual Reality in museums: Exploring the experiences of museum professionals. *Applied Sciences (Basel, Switzerland), 10*(11), 4031. doi:10.3390/app10114031

Sheldon, P., Pollock, A., & Daniele, R. (2017). Social entrepreneurship and tourism: Setting the stage. In P. J. Sheldon & R. Daniele (Eds.), *Social entrepreneurship and tourism: Philosophy and practice* (pp. 1–18). Springer International Publishing. doi:10.1007/978-3-319-46518-0_1

Shen, H., Liu, X., Li, M., & Ji, M. (2018). Development of social enterprises in rural island tourism in China. *Journal of China Tourism Research*, 1–21.

Shinde, K. A. (2007). Case study 6: Visiting sacred sites in India: Religious tourism or pilgrimage. In R. Raj & N. D. Morpeth (Eds.), *Religious Tourism and Pilgrimage Management: An International Perspective* (pp. 184–197). CABI. doi:10.1079/9781845932251.0184

Sigala, M. (2016). Learning with the market: A market approach and framework for developing social entrepreneurship in tourism and hospitality. *International Journal of Contemporary Hospitality Management, 28*(6), 1245–1286. doi:10.1108/IJCHM-06-2014-0285

Sigala, M. (2020). Tourism and COVID-19: Impacts and implications for advancing and resetting industry and research. *Journal of Business Research, 117*, 312–321. doi:10.1016/j.jbusres.2020.06.015 PMID:32546875

Sijabat, R. (2015). The role of social entrepreneurship in enabling economic opportunities for the poor: A synthesis of the literature and empirical works. *International Journal of Business and Social Science, 6*(11), 35–41.

Silva, M. B. (2014). *O segmento low cost na indústria hoteleira em Coimbra: o caso dos hostels.* https://estudogeral.sib.uc.pt/bitstream/10316/27600/3/RelatórioMestrado2014-2.pdf

Silva, J. A. S. (2000). Pensando o planejamento face à intervenção do estado no turismo: A questão do sistema do informações. *Turismo: Visão e Ação, 2*(5), 9–22.

Silva, M. E. M., & Soriane-Sierra, E. J. (2015). Analysis of visitor perception on the revitalization of the seafront: The Calçadão Linear Park model, Itapema - Santa Catarina, Brazil. *Brazilian Journal of Tourism Research., 9*(1), 76–96.

Silva, S. C., Morano, R. P., Santiago, Z. M. P., & Viilarouco, V. (2019). *Hostel: uma forma de hospitalidade aberta.* doi:10.5151/ergodesign2019-2.37

Silva, T. F. F. (2019). *O valor da qualidade de serviço e a satisfação em hóteis de cinco estrelas em Portugal: um estudo exploratório.* Instituto Politécnico do Cávado e do Ave.

Silva, T., & Köhler, A. F. (2015). O mercado de albergues/ hostels do município de São Paulo: Caracterização e avaliação de estabelecimentos e empreendedores. *Revista Iberoamericana de Turismo, 5*(1), 54–78.

Simatupang, J. (2018). How is social entrepreneurship as an asset to implement rural tourism at Muarajambi herritage? *Innovative Issues and Approaches in Social Sciences, 11*(2), 106–115. doi:10.12959/issn.1855-0541.IIASS-2018-no2-art6

Singer, P. (2003). *Globalização e Desemprego: Diagnóstico e Alternativas* (6th ed.). Contexto.

Sirgy, M. J. (2010). Toward a Quality-of-Life Theory of Leisure Travel Satisfaction. *Journal of Travel Research, 49*(2), 246–260. doi:10.1177/0047287509337416

Sirgy, M. J., Kruger, P. S., Lee, D.-J., & Yu, G. B. (2011). How Does a Travel Trip Affect Tourists' Life Satisfaction? *Journal of Travel Research, 50*(3), 261–275. doi:10.1177/0047287510362784

Situmorang, D. B. M., & Mirzantia, I. R. (2012). Social entrepreneurship to develop ecotourism. *Procedia Economics and Finance, 4*, 398–405. doi:10.1016/S2212-5671(12)00354-1

Slemp, G. R., & Vella-Brodrick, D. A. (2013). The Job Crafting Questionnaire: A new scale to measure the extent to which employees engage in job crafting. *International Journal of Wellbeing, 3*(2), 126–146.

Sloan, P., Legrand, W., & Simons-Kaufmann, C. (2014). A survey of social entrepreneurial community-based hospitality and tourism initiatives in developing economies. *Worldwide Hospitality and Tourism Themes, 6*(1), 51–61. doi:10.1108/WHATT-11-2013-0045

Smith, Kwok, Tang, & Nutbeam. (2006). WHO Health Promotion Glossary. Health Promotion International, 21(4), 340–345. doi:10.1093/heapro/dal033

Smith-Godfrey, S. (2016). Defining the blue economy. *Maritime Affairs, 12*(1), 58–64. doi:10.1080/09733159.2016.1175131

Smith, M., & Puczkó, L. (2009). *Health and Wellness Tourism.* Butterworth-Heinemann–Elsevier.

Smith, V. L. (1992). Introduction: The quest in guest. *Annals of Tourism Research*, *19*(1), 1–17. doi:10.1016/0160-7383(92)90103-V

Soares, J. R. R., Pazos, A. P., & Gabriel, L. P. M. C. (2021). Involucrando la comunidad anfitriona en la gestión y desarrollo del turismo. In *Turismo y desarrollo: Contextos Diversos*. Thomson Reuters Aranzadi.

Solomon, M. (2015). *Ritz-Carlton President Herve Humler's leadership, culture and customer service secrets*. Retrieved 6th February 2021, from https://www.forbes.com/sites/micahsolomon/2015/04/21/ritz-carlton-president-herve-humlers-leadership-culture-and-customer-service-secrets/?sh=62d9ccc43b55

Soltani, S., Ghorbanian, M., & Tham, A. (2020). South to South Medical Tourists, the Liminality of Iran? *Journal of Quality Assurance in Hospitality & Tourism*, *22*(1), 55–81. doi:10.1080/1528008X.2020.1747588

Souitaris, V., & Zerbinati, S. (2005). Entrepreneurship in the public sector: A framework of analysis in European Local Governments. *Entrepreneurship and Regional Development*, *17*(1), 43–64. doi:10.1080/0898562042000310723

Sousa Santos, B., & Meneses, M. P. (2009). *Epistemologias do Sul*. Almedina.

Sousa, B. L. (2016). *Avaliação de serviços com base no Servqual: uma aplicação prática*. Universidade dos Açores. http://hdl.handle.net/10400.3/4096

Souza de, J. M. R., Brito, A. S., & Perinotto, A. R. C. (2018). Tradition and religious tourism: The Passion of Christ in Parnaíba-Piauí-Brazil. *Turismo e Sociedade*, *11*(2), 261–288.

Spicker, P. (2000). *The welfare state: a general theory*. Sage Publications.

Standeven, J. (1998). Sport tourism: Joint marketing - A starting point for beneficial synergies. *Journal of Vacation Marketing*, *4*(1), 39–51. doi:10.1177/135676679800400104

Stebbins, R. A. (2001). Serious leisure. *Society*, *38*(4), 53–57. doi:10.100712115-001-1023-8

Stefko, R., Kiralova, A., & Mudrík, M. (2015). Strategic marketing communication in pilgrimage tourism. *Procedia: Social and Behavioral Sciences*, *175*, 423–430. doi:10.1016/j.sbspro.2015.01.1219

Stein, A. C., Hack Neto, E., & Dechechi, E. C. (2018). Reflections on the socioeconomic impacts of tourism in the municipality of Foz do Iguaçu. Turismo: Estudos & Práticas. *Mossoró*, *7*(2), 26–55.

Stiglitz, J.E., Sen, A., & Fitoussi, J.P. (2009). *The Measurement of Economic Performance and Social Progress Revisited*. Paris: OFCE - Centre de recherche en économie de Sciences Po.

Su, L., Tang, B., & Nawijn, J. (2020). Eudaimonic and hedonic well-being pattern changes: Intensity and activity. *Annals of Tourism Research*, *84*(July), 103008. doi:10.1016/j.annals.2020.103008

Suminar, T., Budiartati, E., & Anggraeni, D. (2019). The effectiveness of a women's empowerment model through social entrepreneurship training to strengthen a tourism village program. *International Journal of Innovation. Creativity and Change.*, *5*(5), 324–338.

Swanson, L. A., & Zhang, D. D. (2012). Social Entrepreneurship. In T. Burger-Helmchen (Ed.), *Entrepreneurship - Gender, Geographies and Social Context*. InTech.

Szylvian, K. M. (2004). Transforming Lake Michigan into the "World's Greatest Fishing Hole": The Environmental Politics of Michigan's Great Lakes Sport Fishing, 1965-1985. *Environmental History*, *9*(1), 102–127. doi:10.2307/3985947

Tang, J., & Wang, Y. (2020). Does tourism sports event make residents happier? - Exploring the SWB of Macau residents in the case of Macau Grand Prix. *Journal of Tourism and Cultural Change*, 1–19. Advance online publication. doi:10.1080/14766825.2019.1711103

Tan, W. L., Williams, J., & Tan, T. M. (2005). Defining the 'social' in 'social entrepreneurship': Altruism and entrepreneurship. *The International Entrepreneurship and Management Journal*, *1*(3), 353–365. doi:10.100711365-005-2600-x

Tapia, J., Dieste, M., Royo, E., & Calvo, E. (2020). Factors affecting the choice of medical tourism destination: Spain as a host country. *Journal of Quality Assurance in Hospitality & Tourism*, 1–24. Advance online publication. doi:10.1080/1528008X.2020.1848749

Tarlombani da Silveira, M. A., & Zem, L. C. (2010). Impactos Ambientais do Turismo em Zonas Litorâneas: Um Estudo de Percepção Ambiental no Litoral do Paraná-Brasil. In *VI Latin American Seminar of Physical Geography*. Universidade de Coimbra.

Tatsuyuki-Shishido. (1991). Wakayama Marine City project in Wakanoura Bay—Utilization of the coastal zone through construction of artificial island. *Marine Pollution Bulletin*, *23*(C), 389–395. doi:10.1016/0025-326X(91)90706-X

Tavares, F. A. de O. (2018). *Hostels e Hóteis Boutique: análise de fatores de seleção na cidade de Oporto*. Universidade de Vigo. http://hdl.handle.net/11093/1124

Tavares, F. O., & Brea, J. A. F. (2020). Análise confirmatória à preferência nos hostels na cidade do Porto, em Portugal. *Revista Investigaciones Turísticas, 19*, 104–220. doi:10.14198/INTURI2020.19.05

Tekin, M. (2009). *Girişimcilik*. Günay Ofset Matbaacılık.

Terzidou, M., Scarles, C., & Saunders, M. N. (2018). The complexities of religious tourism motivations: Sacred places, vows and visions. *Annals of Tourism Research*, *70*, 54–65. doi:10.1016/j.annals.2018.02.011

Thal, K. I., & Hudson, S. (2019). A Conceptual Model of Wellness Destination Characteristics That Contribute to Psychological Well-Being. *Journal of Hospitality & Tourism Research (Washington, D.C.)*, *43*(1), 41–57. doi:10.1177/1096348017704498

Thomazi, M. R. (2019). *Hostel: Território de hospedagem marcado pela trama turístico-comunicacional*. Universidade de Caxias do Sul. https://repositorio.ucs.br/11338/4706

Timms, B. F., & Conway, D. (2012). Slow tourism at the Caribbean's geographical margins. *Tourism Geographies, 14*(3), 396–418. doi:10.1080/14616688.2011.610112

Timothy, D. J., & Boyd, S. W. (2006). Heritage tourism in the 21st Century: Valued traditions and new perspectives. *Journal of Heritage Tourism, 1*(1), 1–16. doi:10.1080/17438730608668462

Timothy, D., & Olsen, D. (2006). Management issues for religious heritage attractions. In D. Timothy & D. Olsen (Eds.), *Tourism, Religion and Spiritual Journeys* (pp. 120–134). Routledge. doi:10.4324/9780203001073-18

Tims, M., & Bakker, A. B. (2010). Job crafting: Towards a new model of individual job redesign. *SA Journal of Industrial Psychology, 36*(2), 1–9. doi:10.4102ajip.v36i2.841

Tinoco, M. A. C., & Ribeiro, J. L. D. (2007). Uma nova abordagem para a modelagem das relações entre os determinantes da satisfação dos clientes de serviços. *Produção, 17*(3), 454–470. doi:10.15900103-65132007000300005

Tiryaki, A. (2012). İktisat teorisinde girişimcilik. In Z. Erdoğan (Ed.), *Girişimcilik ve KOBİ'ler*. Ekin Basım Yayın Dağıtım.

Tonini, H., & Dolci, T. S. (2020). Turismo Rural e Novos Mercados para Produtos Alimentares Agroecológicos: Estudo de Caso da Rota Via Orgânica/Rural Tourism and New Markets for Agro-Ecological Food Products: The Case Study of Via Organica Route. *ROSA DOS VENTOS-Turismo e Hospitalidade, 12*(3), 537–554. Advance online publication. doi:10.18226/21789061.v12i3p537

Töppig, S. E., & Moital, M. (2020). An exploration of the techniques, outputs and outcomes of circulation management at exhibitions. *International Journal of Event and Festival Management, 11*(3), 311–335. Advance online publication. doi:10.1108/IJEFM-10-2019-0050

Trentin, F. (2014). *Public Policies in Brazil: Decision Making and the Analysis of Governance Structures in the Inducer Destinations of Armação dos Búzios and Paraty, State of Rio de Janeiro, Brazil* (PhD Thesis). Department of Geography, Faculty of Arts, University of Coimbra.

Tresiana, N., & Duadji, N. (2018). Social entrepreneur model for tourism development in Kiluan Bay. *MIMBAR, 34*(1), 83–92. doi:10.29313/mimbar.v34i1.3141

UFF, Universidade Federal Fluminense. (2021). *Apresentação Experiências Do Brasil Rural*. Retrieved from: http://pesquisaturaf.uff.br/apresentacao-do-projeto/

UN - United Nations. (2015). *Resolution adopted by the General Assembly on 1 September 2015*. 69/315. Draft outcome document of the United Nations summit for the adoption of the post-2015 development agenda. Retrieved from: https://www.unescwa.org/sites/www.unescwa.org/files/un_resolutions/a_res_69_315_e.pdf

UN - United Nations. (2016). *Resolution adopted by the General Assembly on 19 September 2016*. A/RES/71/1, 3 October 2016 (The New York Declaration). Retrieved from: https://www.unescwa.org/sites/www.unescwa.org/files/un_resolutions/a_res_69_315_e.pdf

UN News - United Nations News. (2015). *One in three travel destinations still fully closed to international tourists: UN agency*. Retrieved from: https://news.un.org/en/story/2021/03/1086712

UNCTAD. (2020). *The COVID-19 Pandemic and the Blue Economy: New challenges and prospects for recovery and resilience*. https://unctad.org/system/files/official-document/ditctedinf2020d2_en.pdf

United Nations World Tourism Organization. (n.d.). *Digital Transformation*. Retrieved from https://www.unwto.org/digital-transformation

UNWTO - United Nations World Tourism Organization (n.d.a). *Secretary-General's Policy Brief on Tourism and COVID-19: tourism and covid-19 – unprecedented economic impacts*. Retrieved from https://www.unwto.org/tourism-and-covid-19-unprecedented-economic-impacts

UNWTO - United Nations World Tourism Organization (n.d.b). *Measuring the Sustainability of Tourism: on measuring the sustainability of tourism: MST*. Retrieved from: https://www.unwto.org/standards/measuring-sustainability-tourism

UNWTO - United Nations World Tourism Organization. (2020a). 2020 year of 'tourism and rural development'. Background note.

UNWTO - United Nations World Tourism Organization. (2020b). *Rural tourism competition*. Retrieved from: https://www.unwto.org/unwto-rural-tourism-competition

UNWTO - United Nations World Tourism Organization. (2020c). *UNWTO recommendations on tourism and rural development: A Guide to Making Tourism an Effective Tool for Rural Development*. Retrieved from: https://www.e-unwto.org/doi/epdf/10.18111/9789284422173

UNWTO - United Nations World Tourism Organization. (2021). *Rural Tourism*. Retrieved from: https://www.unwto.org/unwto-rural-tourism

UNWTO - World Tourism Organization. (2013). *Sustainable Tourism for Development*. European Commission, Gran Contract DCI-MULTI-2011/280-663.

UNWTO. (2020). *Tourism Highlights*. Retrieved from https://www.e-unwto.org/doi/epdf/10.18111/9789284422456

UNWTO. (2021). *Recommendations for the Transition to a Green Travel and Tourism Economy*. UNWTO.

Ura, K. (2009). *A proposal for GNH value education in schools*. Gross National Happiness Commission.

Ura, K., & Galay, K. (2004). *Gross National Happiness and Development*. The Centre for Bhutan Studies.

Uysal, M., Sirgy, M. J., Woo, E., & Kim, H. L. (2016). Quality of life (QOL) and well-being research in tourism. *Tourism Management*, *53*, 244–261. doi:10.1016/j.tourman.2015.07.013

Vada, S., Prentice, C., & Hsiao, A. (2019). The role of positive psychology in tourists' behavioural intentions. *Journal of Retailing and Consumer Services*, *51*, 293–303. doi:10.1016/j.jretconser.2019.06.015

Valente, F., Dredge, D., & Lohmann, G. (2015). Leadership and governance in regional tourism. *Journal of Destination Marketing & Management*, *4*(2), 127–136. doi:10.1016/j.jdmm.2015.03.005

van den Heuvel, M., Demerouti, E., & Peeters, M. C. (2015). The job crafting intervention: Effects on job resources, self-efficacy, and affective well-being. *Journal of Occupational and Organizational Psychology*, *88*(3), 511–532. doi:10.1111/joop.12128

Van Praag, B. (1989). *The relativity of the welfare concept.* WIDER Working Paper Series. World Institute for Development Economic Research (UNU-WIDER).

van Wingerden, J., Derks, D., & Bakker, A. B. (2017). The impact of personal resources and job crafting interventions on work engagement and performance. *Human Resource Management*, *56*(1), 51–67. doi:10.1002/hrm.21758

Vasconcelos, M., & de Sampayo, M. T. (2017). Hostels e a nova tendência do turismo urbano. O caso de estudo do centro histórico de Lisboa de 2005 a 2014. In Actas do 4.o Congresso Internacional Casa Nobre - Um património para o futuro (pp. 1195–1205). Academic Press.

Vázquez, R. M. (2020). Nautical Tourism: A Bibliometric Analysis. *Journal of Spatial and Organizational Dynamics*, *8*(4), 320–330.

Vázquez, R. M., García, J. M., & Valenciano, J. D. P. (2021). Analysis and Trends of Global Research on Nautical, Maritime and Marine Tourism. *Journal of Marine Science and Engineering*, *9*(1), 93. doi:10.3390/jmse9010093

Veenhoven, R. (1984). *Conditions of Happiness Reidel.* Springer. doi:10.1007/978-94-009-6432-7

Verhoef, P. C., Lemon, K. N., Parasuraman, A., Roggeveen, A., Tsiros, M., & Schlesinger, L. A. (2009). Customer experience creation: Determinants, dynamics and management strategies. *Journal of Retailing*, *85*(1), 31–41. doi:10.1016/j.jretai.2008.11.001

Viegas Fernandes, J., & Viegas Fernandes, F. M. (2011). *Turismo de saúde e bem-estar no mundo: ética, excelência, segurança e sustentabilidade.* SENAC.

Vieira, E. S., & Gomes, J. A. N. F. (2009). A comparison of Scopus and Web of science for a typical university. *Scientometrics*, *81*(2), 587–600. doi:10.100711192-009-2178-0

Vieira, J., Ardigó, C. M., & Behling, H. P. (2018). Impactos da Volvo Ocean Race – Itajaí Stopover : análise pós- evento da percepção dos residentes da cidade de Itajaí (SC) Impacts of Volvo Ocean Race - Itajaí Stopover : a post-event analysis of the perception of the residents of the city of Itajaí (SC. *Revista Brasileira de Pesquisa em Turismo*, *12*(3), 172–196. doi:10.7784/rbtur.v12i3.1437

Vinhas, R. S. (2008). *Índice de Satisfação e Lealdade do Turista* (Dissertação de Mestrado). Instituto Superior de Estatística e Gestão da Informação da Universidade Nova de Lisboa, Dez.

Vinh, T. T., & Nga, V. T. Q. (2015). The relationship between components of customer-based brand equity for destination: Conceptual framework and preliminary testing for scales. *South East Asia Journal of Contemporary Business, Economics and Law*, *7*(2), 47–53.

Virginio, D. F. (2010). Tourism and Culture: A study on the museum qualification program for tourism. *Electronic Journal of Cultural Tourism*, *4*(1), 65–84.

Voigt, C., Brown, G., & Howat, G. (2011). Wellness tourists: in search of transformation. *Tourism Review, 66*(1/2), 16-30.

Volante, P. J. T. (2011). *O segmento low-cost da indústria hoteleira em Portugal: o caso dos hostels*. https://repositorio.iscte-iul.pt/handle/10071/4068

Vukonic, B. (1996). *Tourism and Religion*. Elsevier Science Ltd.

Waldo, S., & Paulrud, A. (2012). Obstacles to Developing Recreational Fishing Enterprises in Sweden. *Scandinavian Journal of Hospitality and Tourism*, *12*(2), 121–139. doi:10.1080/15022250.2011.633254

Walsham, G. (1995). Interpretive case studies in IS research: Nature and method. *European Journal of Information Systems*, *4*(2), 74–81. doi:10.1057/ejis.1995.9

Wang, C., Duan, Z., & Yu, L. (2016). From nonprofit organization to social enterprise. *International Journal of Contemporary Hospitality Management*, *28*(6), 1287–1306. doi:10.1108/IJCHM-05-2014-0230

Wang, H. (2020). The Development of Marine Sports Tourism in the Context of the Experience Economy. *Journal of Coastal Research*, *112*(sp1), 84–86. doi:10.2112/JCR-SI112-024.1

Wang, R., Dai, M., Ou, Y., & Ma, X. (2021). Residents' happiness of life in rural tourism development. *Journal of Destination Marketing & Management*, *20*, 100612. doi:10.1016/j.jdmm.2021.100612

Waterman, A. S., Schwartz, S. S., & Conti, R. (2008). The implications of two conceptions of happiness (hedonic enjoyment and eudaimonia) for the understanding of intrinsic motivation. *Journal of Happiness Studies*, *9*(1), 41–79. doi:10.100710902-006-9020-7

Watkins, P. C., Woodward, K., Stone, T., & Kolts, R. L. (2003). Gratitude and happiness: Development of a measure of gratitude and relationships with subjective well-being. *Social Behavior and Personality*, *31*(5), 431–452. doi:10.2224bp.2003.31.5.431

WCED. (1987). *Report of the World Commission on Environment and Development: Our common future*. Retrieved from: http://www.un-documents.net/our-common-future.pdf

Weed, M. (2005). Sports Tourism Theory and Method—Concepts, Issues and Epistemologies. *European Sport Management Quarterly*, *5*(3), 229–242. doi:10.1080/16184740500190587

Weed, M. (2009). Progress in sports tourism research? A meta-review and exploration of futures. *Tourism Management, 30*(5), 615–628. doi:10.1016/j.tourman.2009.02.002

Weed, M. (2020). The role of the interface of sport and tourism in the response to the COVID-19 pandemic. *Journal of Sport & Tourism, 24*(2), 79–92. doi:10.1080/14775085.2020.1794351

Weed, M., & Bull, C. (2004). *Sports tourism: Participants, policy and providers*. Elsevier Butterworth-Heinemann., doi:10.4324/9780080474441

Wen, J., Kozak, M., Yang, S., & Liu, F. (2020). COVID-19: Potential effects on Chinese citizens' lifestyle and travel. *Tourism Review, 76*(1), 74–87. doi:10.1108/TR-03-2020-0110

Weppen, J., & Cochrane, J. (2012). Social enterprises in tourism: An exploratory study of operational models and success factor. *Journal of Sustainable Tourism, 20*(3), 497–511. doi:10.1080/09669582.2012.663377

WHO. (2020). *WHO Director-General's opening remarks at the media briefing on COVID-19 - 11 March 2020*. World Health Organisation. https://www.who.int/director-general/speeches/detail/who-director-general-s-opening-remarks-at-the-media-briefing-on-covid-19---11-march-2020

WHO. (n.d.). *Coronavirus disease (COVID-19)*. World Health Organisation.

Williams, L. A., & Bartlett, M. Y. (2015). Warm thanks: Gratitude expression facilitates social affiliation in new relationships via perceived warmth. *Emotion (Washington, D.C.), 15*(1), 1–5. doi:10.1037/emo0000017 PMID:25111881

Williams, R. B., & Slak-Valek, N. (2019). Pokémon GO is serious leisure that increases the touristic engagement, physical activity and sense of happiness of players. *Information Technology & Tourism, 21*(4), 515–533. doi:10.100740558-019-00153-2

Wong, J. O., Gabana, N. T., Brown, J. W., Mcinnis, S., Toth, P., & Gilman, L. (2016). Does gratitude writing improve the mental health of psychotherapy clients? Evidence from a randomized controlled trial. *Psychotherapy Research, 28*(2), 192–202. doi:10.1080/10503307.2016.1169332 PMID:27139595

Wood, A. L., Butler, J. R. A., Sheaves, M., & Wani, J. (2013). Sport fisheries: Opportunities and challenges for diversifying coastal livelihoods in the Pacific. *Marine Policy, 42*, 305–314. doi:10.1016/j.marpol.2013.03.005

Wood, A. M., Froh, J. J., & Geraghty, A. W. A. (2010). Gratitude and well-being: A review and theoretical integration. *Clinical Psychology Review, 30*(7), 890–905. doi:10.1016/j.cpr.2010.03.005 PMID:20451313

World Bank. (2007). World Development Report: Development and the next generation. Washington, DC: Author.

World Bank. (2020). *Poverty and Shared Prosperity 2020: Reversals of Fortune*. The World Bank. doi:10.1596/978-1-4648-1602-4

World Tourism Organization. (2018). *UNWTO/ETC Launch report and health tourism*. Retrieved from: https://www.e-unwto.org/doi/book/10.18111/9789284420209

Wrzesniewski, A., & Dutton, J. E. (2001). Crafting a job: Revisioning employees as active crafters of their work. *Academy of Management Review*, *26*(2), 179–201. doi:10.5465/amr.2001.4378011

Wrzesniewski, A., Dutton, J. E., & Debebe, G. (2003). Interpersonal sensemaking and the meaning of work. In R. M. Kramer & B. M. Staw (Eds.), *Research in organizational behavior. Research in organizational behavior: An annual series of analytical essays and critical reviews* (Vol. 25, pp. 93–135). Elsevier Science Ltd. doi:10.1016/S0191-3085(03)25003-6

Wrzesniewski, A., LoBuglio, N., Dutton, J. E., & Berg, J. M. (2013). Job crafting and cultivating positive meaning and identity in work. In A. B. Bakker (Ed.), *Advances in Positive Organizational Psychology* (Vol. 1, pp. 281–302). Emerald Group Publishing Limited. doi:10.1108/S2046-410X(2013)0000001015

WTO - World Tourism Organization. (1999). *World Code of Tourism Ethics: Preparation for the New Millennium*. General Assembly.

WTO - World Tourism Organization. (2001). *Introdução ao Turismo*. Rocca.

WTTC - World Travel & Tourism Council. (n.d.). *'Safe Travels': Global Protocols & Stamp for the New Normal*. Retrieved from: https://wttc.org/COVID-19/Safe-Travels-Global-Protocols-Stamp

WTTC. (2019). *Mobile payments in travel & tourism: unlocking the potential*. Retrieved from www.wttc.org/-/media/files/reports/2019/mobile-payments-in-travel-tourism.pdf

WTTC. (2019). *World transformed: megatrends and their implications for travel and tourism*. Retrieved from www.wttc.org/-/media/files/reports/2019/megatrends-2019---world-transformed.pdf

WTTC. (2020). *To recovery and beyond: the future of travel & tourism in the wake of Covid-19*. Retrieved from https://wttc.org/Research/To-Recovery-Beyond

Wu, H. C., Ai, C. H., & Cheng, C. C. (2019). Virtual reality experiences, attachment and experiential outcomes in tourism. *Tourism Review*, *75*(3), 481–495. doi:10.1108/TR-06-2019-0205

Wulleman, M., & Hudon, M. (2015). *Models of social entrepreneurship: empirical evidence from Mexico*. CEB Working Paper, 15/024, 1-41.

Wyman, O. (2020). *To Recovery & Beyond: The Future of Travel & Tourism in the Wake of COVID-19*. https://wttc.org/Research/To-Recovery-Beyond

Xia, Y. U., Zheng, Y., Xiaofeng, S. U., & Zheng, Q. (2020). A Study on the Correlation among Social Interaction, Career Self-Efficacy and Well-Being of Employees in Rural Tourism. *Revista de Cercetare si Interventie Sociala*, *70*, 1.

Yadav, A. (2020, June). *How job crafting is shaping the future of hospitality industry: Through the lens of industry 4.0*. Paper presented at the Travel and Tourism Research Association International Conference, Victoria, Canada.

Yamamura, E., & Tsutsui, Y. (2020). The impact of postponing 2020 tokyo olympics on the happiness of O-MO-TE-NA-SHI Workers in tourism: A consequence of COVID-19. *Sustainability (Switzerland), 12*(19), 1–16. doi:10.3390u12198168

Yamauchi, D. A. M., Reis, B. C. M., & Rodrigues, M. D. (2019). Análise da qualidade dos serviços de hostels em Ouro Preto - Minas Gerais. *Revista Eletrônica de Gestão e Serviços, 10*(1), 2599–2623. doi:10.15603/2177-7284/regs.v10n1p2599-2626

Yang, T., Lai, I., Fan, Z. B., & Mo, Q. M. (2021). The impact of a 360° virtual tour on the reduction of psychological stress caused by COVID-19. *Technology in Society, 64*, 101514. doi:10.1016/j.techsoc.2020.101514 PMID:33424061

Yang, Y., Gong, Y., & Li, B. (2020). The Development of Marine Sports Tourism Based on Experience Economy. *Journal of Coastal Research, 112*(sp1), 106–108. doi:10.2112/JCR-SI112-031.1

Yu, G. B., Sirgy, M. J., & Bosnjak, M. (2020). The Effects of Holiday Leisure Travel on Subjective Well-Being: The Moderating Role of Experience Sharing. *Journal of Travel Research*. Advance online publication. doi:10.1177/0047287520966381

Yukl, G. (2002). *Leadership in organizations* (5th ed.). Prentice Hall International Inc.

Yüksel & Oğuz Duran. (2012). Turkish Adaptation of the Gratitude Questionnaire. *Eurasian Journal of Educational Research, 46*, 199–216.

Yurcu, G., & Akinci, Z. (2017). influence of organizational citizenship behavior on hotel employees' job satisfaction and subjective well-being. *Advances in Hospitality and Tourism Research-AHTR, 5*(1), 57–83.

Yu, Y. (2020). Research on Current Situation and Development of Marine Leisure Sports in Shandong Peninsula. *Journal of Coastal Research, 115*(sp1), 123–126. doi:10.2112/JCR-SI115-037.1

Zahra, S. A., Gedajlovic, E., Neubaum, D. O., & Shulman, J. M. (2009). A typology of social entrepreneurs: Motives, search processes and ethical challenges. *Journal of Business Venturing, 24*(5), 519–532. doi:10.1016/j.jbusvent.2008.04.007

Zebryte, I., & Jorquera, H. (2017). Chilean tourism sector "B corporations": Evidence of social entrepreneurship and innovation. *International Journal of Entrepreneurial Behaviour & Research, 23*(6), 866–879. doi:10.1108/IJEBR-07-2017-0218

Zeng, B. (2018). How can social enterprises contribute to sustainable pro-poor tourism development? *Zhongguo Renkou Ziyuan Yu Huanjing, 16*(2), 1–12. doi:10.1080/10042857.2018.1466955

Zhang, J. (2020). The Development of Marine Sports Tourism Resources and Product Portfolio under the Background of the Belt and Road. *Journal of Coastal Research*, *112*(sp1), 81–83. doi:10.2112/JCR-SI112-023.1

Zhang, R. (2020). Research on the Development Strategy of China's Marine Sports Tourism from the Perspective of Experience Economy. *Journal of Coastal Research*, *115*(sp1), 75–77. doi:10.2112/JCR-SI115-022.1

Zhang, T., Kandampully, J., & Choi, H. Y. (2014). The role of employee wellness programme in the hospitality industry: A review of concepts, research, and practice. *Research in Hospitality Management*, *4*(1-2), 45–54. doi:10.1080/22243534.2014.11828313

Zhang, Z., & Ding, Z. (2020). The systematic spatial characteristics of sports tourism destinations in the core area of marine economy. *Journal of Coastal Research*, *112*(sp1), 109–111. doi:10.2112/JCR-SI112-032.1

Zhao, P., & Lv, J. (2020). An Empirical Study on the Development of Marine Sports Tourism in Northern China. *Journal of Coastal Research*, *106*(sp1), 105–108. doi:10.2112/SI106-027.1

Zmyślony, P., & Kowalczyk-Anioł, J. (2019). Urban tourism hypertrophy: Who should deal with it? The case of Krakow (Poland). *International Journal of Tourism Cities*, *5*(2), 247–269. doi:10.1108/IJTC-07-2018-0051

Related References

To continue our tradition of advancing information science and technology research, we have compiled a list of recommended IGI Global readings. These references will provide additional information and guidance to further enrich your knowledge and assist you with your own research and future publications.

Abtahi, M. S., Behboudi, L., & Hasanabad, H. M. (2017). Factors Affecting Internet Advertising Adoption in Ad Agencies. *International Journal of Innovation in the Digital Economy*, *8*(4), 18–29. doi:10.4018/IJIDE.2017100102

Agrawal, S. (2017). The Impact of Emerging Technologies and Social Media on Different Business(es): Marketing and Management. In O. Rishi & A. Sharma (Eds.), *Maximizing Business Performance and Efficiency Through Intelligent Systems* (pp. 37–49). Hershey, PA: IGI Global. doi:10.4018/978-1-5225-2234-8.ch002

Alnoukari, M., Razouk, R., & Hanano, A. (2016). BSC-SI: A Framework for Integrating Strategic Intelligence in Corporate Strategic Management. *International Journal of Social and Organizational Dynamics in IT*, *5*(2), 1–14. doi:10.4018/IJSODIT.2016070101

Alnoukari, M., Razouk, R., & Hanano, A. (2016). BSC-SI, A Framework for Integrating Strategic Intelligence in Corporate Strategic Management. *International Journal of Strategic Information Technology and Applications*, *7*(1), 32–44. doi:10.4018/IJSITA.2016010103

Altındağ, E. (2016). Current Approaches in Change Management. In A. Goksoy (Ed.), *Organizational Change Management Strategies in Modern Business* (pp. 24–51). Hershey, PA: IGI Global. doi:10.4018/978-1-4666-9533-7.ch002

Alvarez-Dionisi, L. E., Turner, R., & Mittra, M. (2016). Global Project Management Trends. *International Journal of Information Technology Project Management, 7*(3), 54–73. doi:10.4018/IJITPM.2016070104

Anantharaman, R. N., Rajeswari, K. S., Angusamy, A., & Kuppusamy, J. (2017). Role of Self-Efficacy and Collective Efficacy as Moderators of Occupational Stress Among Software Development Professionals. *International Journal of Human Capital and Information Technology Professionals, 8*(2), 45–58. doi:10.4018/IJHCITP.2017040103

Aninze, F., El-Gohary, H., & Hussain, J. (2018). The Role of Microfinance to Empower Women: The Case of Developing Countries. *International Journal of Customer Relationship Marketing and Management, 9*(1), 54–78. doi:10.4018/IJCRMM.2018010104

Arsenijević, O. M., Orčić, D., & Kastratović, E. (2017). Development of an Optimization Tool for Intangibles in SMEs: A Case Study from Serbia with a Pilot Research in the Prestige by Milka Company. In M. Vemić (Ed.), *Optimal Management Strategies in Small and Medium Enterprises* (pp. 320–347). Hershey, PA: IGI Global. doi:10.4018/978-1-5225-1949-2.ch015

Aryanto, V. D., Wismantoro, Y., & Widyatmoko, K. (2018). Implementing Eco-Innovation by Utilizing the Internet to Enhance Firm's Marketing Performance: Study of Green Batik Small and Medium Enterprises in Indonesia. *International Journal of E-Business Research, 14*(1), 21–36. doi:10.4018/IJEBR.2018010102

Atiku, S. O., & Fields, Z. (2017). Multicultural Orientations for 21st Century Global Leadership. In N. Baporikar (Ed.), *Management Education for Global Leadership* (pp. 28–51). Hershey, PA: IGI Global. doi:10.4018/978-1-5225-1013-0.ch002

Atiku, S. O., & Fields, Z. (2018). Organisational Learning Dimensions and Talent Retention Strategies for the Service Industries. In N. Baporikar (Ed.), *Global Practices in Knowledge Management for Societal and Organizational Development* (pp. 358–381). Hershey, PA: IGI Global. doi:10.4018/978-1-5225-3009-1.ch017

Ávila, L., & Teixeira, L. (2018). The Main Concepts Behind the Dematerialization of Business Processes. In M. Khosrow-Pour, D.B.A. (Ed.), Encyclopedia of Information Science and Technology, Fourth Edition (pp. 888-898). Hershey, PA: IGI Global. doi:10.4018/978-1-5225-2255-3.ch076

Bartens, Y., Chunpir, H. I., Schulte, F., & Voß, S. (2017). Business/IT Alignment in Two-Sided Markets: A COBIT 5 Analysis for Media Streaming Business Models. In S. De Haes & W. Van Grembergen (Eds.), *Strategic IT Governance and Alignment in Business Settings* (pp. 82–111). Hershey, PA: IGI Global. doi:10.4018/978-1-5225-0861-8.ch004

Bashayreh, A. M. (2018). Organizational Culture and Organizational Performance. In W. Lee & F. Sabetzadeh (Eds.), *Contemporary Knowledge and Systems Science* (pp. 50–69). Hershey, PA: IGI Global. doi:10.4018/978-1-5225-5655-8.ch003

Bedford, D. A. (2018). Sustainable Knowledge Management Strategies: Aligning Business Capabilities and Knowledge Management Goals. In N. Baporikar (Ed.), *Global Practices in Knowledge Management for Societal and Organizational Development* (pp. 46–73). Hershey, PA: IGI Global. doi:10.4018/978-1-5225-3009-1.ch003

Benmoussa, F., Nakara, W. A., & Jaouen, A. (2016). The Use of Social Media by SMEs in the Tourism Industry. In I. Lee (Ed.), *Encyclopedia of E-Commerce Development, Implementation, and Management* (pp. 2159–2170). Hershey, PA: IGI Global. doi:10.4018/978-1-4666-9787-4.ch155

Berger, R. (2016). Indigenous Management and Bottom of Pyramid Countries: The Role of National Institutions. In U. Aung & P. Ordoñez de Pablos (Eds.), *Managerial Strategies and Practice in the Asian Business Sector* (pp. 107–123). Hershey, PA: IGI Global. doi:10.4018/978-1-4666-9758-4.ch007

Bharwani, S., & Musunuri, D. (2018). Reflection as a Process From Theory to Practice. In M. Khosrow-Pour, D.B.A. (Ed.), Encyclopedia of Information Science and Technology, Fourth Edition (pp. 1529-1539). Hershey, PA: IGI Global. doi:10.4018/978-1-5225-2255-3.ch132

Bhatt, G. D., Wang, Z., & Rodger, J. A. (2017). Information Systems Capabilities and Their Effects on Competitive Advantages: A Study of Chinese Companies. *Information Resources Management Journal, 30*(3), 41–57. doi:10.4018/IRMJ.2017070103

Bhushan, M., & Yadav, A. (2017). Concept of Cloud Computing in ESB. In R. Bhadoria, N. Chaudhari, G. Tomar, & S. Singh (Eds.), *Exploring Enterprise Service Bus in the Service-Oriented Architecture Paradigm* (pp. 116–127). Hershey, PA: IGI Global. doi:10.4018/978-1-5225-2157-0.ch008

Bhushan, S. (2017). System Dynamics Base-Model of Humanitarian Supply Chain (HSCM) in Disaster Prone Eco-Communities of India: A Discussion on Simulation and Scenario Results. *International Journal of System Dynamics Applications, 6*(3), 20–37. doi:10.4018/IJSDA.2017070102

Biswas, A., & De, A. K. (2017). On Development of a Fuzzy Stochastic Programming Model with Its Application to Business Management. In S. Trivedi, S. Dey, A. Kumar, & T. Panda (Eds.), *Handbook of Research on Advanced Data Mining Techniques and Applications for Business Intelligence* (pp. 353–378). Hershey, PA: IGI Global. doi:10.4018/978-1-5225-2031-3.ch021

Bücker, J., & Ernste, K. (2018). Use of Brand Heroes in Strategic Reputation Management: The Case of Bacardi, Adidas, and Daimler. In A. Erdemir (Ed.), *Reputation Management Techniques in Public Relations* (pp. 126–150). Hershey, PA: IGI Global. doi:10.4018/978-1-5225-3619-2.ch007

Bureš, V. (2018). Industry 4.0 From the Systems Engineering Perspective: Alternative Holistic Framework Development. In R. Brunet-Thornton & F. Martinez (Eds.), *Analyzing the Impacts of Industry 4.0 in Modern Business Environments* (pp. 199–223). Hershey, PA: IGI Global. doi:10.4018/978-1-5225-3468-6.ch011

Buzady, Z. (2017). Resolving the Magic Cube of Effective Case Teaching: Benchmarking Case Teaching Practices in Emerging Markets – Insights from the Central European University Business School, Hungary. In D. Latusek (Ed.), *Case Studies as a Teaching Tool in Management Education* (pp. 79–103). Hershey, PA: IGI Global. doi:10.4018/978-1-5225-0770-3.ch005

Campatelli, G., Richter, A., & Stocker, A. (2016). Participative Knowledge Management to Empower Manufacturing Workers. *International Journal of Knowledge Management, 12*(4), 37–50. doi:10.4018/IJKM.2016100103

Căpusneanu, S., & Topor, D. I. (2018). Business Ethics and Cost Management in SMEs: Theories of Business Ethics and Cost Management Ethos. In I. Oncioiu (Ed.), *Ethics and Decision-Making for Sustainable Business Practices* (pp. 109–127). Hershey, PA: IGI Global. doi:10.4018/978-1-5225-3773-1.ch007

Carneiro, A. (2016). Maturity in Health Organization Information Systems: Metrics and Privacy Perspectives. *International Journal of Privacy and Health Information Management, 4*(2), 1–18. doi:10.4018/IJPHIM.2016070101

Chan, R. L., Mo, P. L., & Moon, K. K. (2018). Strategic and Tactical Measures in Managing Enterprise Risks: A Study of the Textile and Apparel Industry. In K. Strang, M. Korstanje, & N. Vajjhala (Eds.), *Research, Practices, and Innovations in Global Risk and Contingency Management* (pp. 1–19). Hershey, PA: IGI Global. doi:10.4018/978-1-5225-4754-9.ch001

Chandan, H. C. (2016). Motivations and Challenges of Female Entrepreneurship in Developed and Developing Economies. In N. Baporikar (Ed.), *Handbook of Research on Entrepreneurship in the Contemporary Knowledge-Based Global Economy* (pp. 260–286). Hershey, PA: IGI Global. doi:10.4018/978-1-4666-8798-1.ch012

Charlier, S. D., Burke-Smalley, L. A., & Fisher, S. L. (2018). Undergraduate Programs in the U.S: A Contextual and Content-Based Analysis. In J. Mendy (Ed.), *Teaching Human Resources and Organizational Behavior at the College Level* (pp. 26–57). Hershey, PA: IGI Global. doi:10.4018/978-1-5225-2820-3.ch002

Chaudhuri, S. (2016). Application of Web-Based Geographical Information System (GIS) in E-Business. In U. Panwar, R. Kumar, & N. Ray (Eds.), *Handbook of Research on Promotional Strategies and Consumer Influence in the Service Sector* (pp. 389–405). Hershey, PA: IGI Global. doi:10.4018/978-1-5225-0143-5.ch023

Choudhuri, P. S. (2016). An Empirical Study on the Quality of Services Offered by the Private Life Insurers in Burdwan. In U. Panwar, R. Kumar, & N. Ray (Eds.), *Handbook of Research on Promotional Strategies and Consumer Influence in the Service Sector* (pp. 31–55). Hershey, PA: IGI Global. doi:10.4018/978-1-5225-0143-5.ch002

Dahlberg, T., Kivijärvi, H., & Saarinen, T. (2017). IT Investment Consistency and Other Factors Influencing the Success of IT Performance. In S. De Haes & W. Van Grembergen (Eds.), *Strategic IT Governance and Alignment in Business Settings* (pp. 176–208). Hershey, PA: IGI Global. doi:10.4018/978-1-5225-0861-8.ch007

Damnjanović, A. M. (2017). Knowledge Management Optimization through IT and E-Business Utilization: A Qualitative Study on Serbian SMEs. In M. Vemić (Ed.), *Optimal Management Strategies in Small and Medium Enterprises* (pp. 249–267). Hershey, PA: IGI Global. doi:10.4018/978-1-5225-1949-2.ch012

Daneshpour, H. (2017). Integrating Sustainable Development into Project Portfolio Management through Application of Open Innovation. In M. Vemić (Ed.), *Optimal Management Strategies in Small and Medium Enterprises* (pp. 370–387). Hershey, PA: IGI Global. doi:10.4018/978-1-5225-1949-2.ch017

Daniel, A. D., & Reis de Castro, V. (2018). Entrepreneurship Education: How to Measure the Impact on Nascent Entrepreneurs. In A. Carrizo Moreira, J. Guilherme Leitão Dantas, & F. Manuel Valente (Eds.), *Nascent Entrepreneurship and Successful New Venture Creation* (pp. 85–110). Hershey, PA: IGI Global. doi:10.4018/978-1-5225-2936-1.ch004

David, F., van der Sijde, P., & van den Besselaar, P. (2016). Enterpreneurial Incentives, Obstacles, and Management in University-Business Co-Operation: The Case of Indonesia. In J. Saiz-Álvarez (Ed.), *Handbook of Research on Social Entrepreneurship and Solidarity Economics* (pp. 499–518). Hershey, PA: IGI Global. doi:10.4018/978-1-5225-0097-1.ch024

David, R., Swami, B. N., & Tangirala, S. (2018). Ethics Impact on Knowledge Management in Organizational Development: A Case Study. In N. Baporikar (Ed.), *Global Practices in Knowledge Management for Societal and Organizational Development* (pp. 19–45). Hershey, PA: IGI Global. doi:10.4018/978-1-5225-3009-1.ch002

Delias, P., & Lakiotaki, K. (2018). Discovering Process Horizontal Boundaries to Facilitate Process Comprehension. *International Journal of Operations Research and Information Systems*, *9*(2), 1–31. doi:10.4018/IJORIS.2018040101

Denholm, J., & Lee-Davies, L. (2018). Success Factors for Games in Business and Project Management. In *Enhancing Education and Training Initiatives Through Serious Games* (pp. 34–68). Hershey, PA: IGI Global. doi:10.4018/978-1-5225-3689-5.ch002

Deshpande, M. (2017). Best Practices in Management Institutions for Global Leadership: Policy Aspects. In N. Baporikar (Ed.), *Management Education for Global Leadership* (pp. 1–27). Hershey, PA: IGI Global. doi:10.4018/978-1-5225-1013-0.ch001

Deshpande, M. (2018). Policy Perspectives for SMEs Knowledge Management. In N. Baporikar (Ed.), *Knowledge Integration Strategies for Entrepreneurship and Sustainability* (pp. 23–46). Hershey, PA: IGI Global. doi:10.4018/978-1-5225-5115-7.ch002

Dezdar, S. (2017). ERP Implementation Projects in Asian Countries: A Comparative Study on Iran and China. *International Journal of Information Technology Project Management*, *8*(3), 52–68. doi:10.4018/IJITPM.2017070104

Domingos, D., Martinho, R., & Varajão, J. (2016). Controlled Flexibility in Healthcare Processes: A BPMN-Extension Approach. In M. Cruz-Cunha, I. Miranda, R. Martinho, & R. Rijo (Eds.), *Encyclopedia of E-Health and Telemedicine* (pp. 521–535). Hershey, PA: IGI Global. doi:10.4018/978-1-4666-9978-6.ch040

Domingos, D., Respício, A., & Martinho, R. (2017). Reliability of IoT-Aware BPMN Healthcare Processes. In C. Reis & M. Maximiano (Eds.), *Internet of Things and Advanced Application in Healthcare* (pp. 214–248). Hershey, PA: IGI Global. doi:10.4018/978-1-5225-1820-4.ch008

Dosumu, O., Hussain, J., & El-Gohary, H. (2017). An Exploratory Study of the Impact of Government Policies on the Development of Small and Medium Enterprises in Developing Countries: The Case of Nigeria. *International Journal of Customer Relationship Marketing and Management*, *8*(4), 51–62. doi:10.4018/IJCRMM.2017100104

Durst, S., Bruns, G., & Edvardsson, I. R. (2017). Retaining Knowledge in Smaller Building and Construction Firms. *International Journal of Knowledge and Systems Science*, *8*(3), 1–12. doi:10.4018/IJKSS.2017070101

Edvardsson, I. R., & Durst, S. (2017). Outsourcing, Knowledge, and Learning: A Critical Review. *International Journal of Knowledge-Based Organizations*, *7*(2), 13–26. doi:10.4018/IJKBO.2017040102

Edwards, J. S. (2018). Integrating Knowledge Management and Business Processes. In M. Khosrow-Pour, D.B.A. (Ed.), Encyclopedia of Information Science and Technology, Fourth Edition (pp. 5046-5055). Hershey, PA: IGI Global. doi:10.4018/978-1-5225-2255-3.ch437

Ejiogu, A. O. (2018). Economics of Farm Management. In *Agricultural Finance and Opportunities for Investment and Expansion* (pp. 56–72). Hershey, PA: IGI Global. doi:10.4018/978-1-5225-3059-6.ch003

Ekanem, I., & Abiade, G. E. (2018). Factors Influencing the Use of E-Commerce by Small Enterprises in Nigeria. *International Journal of ICT Research in Africa and the Middle East*, *7*(1), 37–53. doi:10.4018/IJICTRAME.2018010103

Ekanem, I., & Alrossais, L. A. (2017). Succession Challenges Facing Family Businesses in Saudi Arabia. In P. Zgheib (Ed.), *Entrepreneurship and Business Innovation in the Middle East* (pp. 122–146). Hershey, PA: IGI Global. doi:10.4018/978-1-5225-2066-5.ch007

El Faquih, L., & Fredj, M. (2017). Ontology-Based Framework for Quality in Configurable Process Models. *Journal of Electronic Commerce in Organizations*, *15*(2), 48–60. doi:10.4018/JECO.2017040104

El-Gohary, H., & El-Gohary, Z. (2016). An Attempt to Explore Electronic Marketing Adoption and Implementation Aspects in Developing Countries: The Case of Egypt. *International Journal of Customer Relationship Marketing and Management*, *7*(4), 1–26. doi:10.4018/IJCRMM.2016100101

Entico, G. J. (2016). Knowledge Management and the Medical Health Librarians: A Perception Study. In J. Yap, M. Perez, M. Ayson, & G. Entico (Eds.), *Special Library Administration, Standardization and Technological Integration* (pp. 52–77). Hershey, PA: IGI Global. doi:10.4018/978-1-4666-9542-9.ch003

Faisal, M. N., & Talib, F. (2017). Building Ambidextrous Supply Chains in SMEs: How to Tackle the Barriers? *International Journal of Information Systems and Supply Chain Management, 10*(4), 80–100. doi:10.4018/IJISSCM.2017100105

Fernandes, T. M., Gomes, J., & Romão, M. (2017). Investments in E-Government: A Benefit Management Case Study. *International Journal of Electronic Government Research, 13*(3), 1–17. doi:10.4018/IJEGR.2017070101

Fouda, F. A. (2016). A Suggested Curriculum in Career Education to Develop Business Secondary Schools Students' Career Knowledge Management Domains and Professional Thinking. *International Journal of Technology Diffusion, 7*(2), 42–62. doi:10.4018/IJTD.2016040103

Gallardo-Vázquez, D., & Pajuelo-Moreno, M. L. (2016). How Spanish Universities are Promoting Entrepreneurship through Your Own Lines of Teaching and Research? In L. Carvalho (Ed.), *Handbook of Research on Entrepreneurial Success and its Impact on Regional Development* (pp. 431–454). Hershey, PA: IGI Global. doi:10.4018/978-1-4666-9567-2.ch019

Gao, S. S., Oreal, S., & Zhang, J. (2018). Contemporary Financial Risk Management Perceptions and Practices of Small-Sized Chinese Businesses. In I. Management Association (Ed.), Global Business Expansion: Concepts, Methodologies, Tools, and Applications (pp. 917-931). Hershey, PA: IGI Global. doi:10.4018/978-1-5225-5481-3.ch041

Garg, R., & Berning, S. C. (2017). Indigenous Chinese Management Philosophies: Key Concepts and Relevance for Modern Chinese Firms. In B. Christiansen & G. Koc (Eds.), *Transcontinental Strategies for Industrial Development and Economic Growth* (pp. 43–57). Hershey, PA: IGI Global. doi:10.4018/978-1-5225-2160-0.ch003

Gencer, Y. G. (2017). Supply Chain Management in Retailing Business. In U. Akkucuk (Ed.), *Ethics and Sustainability in Global Supply Chain Management* (pp. 197–210). Hershey, PA: IGI Global. doi:10.4018/978-1-5225-2036-8.ch011

Giacosa, E. (2016). Innovation in Luxury Fashion Businesses as a Means for the Regional Development. In L. Carvalho (Ed.), *Handbook of Research on Entrepreneurial Success and its Impact on Regional Development* (pp. 206–222). Hershey, PA: IGI Global. doi:10.4018/978-1-4666-9567-2.ch010

Giacosa, E. (2018). The Increasing of the Regional Development Thanks to the Luxury Business Innovation. In L. Carvalho (Ed.), *Handbook of Research on Entrepreneurial Ecosystems and Social Dynamics in a Globalized World* (pp. 260–273). Hershey, PA: IGI Global. doi:10.4018/978-1-5225-3525-6.ch011

Gianni, M., & Gotzamani, K. (2016). Integrated Management Systems and Information Management Systems: Common Threads. In P. Papajorgji, F. Pinet, A. Guimarães, & J. Papathanasiou (Eds.), *Automated Enterprise Systems for Maximizing Business Performance* (pp. 195–214). Hershey, PA: IGI Global. doi:10.4018/978-1-4666-8841-4.ch011

Gianni, M., Gotzamani, K., & Linden, I. (2016). How a BI-wise Responsible Integrated Management System May Support Food Traceability. *International Journal of Decision Support System Technology*, *8*(2), 1–17. doi:10.4018/IJDSST.2016040101

Glykas, M., & George, J. (2017). Quality and Process Management Systems in the UAE Maritime Industry. *International Journal of Productivity Management and Assessment Technologies*, *5*(1), 20–39. doi:10.4018/IJPMAT.2017010102

Glykas, M., Valiris, G., Kokkinaki, A., & Koutsoukou, Z. (2018). Banking Business Process Management Implementation. *International Journal of Productivity Management and Assessment Technologies*, *6*(1), 50–69. doi:10.4018/IJPMAT.2018010104

Gomes, J., & Romão, M. (2017). The Balanced Scorecard: Keeping Updated and Aligned with Today's Business Trends. *International Journal of Productivity Management and Assessment Technologies*, *5*(2), 1–15. doi:10.4018/IJPMAT.2017070101

Gomes, J., & Romão, M. (2017). Aligning Information Systems and Technology with Benefit Management and Balanced Scorecard. In S. De Haes & W. Van Grembergen (Eds.), *Strategic IT Governance and Alignment in Business Settings* (pp. 112–131). Hershey, PA: IGI Global. doi:10.4018/978-1-5225-0861-8.ch005

Grefen, P., & Turetken, O. (2017). Advanced Business Process Management in Networked E-Business Scenarios. *International Journal of E-Business Research*, *13*(4), 70–104. doi:10.4018/IJEBR.2017100105

Haider, A., & Saetang, S. (2017). Strategic IT Alignment in Service Sector. In S. Rozenes & Y. Cohen (Eds.), *Handbook of Research on Strategic Alliances and Value Co-Creation in the Service Industry* (pp. 231–258). Hershey, PA: IGI Global. doi:10.4018/978-1-5225-2084-9.ch012

Haider, A., & Tang, S. S. (2016). Maximising Value Through IT and Business Alignment: A Case of IT Governance Institutionalisation at a Thai Bank. *International Journal of Technology Diffusion*, *7*(3), 33–58. doi:10.4018/IJTD.2016070104

Hajilari, A. B., Ghadaksaz, M., & Fasghandis, G. S. (2017). Assessing Organizational Readiness for Implementing ERP System Using Fuzzy Expert System Approach. *International Journal of Enterprise Information Systems*, *13*(1), 67–85. doi:10.4018/IJEIS.2017010105

Haldorai, A., Ramu, A., & Murugan, S. (2018). Social Aware Cognitive Radio Networks: Effectiveness of Social Networks as a Strategic Tool for Organizational Business Management. In H. Bansal, G. Shrivastava, G. Nguyen, & L. Stanciu (Eds.), *Social Network Analytics for Contemporary Business Organizations* (pp. 188–202). Hershey, PA: IGI Global. doi:10.4018/978-1-5225-5097-6.ch010

Hall, O. P. Jr. (2017). Social Media Driven Management Education. *International Journal of Knowledge-Based Organizations*, *7*(2), 43–59. doi:10.4018/IJKBO.2017040104

Hanifah, H., Halim, H. A., Ahmad, N. H., & Vafaei-Zadeh, A. (2017). Innovation Culture as a Mediator Between Specific Human Capital and Innovation Performance Among Bumiputera SMEs in Malaysia. In N. Ahmad, T. Ramayah, H. Halim, & S. Rahman (Eds.), *Handbook of Research on Small and Medium Enterprises in Developing Countries* (pp. 261–279). Hershey, PA: IGI Global. doi:10.4018/978-1-5225-2165-5.ch012

Hartlieb, S., & Silvius, G. (2017). Handling Uncertainty in Project Management and Business Development: Similarities and Differences. In Y. Raydugin (Ed.), *Handbook of Research on Leveraging Risk and Uncertainties for Effective Project Management* (pp. 337–362). Hershey, PA: IGI Global. doi:10.4018/978-1-5225-1790-0.ch016

Hass, K. B. (2017). Living on the Edge: Managing Project Complexity. In Y. Raydugin (Ed.), *Handbook of Research on Leveraging Risk and Uncertainties for Effective Project Management* (pp. 177–201). Hershey, PA: IGI Global. doi:10.4018/978-1-5225-1790-0.ch009

Hassan, A., & Privitera, D. S. (2016). Google AdSense as a Mobile Technology in Education. In J. Holland (Ed.), *Wearable Technology and Mobile Innovations for Next-Generation Education* (pp. 200–223). Hershey, PA: IGI Global. doi:10.4018/978-1-5225-0069-8.ch011

Hassan, A., & Rahimi, R. (2016). Consuming "Innovation" in Tourism: Augmented Reality as an Innovation Tool in Digital Tourism Marketing. In N. Pappas & I. Bregoli (Eds.), *Global Dynamics in Travel, Tourism, and Hospitality* (pp. 130–147). Hershey, PA: IGI Global. doi:10.4018/978-1-5225-0201-2.ch008

Hawking, P., & Carmine Sellitto, C. (2017). Developing an Effective Strategy for Organizational Business Intelligence. In M. Tavana (Ed.), *Enterprise Information Systems and the Digitalization of Business Functions* (pp. 222–237). Hershey, PA: IGI Global. doi:10.4018/978-1-5225-2382-6.ch010

Hawking, P., & Sellitto, C. (2017). A Fast-Moving Consumer Goods Company and Business Intelligence Strategy Development. *International Journal of Enterprise Information Systems, 13*(2), 22–33. doi:10.4018/IJEIS.2017040102

Hawking, P., & Sellitto, C. (2017). Business Intelligence Strategy: Two Case Studies. *International Journal of Business Intelligence Research, 8*(2), 17–30. doi:10.4018/IJBIR.2017070102

Haynes, J. D., Arockiasamy, S., Al Rashdi, M., & Al Rashdi, S. (2016). Business and E Business Strategies for Coopetition and Thematic Management as a Sustained Basis for Ethics and Social Responsibility in Emerging Markets. In M. Al-Shammari & H. Masri (Eds.), *Ethical and Social Perspectives on Global Business Interaction in Emerging Markets* (pp. 25–39). Hershey, PA: IGI Global. doi:10.4018/978-1-4666-9864-2.ch002

Hee, W. J., Jalleh, G., Lai, H., & Lin, C. (2017). E-Commerce and IT Projects: Evaluation and Management Issues in Australian and Taiwanese Hospitals. *International Journal of Public Health Management and Ethics, 2*(1), 69–90. doi:10.4018/IJPHME.2017010104

Hernandez, A. A. (2018). Exploring the Factors to Green IT Adoption of SMEs in the Philippines. *Journal of Cases on Information Technology, 20*(2), 49–66. doi:10.4018/JCIT.2018040104

Hernandez, A. A., & Ona, S. E. (2016). Green IT Adoption: Lessons from the Philippines Business Process Outsourcing Industry. *International Journal of Social Ecology and Sustainable Development, 7*(1), 1–34. doi:10.4018/IJSESD.2016010101

Hollman, A., Bickford, S., & Hollman, T. (2017). Cyber InSecurity: A Post-Mortem Attempt to Assess Cyber Problems from IT and Business Management Perspectives. *Journal of Cases on Information Technology, 19*(3), 42–70. doi:10.4018/JCIT.2017070104

Igbinakhase, I. (2017). Responsible and Sustainable Management Practices in Developing and Developed Business Environments. In Z. Fields (Ed.), *Collective Creativity for Responsible and Sustainable Business Practice* (pp. 180–207). Hershey, PA: IGI Global. doi:10.4018/978-1-5225-1823-5.ch010

Ilahi, L., Ghannouchi, S. A., & Martinho, R. (2016). A Business Process Management Approach to Home Healthcare Processes: On the Gap between Intention and Reality. In M. Cruz-Cunha, I. Miranda, R. Martinho, & R. Rijo (Eds.), *Encyclopedia of E-Health and Telemedicine* (pp. 439–457). Hershey, PA: IGI Global. doi:10.4018/978-1-4666-9978-6.ch035

Iwata, J. J., & Hoskins, R. G. (2017). Managing Indigenous Knowledge in Tanzania: A Business Perspective. In P. Jain & N. Mnjama (Eds.), *Managing Knowledge Resources and Records in Modern Organizations* (pp. 198–214). Hershey, PA: IGI Global. doi:10.4018/978-1-5225-1965-2.ch012

Jabeen, F., Ahmad, S. Z., & Alkaabi, S. (2016). The Internationalization Decision-Making of United Arab Emirates Family Businesses. In N. Zakaria, A. Abdul-Talib, & N. Osman (Eds.), *Handbook of Research on Impacts of International Business and Political Affairs on the Global Economy* (pp. 1–22). Hershey, PA: IGI Global. doi:10.4018/978-1-4666-9806-2.ch001

Jain, P. (2017). Ethical and Legal Issues in Knowledge Management Life-Cycle in Business. In P. Jain & N. Mnjama (Eds.), *Managing Knowledge Resources and Records in Modern Organizations* (pp. 82–101). Hershey, PA: IGI Global. doi:10.4018/978-1-5225-1965-2.ch006

Jamali, D., Abdallah, H., & Matar, F. (2016). Opportunities and Challenges for CSR Mainstreaming in Business Schools. *International Journal of Technology and Educational Marketing*, *6*(2), 1–29. doi:10.4018/IJTEM.2016070101

James, S., & Hauli, E. (2017). Holistic Management Education at Tanzanian Rural Development Planning Institute. In N. Baporikar (Ed.), *Management Education for Global Leadership* (pp. 112–136). Hershey, PA: IGI Global. doi:10.4018/978-1-5225-1013-0.ch006

Janošková, M., Csikósová, A., & Čulková, K. (2018). Measurement of Company Performance as Part of Its Strategic Management. In R. Leon (Ed.), *Managerial Strategies for Business Sustainability During Turbulent Times* (pp. 309–335). Hershey, PA: IGI Global. doi:10.4018/978-1-5225-2716-9.ch017

Jean-Vasile, A., & Alecu, A. (2017). Theoretical and Practical Approaches in Understanding the Influences of Cost-Productivity-Profit Trinomial in Contemporary Enterprises. In A. Jean Vasile & D. Nicolò (Eds.), *Sustainable Entrepreneurship and Investments in the Green Economy* (pp. 28–62). Hershey, PA: IGI Global. doi:10.4018/978-1-5225-2075-7.ch002

Jha, D. G. (2016). Preparing for Information Technology Driven Changes. In S. Tiwari & L. Nafees (Eds.), *Innovative Management Education Pedagogies for Preparing Next-Generation Leaders* (pp. 258–274). Hershey, PA: IGI Global. doi:10.4018/978-1-4666-9691-4.ch015

Joia, L. A., & Correia, J. C. (2018). CIO Competencies From the IT Professional Perspective: Insights From Brazil. *Journal of Global Information Management, 26*(2), 74–103. doi:10.4018/JGIM.2018040104

Juma, A., & Mzera, N. (2017). Knowledge Management and Records Management and Competitive Advantage in Business. In P. Jain & N. Mnjama (Eds.), *Managing Knowledge Resources and Records in Modern Organizations* (pp. 15–28). Hershey, PA: IGI Global. doi:10.4018/978-1-5225-1965-2.ch002

K., I., & A, V. (2018). Monitoring and Auditing in the Cloud. In K. Munir (Ed.), *Cloud Computing Technologies for Green Enterprises* (pp. 318-350). Hershey, PA: IGI Global. doi:10.4018/978-1-5225-3038-1.ch013

Kabra, G., Ghosh, V., & Ramesh, A. (2018). Enterprise Integrated Business Process Management and Business Intelligence Framework for Business Process Sustainability. In A. Paul, D. Bhattacharyya, & S. Anand (Eds.), *Green Initiatives for Business Sustainability and Value Creation* (pp. 228–238). Hershey, PA: IGI Global. doi:10.4018/978-1-5225-2662-9.ch010

Kaoud, M. (2017). Investigation of Customer Knowledge Management: A Case Study Research. *International Journal of Service Science, Management, Engineering, and Technology, 8*(2), 12–22. doi:10.4018/IJSSMET.2017040102

Kara, M. E., & Fırat, S. Ü. (2016). Sustainability, Risk, and Business Intelligence in Supply Chains. In M. Erdoğdu, T. Arun, & I. Ahmad (Eds.), *Handbook of Research on Green Economic Development Initiatives and Strategies* (pp. 501–538). Hershey, PA: IGI Global. doi:10.4018/978-1-5225-0440-5.ch022

Katuu, S. (2018). A Comparative Assessment of Enterprise Content Management Maturity Models. In N. Gwangwava & M. Mutingi (Eds.), *E-Manufacturing and E-Service Strategies in Contemporary Organizations* (pp. 93–118). Hershey, PA: IGI Global. doi:10.4018/978-1-5225-3628-4.ch005

Khan, M. A. (2016). MNEs Management Strategies in Developing Countries: Establishing the Context. In M. Khan (Ed.), *Multinational Enterprise Management Strategies in Developing Countries* (pp. 1–33). Hershey, PA: IGI Global. doi:10.4018/978-1-5225-0276-0.ch001

Khan, M. A. (2016). Operational Approaches in Organizational Structure: A Case for MNEs in Developing Countries. In M. Khan (Ed.), *Multinational Enterprise Management Strategies in Developing Countries* (pp. 129–151). Hershey, PA: IGI Global. doi:10.4018/978-1-5225-0276-0.ch007

Kinnunen, S., Ylä-Kujala, A., Marttonen-Arola, S., Kärri, T., & Baglee, D. (2018). Internet of Things in Asset Management: Insights from Industrial Professionals and Academia. *International Journal of Service Science, Management, Engineering, and Technology*, *9*(2), 104–119. doi:10.4018/IJSSMET.2018040105

Klein, A. Z., Sabino de Freitas, A., Machado, L., Freitas, J. C. Jr, Graziola, P. G. Jr, & Schlemmer, E. (2017). Virtual Worlds Applications for Management Education. In L. Tomei (Ed.), *Exploring the New Era of Technology-Infused Education* (pp. 279–299). Hershey, PA: IGI Global. doi:10.4018/978-1-5225-1709-2.ch017

Kożuch, B., & Jabłoński, A. (2017). Adopting the Concept of Business Models in Public Management. In M. Lewandowski & B. Kożuch (Eds.), *Public Sector Entrepreneurship and the Integration of Innovative Business Models* (pp. 10–46). Hershey, PA: IGI Global. doi:10.4018/978-1-5225-2215-7.ch002

Kumar, J., Adhikary, A., & Jha, A. (2017). Small Active Investors' Perceptions and Preferences Towards Tax Saving Mutual Fund Schemes in Eastern India: An Empirical Note. *International Journal of Asian Business and Information Management*, *8*(2), 35–45. doi:10.4018/IJABIM.2017040103

Lassoued, Y., Bouzguenda, L., & Mahmoud, T. (2016). Context-Aware Business Process Versions Management. *International Journal of e-Collaboration*, *12*(3), 7–33. doi:10.4018/IJeC.2016070102

Lavassani, K. M., & Movahedi, B. (2017). Applications Driven Information Systems: Beyond Networks toward Business Ecosystems. *International Journal of Innovation in the Digital Economy*, *8*(1), 61–75. doi:10.4018/IJIDE.2017010104

Lazzareschi, V. H., & Brito, M. S. (2017). Strategic Information Management: Proposal of Business Project Model. In G. Jamil, A. Soares, & C. Pessoa (Eds.), *Handbook of Research on Information Management for Effective Logistics and Supply Chains* (pp. 59–88). Hershey, PA: IGI Global. doi:10.4018/978-1-5225-0973-8.ch004

Lederer, M., Kurz, M., & Lazarov, P. (2017). Usage and Suitability of Methods for Strategic Business Process Initiatives: A Multi Case Study Research. *International Journal of Productivity Management and Assessment Technologies*, *5*(1), 40–51. doi:10.4018/IJPMAT.2017010103

Lee, I. (2017). A Social Enterprise Business Model and a Case Study of Pacific Community Ventures (PCV). In V. Potocan, M. Ünğan, & Z. Nedelko (Eds.), *Handbook of Research on Managerial Solutions in Non-Profit Organizations* (pp. 182–204). Hershey, PA: IGI Global. doi:10.4018/978-1-5225-0731-4.ch009

Lee, L. J., & Leu, J. (2016). Exploring the Effectiveness of IT Application and Value Method in the Innovation Performance of Enterprise. *International Journal of Enterprise Information Systems*, *12*(2), 47–65. doi:10.4018/IJEIS.2016040104

Lee, Y. (2016). Alignment Effect of Entrepreneurial Orientation and Marketing Orientation on Firm Performance. *International Journal of Customer Relationship Marketing and Management*, *7*(4), 58–69. doi:10.4018/IJCRMM.2016100104

Leon, L. A., Seal, K. C., Przasnyski, Z. H., & Wiedenman, I. (2017). Skills and Competencies Required for Jobs in Business Analytics: A Content Analysis of Job Advertisements Using Text Mining. *International Journal of Business Intelligence Research*, *8*(1), 1–25. doi:10.4018/IJBIR.2017010101

Leu, J., Lee, L. J., & Krischke, A. (2016). Value Engineering-Based Method for Implementing the ISO14001 System in the Green Supply Chains. *International Journal of Strategic Decision Sciences*, *7*(4), 1–20. doi:10.4018/IJSDS.2016100101

Levy, C. L., & Elias, N. I. (2017). SOHO Users' Perceptions of Reliability and Continuity of Cloud-Based Services. In M. Moore (Ed.), *Cybersecurity Breaches and Issues Surrounding Online Threat Protection* (pp. 248–287). Hershey, PA: IGI Global. doi:10.4018/978-1-5225-1941-6.ch011

Levy, M. (2018). Change Management Serving Knowledge Management and Organizational Development: Reflections and Review. In N. Baporikar (Ed.), *Global Practices in Knowledge Management for Societal and Organizational Development* (pp. 256–270). Hershey, PA: IGI Global. doi:10.4018/978-1-5225-3009-1.ch012

Lewandowski, M. (2017). Public Organizations and Business Model Innovation: The Role of Public Service Design. In M. Lewandowski & B. Kożuch (Eds.), *Public Sector Entrepreneurship and the Integration of Innovative Business Models* (pp. 47–72). Hershey, PA: IGI Global. doi:10.4018/978-1-5225-2215-7.ch003

Lhannaoui, H., Kabbaj, M. I., & Bakkoury, Z. (2017). A Survey of Risk-Aware Business Process Modelling. *International Journal of Risk and Contingency Management*, *6*(3), 14–26. doi:10.4018/IJRCM.2017070102

Li, J., Sun, W., Jiang, W., Yang, H., & Zhang, L. (2017). How the Nature of Exogenous Shocks and Crises Impact Company Performance?: The Effects of Industry Characteristics. *International Journal of Risk and Contingency Management*, *6*(4), 40–55. doi:10.4018/IJRCM.2017100103

Lu, C., & Liu, S. (2016). Cultural Tourism O2O Business Model Innovation-A Case Study of CTrip. *Journal of Electronic Commerce in Organizations*, *14*(2), 16–31. doi:10.4018/JECO.2016040102

Machen, B., Hosseini, M. R., Wood, A., & Bakhshi, J. (2016). An Investigation into using SAP-PS as a Multidimensional Project Control System (MPCS). *International Journal of Enterprise Information Systems*, *12*(2), 66–81. doi:10.4018/IJEIS.2016040105

Malega, P. (2017). Small and Medium Enterprises in the Slovak Republic: Status and Competitiveness of SMEs in the Global Markets and Possibilities of Optimization. In M. Vemić (Ed.), *Optimal Management Strategies in Small and Medium Enterprises* (pp. 102–124). Hershey, PA: IGI Global. doi:10.4018/978-1-5225-1949-2.ch006

Malewska, K. M. (2017). Intuition in Decision-Making on the Example of a Non-Profit Organization. In V. Potocan, M. Üngan, & Z. Nedelko (Eds.), *Handbook of Research on Managerial Solutions in Non-Profit Organizations* (pp. 378–399). Hershey, PA: IGI Global. doi:10.4018/978-1-5225-0731-4.ch018

Maroofi, F. (2017). Entrepreneurial Orientation and Organizational Learning Ability Analysis for Innovation and Firm Performance. In N. Baporikar (Ed.), *Innovation and Shifting Perspectives in Management Education* (pp. 144–165). Hershey, PA: IGI Global. doi:10.4018/978-1-5225-1019-2.ch007

Martins, P. V., & Zacarias, M. (2017). A Web-based Tool for Business Process Improvement. *International Journal of Web Portals*, *9*(2), 68–84. doi:10.4018/IJWP.2017070104

Matthies, B., & Coners, A. (2017). Exploring the Conceptual Nature of e-Business Projects. *Journal of Electronic Commerce in Organizations*, *15*(3), 33–63. doi:10.4018/JECO.2017070103

McKee, J. (2018). Architecture as a Tool to Solve Business Planning Problems. In M. Khosrow-Pour, D.B.A. (Ed.), Encyclopedia of Information Science and Technology, Fourth Edition (pp. 573-586). Hershey, PA: IGI Global. doi:10.4018/978-1-5225-2255-3.ch050

McMurray, A. J., Cross, J., & Caponecchia, C. (2018). The Risk Management Profession in Australia: Business Continuity Plan Practices. In N. Bajgoric (Ed.), *Always-On Enterprise Information Systems for Modern Organizations* (pp. 112–129). Hershey, PA: IGI Global. doi:10.4018/978-1-5225-3704-5.ch006

Meddah, I. H., & Belkadi, K. (2018). Mining Patterns Using Business Process Management. In R. Hamou (Ed.), *Handbook of Research on Biomimicry in Information Retrieval and Knowledge Management* (pp. 78–89). Hershey, PA: IGI Global. doi:10.4018/978-1-5225-3004-6.ch005

Mendes, L. (2017). TQM and Knowledge Management: An Integrated Approach Towards Tacit Knowledge Management. In D. Jaziri-Bouagina & G. Jamil (Eds.), *Handbook of Research on Tacit Knowledge Management for Organizational Success* (pp. 236–263). Hershey, PA: IGI Global. doi:10.4018/978-1-5225-2394-9.ch009

Mnjama, N. M. (2017). Preservation of Recorded Information in Public and Private Sector Organizations. In P. Jain & N. Mnjama (Eds.), *Managing Knowledge Resources and Records in Modern Organizations* (pp. 149–167). Hershey, PA: IGI Global. doi:10.4018/978-1-5225-1965-2.ch009

Mokoqama, M., & Fields, Z. (2017). Principles of Responsible Management Education (PRME): Call for Responsible Management Education. In Z. Fields (Ed.), *Collective Creativity for Responsible and Sustainable Business Practice* (pp. 229–241). Hershey, PA: IGI Global. doi:10.4018/978-1-5225-1823-5.ch012

Muniapan, B. (2017). Philosophy and Management: The Relevance of Vedanta in Management. In P. Ordóñez de Pablos (Ed.), *Managerial Strategies and Solutions for Business Success in Asia* (pp. 124–139). Hershey, PA: IGI Global. doi:10.4018/978-1-5225-1886-0.ch007

Muniapan, B., Gregory, M. L., & Ling, L. A. (2016). Marketing Education in Sarawak: Looking at It from the Employers' Viewpoint. In B. Smith & A. Porath (Eds.), *Global Perspectives on Contemporary Marketing Education* (pp. 112–130). Hershey, PA: IGI Global. doi:10.4018/978-1-4666-9784-3.ch008

Murad, S. E., & Dowaji, S. (2017). Using Value-Based Approach for Managing Cloud-Based Services. In A. Turuk, B. Sahoo, & S. Addya (Eds.), *Resource Management and Efficiency in Cloud Computing Environments* (pp. 33–60). Hershey, PA: IGI Global. doi:10.4018/978-1-5225-1721-4.ch002

Mutahar, A. M., Daud, N. M., Thurasamy, R., Isaac, O., & Abdulsalam, R. (2018). The Mediating of Perceived Usefulness and Perceived Ease of Use: The Case of Mobile Banking in Yemen. *International Journal of Technology Diffusion, 9*(2), 21–40. doi:10.4018/IJTD.2018040102

Naidoo, V. (2017). E-Learning and Management Education at African Universities. In N. Baporikar (Ed.), *Management Education for Global Leadership* (pp. 181–201). Hershey, PA: IGI Global. doi:10.4018/978-1-5225-1013-0.ch009

Naidoo, V., & Igbinakhase, I. (2018). Opportunities and Challenges of Knowledge Retention in SMEs. In N. Baporikar (Ed.), *Knowledge Integration Strategies for Entrepreneurship and Sustainability* (pp. 70–94). Hershey, PA: IGI Global. doi:10.4018/978-1-5225-5115-7.ch004

Nayak, S., & Prabhu, N. (2017). Paradigm Shift in Management Education: Need for a Cross Functional Perspective. In N. Baporikar (Ed.), *Management Education for Global Leadership* (pp. 241–255). Hershey, PA: IGI Global. doi:10.4018/978-1-5225-1013-0.ch012

Ndede-Amadi, A. A. (2016). Student Interest in the IS Specialization as Predictor of the Success Potential of New Information Systems Programmes within the Schools of Business in Kenyan Public Universities. *International Journal of Information Systems and Social Change, 7*(2), 63–79. doi:10.4018/IJISSC.2016040104

Nedelko, Z., & Potocan, V. (2016). Management Practices for Processes Optimization: Case of Slovenia. In G. Alor-Hernández, C. Sánchez-Ramírez, & J. García-Alcaraz (Eds.), *Handbook of Research on Managerial Strategies for Achieving Optimal Performance in Industrial Processes* (pp. 545–561). Hershey, PA: IGI Global. doi:10.4018/978-1-5225-0130-5.ch025

Nedelko, Z., & Potocan, V. (2017). Management Solutions in Non-Profit Organizations: Case of Slovenia. In V. Potocan, M. Üngan, & Z. Nedelko (Eds.), *Handbook of Research on Managerial Solutions in Non-Profit Organizations* (pp. 1–22). Hershey, PA: IGI Global. doi:10.4018/978-1-5225-0731-4.ch001

Nedelko, Z., & Potocan, V. (2017). Priority of Management Tools Utilization among Managers: International Comparison. In V. Wang (Ed.), *Encyclopedia of Strategic Leadership and Management* (pp. 1083–1094). Hershey, PA: IGI Global. doi:10.4018/978-1-5225-1049-9.ch075

Nedelko, Z., Raudeliūnienė, J., & Črešnar, R. (2018). Knowledge Dynamics in Supply Chain Management. In N. Baporikar (Ed.), *Knowledge Integration Strategies for Entrepreneurship and Sustainability* (pp. 150–166). Hershey, PA: IGI Global. doi:10.4018/978-1-5225-5115-7.ch008

Nguyen, H. T., & Hipsher, S. A. (2018). Innovation and Creativity Used by Private Sector Firms in a Resources-Constrained Environment. In S. Hipsher (Ed.), *Examining the Private Sector's Role in Wealth Creation and Poverty Reduction* (pp. 219–238). Hershey, PA: IGI Global. doi:10.4018/978-1-5225-3117-3.ch010

Nycz, M., & Półkowski, Z. (2016). Business Intelligence as a Modern IT Supporting Management of Local Government Units in Poland. *International Journal of Knowledge and Systems Science, 7*(4), 1–18. doi:10.4018/IJKSS.2016100101

Obaji, N. O., Senin, A. A., & Olugu, M. U. (2016). Supportive Government Policy as a Mechanism for Business Incubation Performance in Nigeria. *International Journal of Information Systems and Social Change, 7*(4), 52–66. doi:10.4018/IJISSC.2016100103

Obicci, P. A. (2017). Risk Sharing in a Partnership. In *Risk Management Strategies in Public-Private Partnerships* (pp. 115–152). Hershey, PA: IGI Global. doi:10.4018/978-1-5225-2503-5.ch004

Obidallah, W. J., & Raahemi, B. (2017). Managing Changes in Service Oriented Virtual Organizations: A Structural and Procedural Framework to Facilitate the Process of Change. *Journal of Electronic Commerce in Organizations, 15*(1), 59–83. doi:10.4018/JECO.2017010104

Ojasalo, J., & Ojasalo, K. (2016). Service Logic Business Model Canvas for Lean Development of SMEs and Start-Ups. In N. Baporikar (Ed.), *Handbook of Research on Entrepreneurship in the Contemporary Knowledge-Based Global Economy* (pp. 217–243). Hershey, PA: IGI Global. doi:10.4018/978-1-4666-8798-1.ch010

Ojo, O. (2017). Impact of Innovation on the Entrepreneurial Success in Selected Business Enterprises in South-West Nigeria. *International Journal of Innovation in the Digital Economy, 8*(2), 29–38. doi:10.4018/IJIDE.2017040103

Okdinawati, L., Simatupang, T. M., & Sunitiyoso, Y. (2017). Multi-Agent Reinforcement Learning for Value Co-Creation of Collaborative Transportation Management (CTM). *International Journal of Information Systems and Supply Chain Management, 10*(3), 84–95. doi:10.4018/IJISSCM.2017070105

Ortner, E., Mevius, M., Wiedmann, P., & Kurz, F. (2016). Design of Interactional Decision Support Applications for E-Participation in Smart Cities. *International Journal of Electronic Government Research, 12*(2), 18–38. doi:10.4018/IJEGR.2016040102

Pal, K. (2018). Building High Quality Big Data-Based Applications in Supply Chains. In A. Kumar & S. Saurav (Eds.), *Supply Chain Management Strategies and Risk Assessment in Retail Environments* (pp. 1–24). Hershey, PA: IGI Global. doi:10.4018/978-1-5225-3056-5.ch001

Palos-Sanchez, P. R., & Correia, M. B. (2018). Perspectives of the Adoption of Cloud Computing in the Tourism Sector. In J. Rodrigues, C. Ramos, P. Cardoso, & C. Henriques (Eds.), *Handbook of Research on Technological Developments for Cultural Heritage and eTourism Applications* (pp. 377–400). Hershey, PA: IGI Global. doi:10.4018/978-1-5225-2927-9.ch018

Parry, V. K., & Lind, M. L. (2016). Alignment of Business Strategy and Information Technology Considering Information Technology Governance, Project Portfolio Control, and Risk Management. *International Journal of Information Technology Project Management*, *7*(4), 21–37. doi:10.4018/IJITPM.2016100102

Pashkova, N., Trujillo-Barrera, A., Apostolakis, G., Van Dijk, G., Drakos, P. D., & Baourakis, G. (2016). Business Management Models of Microfinance Institutions (MFIs) in Africa: A Study into Their Enabling Environments. *International Journal of Food and Beverage Manufacturing and Business Models*, *1*(2), 63–82. doi:10.4018/IJFBMBM.2016070105

Patiño, B. E. (2017). New Generation Management by Convergence and Individual Identity: A Systemic and Human-Oriented Approach. In N. Baporikar (Ed.), *Innovation and Shifting Perspectives in Management Education* (pp. 119–143). Hershey, PA: IGI Global. doi:10.4018/978-1-5225-1019-2.ch006

Pawliczek, A., & Rössler, M. (2017). Knowledge of Management Tools and Systems in SMEs: Knowledge Transfer in Management. In A. Bencsik (Ed.), *Knowledge Management Initiatives and Strategies in Small and Medium Enterprises* (pp. 180–203). Hershey, PA: IGI Global. doi:10.4018/978-1-5225-1642-2.ch009

Pejic-Bach, M., Omazic, M. A., Aleksic, A., & Zoroja, J. (2018). Knowledge-Based Decision Making: A Multi-Case Analysis. In R. Leon (Ed.), *Managerial Strategies for Business Sustainability During Turbulent Times* (pp. 160–184). Hershey, PA: IGI Global. doi:10.4018/978-1-5225-2716-9.ch009

Perano, M., Hysa, X., & Calabrese, M. (2018). Strategic Planning, Cultural Context, and Business Continuity Management: Business Cases in the City of Shkoder. In A. Presenza & L. Sheehan (Eds.), *Geopolitics and Strategic Management in the Global Economy* (pp. 57–77). Hershey, PA: IGI Global. doi:10.4018/978-1-5225-2673-5.ch004

Pereira, R., Mira da Silva, M., & Lapão, L. V. (2017). IT Governance Maturity Patterns in Portuguese Healthcare. In S. De Haes & W. Van Grembergen (Eds.), *Strategic IT Governance and Alignment in Business Settings* (pp. 24–52). Hershey, PA: IGI Global. doi:10.4018/978-1-5225-0861-8.ch002

Perez-Uribe, R., & Ocampo-Guzman, D. (2016). Conflict within Colombian Family Owned SMEs: An Explosive Blend between Feelings and Business. In J. Saiz-Álvarez (Ed.), *Handbook of Research on Social Entrepreneurship and Solidarity Economics* (pp. 329–354). Hershey, PA: IGI Global. doi:10.4018/978-1-5225-0097-1.ch017

Pérez-Uribe, R. I., Torres, D. A., Jurado, S. P., & Prada, D. M. (2018). Cloud Tools for the Development of Project Management in SMEs. In R. Perez-Uribe, C. Salcedo-Perez, & D. Ocampo-Guzman (Eds.), *Handbook of Research on Intrapreneurship and Organizational Sustainability in SMEs* (pp. 95–120). Hershey, PA: IGI Global. doi:10.4018/978-1-5225-3543-0.ch005

Petrisor, I., & Cozmiuc, D. (2017). Global Supply Chain Management Organization at Siemens in the Advent of Industry 4.0. In L. Saglietto & C. Cezanne (Eds.), *Global Intermediation and Logistics Service Providers* (pp. 123–142). Hershey, PA: IGI Global. doi:10.4018/978-1-5225-2133-4.ch007

Pierce, J. M., Velliaris, D. M., & Edwards, J. (2017). A Living Case Study: A Journey Not a Destination. In N. Silton (Ed.), *Exploring the Benefits of Creativity in Education, Media, and the Arts* (pp. 158–178). Hershey, PA: IGI Global. doi:10.4018/978-1-5225-0504-4.ch008

Radosavljevic, M., & Andjelkovic, A. (2017). Multi-Criteria Decision Making Approach for Choosing Business Process for the Improvement: Upgrading of the Six Sigma Methodology. In J. Stanković, P. Delias, S. Marinković, & S. Rochhia (Eds.), *Tools and Techniques for Economic Decision Analysis* (pp. 225–247). Hershey, PA: IGI Global. doi:10.4018/978-1-5225-0959-2.ch011

Radovic, V. M. (2017). Corporate Sustainability and Responsibility and Disaster Risk Reduction: A Serbian Overview. In M. Camilleri (Ed.), *CSR 2.0 and the New Era of Corporate Citizenship* (pp. 147–164). Hershey, PA: IGI Global. doi:10.4018/978-1-5225-1842-6.ch008

Raghunath, K. M., Devi, S. L., & Patro, C. S. (2018). Impact of Risk Assessment Models on Risk Factors: A Holistic Outlook. In K. Strang, M. Korstanje, & N. Vajjhala (Eds.), *Research, Practices, and Innovations in Global Risk and Contingency Management* (pp. 134–153). Hershey, PA: IGI Global. doi:10.4018/978-1-5225-4754-9.ch008

Raman, A., & Goyal, D. P. (2017). Extending IMPLEMENT Framework for Enterprise Information Systems Implementation to Information System Innovation. In M. Tavana (Ed.), *Enterprise Information Systems and the Digitalization of Business Functions* (pp. 137–177). Hershey, PA: IGI Global. doi:10.4018/978-1-5225-2382-6.ch007

Rao, Y., & Zhang, Y. (2017). The Construction and Development of Academic Library Digital Special Subject Databases. In L. Ruan, Q. Zhu, & Y. Ye (Eds.), *Academic Library Development and Administration in China* (pp. 163–183). Hershey, PA: IGI Global. doi:10.4018/978-1-5225-0550-1.ch010

Ravasan, A. Z., Mohammadi, M. M., & Hamidi, H. (2018). An Investigation Into the Critical Success Factors of Implementing Information Technology Service Management Frameworks. In K. Jakobs (Ed.), *Corporate and Global Standardization Initiatives in Contemporary Society* (pp. 200–218). Hershey, PA: IGI Global. doi:10.4018/978-1-5225-5320-5.ch009

Renna, P., Izzo, C., & Romaniello, T. (2016). The Business Process Management Systems to Support Continuous Improvements. In W. Nuninger & J. Châtelet (Eds.), *Handbook of Research on Quality Assurance and Value Management in Higher Education* (pp. 237–256). Hershey, PA: IGI Global. doi:10.4018/978-1-5225-0024-7.ch009

Rezaie, S., Mirabedini, S. J., & Abtahi, A. (2018). Designing a Model for Implementation of Business Intelligence in the Banking Industry. *International Journal of Enterprise Information Systems, 14*(1), 77–103. doi:10.4018/IJEIS.2018010105

Riccò, R. (2016). Diversity Management: Bringing Equality, Equity, and Inclusion in the Workplace. In J. Prescott (Ed.), *Handbook of Research on Race, Gender, and the Fight for Equality* (pp. 335–359). Hershey, PA: IGI Global. doi:10.4018/978-1-5225-0047-6.ch015

Romano, L., Grimaldi, R., & Colasuonno, F. S. (2017). Demand Management as a Success Factor in Project Portfolio Management. In L. Romano (Ed.), *Project Portfolio Management Strategies for Effective Organizational Operations* (pp. 202–219). Hershey, PA: IGI Global. doi:10.4018/978-1-5225-2151-8.ch008

Rostek, K. B. (2016). Risk Management: Role and Importance in Business Organization. In D. Jakóbczak (Ed.), *Analyzing Risk through Probabilistic Modeling in Operations Research* (pp. 149–178). Hershey, PA: IGI Global. doi:10.4018/978-1-4666-9458-3.ch007

Rouhani, S., & Savoji, S. R. (2016). A Success Assessment Model for BI Tools Implementation: An Empirical Study of Banking Industry. *International Journal of Business Intelligence Research, 7*(1), 25–44. doi:10.4018/IJBIR.2016010103

Ruan, Z. (2016). A Corpus-Based Functional Analysis of Complex Nominal Groups in Written Business Discourse: The Case of "Business". *International Journal of Computer-Assisted Language Learning and Teaching*, *6*(2), 74–90. doi:10.4018/IJCALLT.2016040105

Ruhi, U. (2018). Towards an Interdisciplinary Socio-Technical Definition of Virtual Communities. In M. Khosrow-Pour, D.B.A. (Ed.), Encyclopedia of Information Science and Technology, Fourth Edition (pp. 4278-4295). Hershey, PA: IGI Global. doi:10.4018/978-1-5225-2255-3.ch371

Ryan, J., Doster, B., Daily, S., & Lewis, C. (2016). A Case Study Perspective for Balanced Perioperative Workflow Achievement through Data-Driven Process Improvement. *International Journal of Healthcare Information Systems and Informatics*, *11*(3), 19–41. doi:10.4018/IJHISI.2016070102

Safari, M. R., & Jiang, Q. (2018). The Theory and Practice of IT Governance Maturity and Strategies Alignment: Evidence From Banking Industry. *Journal of Global Information Management*, *26*(2), 127–146. doi:10.4018/JGIM.2018040106

Sahoo, J., Pati, B., & Mohanty, B. (2017). Knowledge Management as an Academic Discipline: An Assessment. In B. Gunjal (Ed.), *Managing Knowledge and Scholarly Assets in Academic Libraries* (pp. 99–126). Hershey, PA: IGI Global. doi:10.4018/978-1-5225-1741-2.ch005

Saini, D. (2017). Relevance of Teaching Values and Ethics in Management Education. In N. Baporikar (Ed.), *Management Education for Global Leadership* (pp. 90–111). Hershey, PA: IGI Global. doi:10.4018/978-1-5225-1013-0.ch005

Sambhanthan, A. (2017). Assessing and Benchmarking Sustainability in Organisations: An Integrated Conceptual Model. *International Journal of Systems and Service-Oriented Engineering*, *7*(4), 22–43. doi:10.4018/IJSSOE.2017100102

Sambhanthan, A., & Potdar, V. (2017). A Study of the Parameters Impacting Sustainability in Information Technology Organizations. *International Journal of Knowledge-Based Organizations*, *7*(3), 27–39. doi:10.4018/IJKBO.2017070103

Sánchez-Fernández, M. D., & Manríquez, M. R. (2018). The Entrepreneurial Spirit Based on Social Values: The Digital Generation. In P. Isaias & L. Carvalho (Eds.), *User Innovation and the Entrepreneurship Phenomenon in the Digital Economy* (pp. 173–193). Hershey, PA: IGI Global. doi:10.4018/978-1-5225-2826-5.ch009

Sanchez-Ruiz, L., & Blanco, B. (2017). Process Management for SMEs: Barriers, Enablers, and Benefits. In M. Vemić (Ed.), *Optimal Management Strategies in Small and Medium Enterprises* (pp. 293–319). Hershey, PA: IGI Global. doi:10.4018/978-1-5225-1949-2.ch014

Sanz, L. F., Gómez-Pérez, J., & Castillo-Martinez, A. (2018). Analysis of the European ICT Competence Frameworks. In V. Ahuja & S. Rathore (Eds.), *Multidisciplinary Perspectives on Human Capital and Information Technology Professionals* (pp. 225–245). Hershey, PA: IGI Global. doi:10.4018/978-1-5225-5297-0.ch012

Sarvepalli, A., & Godin, J. (2017). Business Process Management in the Classroom. *Journal of Cases on Information Technology*, *19*(2), 17–28. doi:10.4018/JCIT.2017040102

Satpathy, B., & Muniapan, B. (2016). Ancient Wisdom for Transformational Leadership and Its Insights from the Bhagavad-Gita. In U. Aung & P. Ordoñez de Pablos (Eds.), *Managerial Strategies and Practice in the Asian Business Sector* (pp. 1–10). Hershey, PA: IGI Global. doi:10.4018/978-1-4666-9758-4.ch001

Saygili, E. E., Ozturkoglu, Y., & Kocakulah, M. C. (2017). End Users' Perceptions of Critical Success Factors in ERP Applications. *International Journal of Enterprise Information Systems*, *13*(4), 58–75. doi:10.4018/IJEIS.2017100104

Saygili, E. E., & Saygili, A. T. (2017). Contemporary Issues in Enterprise Information Systems: A Critical Review of CSFs in ERP Implementations. In M. Tavana (Ed.), *Enterprise Information Systems and the Digitalization of Business Functions* (pp. 120–136). Hershey, PA: IGI Global. doi:10.4018/978-1-5225-2382-6.ch006

Seidenstricker, S., & Antonino, A. (2018). Business Model Innovation-Oriented Technology Management for Emergent Technologies. In M. Khosrow-Pour, D.B.A. (Ed.), Encyclopedia of Information Science and Technology, Fourth Edition (pp. 4560-4569). Hershey, PA: IGI Global. doi:10.4018/978-1-5225-2255-3.ch396

Senaratne, S., & Gunarathne, A. D. (2017). Excellence Perspective for Management Education from a Global Accountants' Hub in Asia. In N. Baporikar (Ed.), *Management Education for Global Leadership* (pp. 158–180). Hershey, PA: IGI Global. doi:10.4018/978-1-5225-1013-0.ch008

Sensuse, D. I., & Cahyaningsih, E. (2018). Knowledge Management Models: A Summative Review. *International Journal of Information Systems in the Service Sector*, *10*(1), 71–100. doi:10.4018/IJISSS.2018010105

Sensuse, D. I., Wibowo, W. C., & Cahyaningsih, E. (2016). Indonesian Government Knowledge Management Model: A Theoretical Model. *Information Resources Management Journal, 29*(1), 91–108. doi:10.4018/irmj.2016010106

Seth, M., Goyal, D., & Kiran, R. (2017). Diminution of Impediments in Implementation of Supply Chain Management Information System for Enhancing its Effectiveness in Indian Automobile Industry. *Journal of Global Information Management, 25*(3), 1–20. doi:10.4018/JGIM.2017070101

Seyal, A. H., & Rahman, M. N. (2017). Investigating Impact of Inter-Organizational Factors in Measuring ERP Systems Success: Bruneian Perspectives. In M. Tavana (Ed.), *Enterprise Information Systems and the Digitalization of Business Functions* (pp. 178–204). Hershey, PA: IGI Global. doi:10.4018/978-1-5225-2382-6.ch008

Shaikh, A. A., & Karjaluoto, H. (2016). On Some Misconceptions Concerning Digital Banking and Alternative Delivery Channels. *International Journal of E-Business Research, 12*(3), 1–16. doi:10.4018/IJEBR.2016070101

Shams, S. M. (2016). Stakeholder Relationship Management in Online Business and Competitive Value Propositions: Evidence from the Sports Industry. *International Journal of Online Marketing, 6*(2), 1–17. doi:10.4018/IJOM.2016040101

Shamsuzzoha, A. (2016). Management of Risk and Resilience within Collaborative Business Network. In R. Addo-Tenkorang, J. Kantola, P. Helo, & A. Shamsuzzoha (Eds.), *Supply Chain Strategies and the Engineer-to-Order Approach* (pp. 143–159). Hershey, PA: IGI Global. doi:10.4018/978-1-5225-0021-6.ch008

Shaqrah, A. A. (2018). Analyzing Business Intelligence Systems Based on 7s Model of McKinsey. *International Journal of Business Intelligence Research, 9*(1), 53–63. doi:10.4018/IJBIR.2018010104

Sharma, A. J. (2017). Enhancing Sustainability through Experiential Learning in Management Education. In N. Baporikar (Ed.), *Management Education for Global Leadership* (pp. 256–274). Hershey, PA: IGI Global. doi:10.4018/978-1-5225-1013-0.ch013

Shetty, K. P. (2017). Responsible Global Leadership: Ethical Challenges in Management Education. In N. Baporikar (Ed.), *Innovation and Shifting Perspectives in Management Education* (pp. 194–223). Hershey, PA: IGI Global. doi:10.4018/978-1-5225-1019-2.ch009

Sinthupundaja, J., & Kohda, Y. (2017). Effects of Corporate Social Responsibility and Creating Shared Value on Sustainability. *International Journal of Sustainable Entrepreneurship and Corporate Social Responsibility*, *2*(1), 27–38. doi:10.4018/IJSECSR.2017010103

Škarica, I., & Hrgović, A. V. (2018). Implementation of Total Quality Management Principles in Public Health Institutes in the Republic of Croatia. *International Journal of Productivity Management and Assessment Technologies*, *6*(1), 1–16. doi:10.4018/IJPMAT.2018010101

Smuts, H., Kotzé, P., Van der Merwe, A., & Loock, M. (2017). Framework for Managing Shared Knowledge in an Information Systems Outsourcing Context. *International Journal of Knowledge Management*, *13*(4), 1–30. doi:10.4018/IJKM.2017100101

Soares, E. R., & Zaidan, F. H. (2016). Information Architecture and Business Modeling in Modern Organizations of Information Technology: Professional Career Plan in Organizations IT. In G. Jamil, J. Poças Rascão, F. Ribeiro, & A. Malheiro da Silva (Eds.), *Handbook of Research on Information Architecture and Management in Modern Organizations* (pp. 439–457). Hershey, PA: IGI Global. doi:10.4018/978-1-4666-8637-3.ch020

Sousa, M. J., Cruz, R., Dias, I., & Caracol, C. (2017). Information Management Systems in the Supply Chain. In G. Jamil, A. Soares, & C. Pessoa (Eds.), *Handbook of Research on Information Management for Effective Logistics and Supply Chains* (pp. 469–485). Hershey, PA: IGI Global. doi:10.4018/978-1-5225-0973-8.ch025

Spremic, M., Turulja, L., & Bajgoric, N. (2018). Two Approaches in Assessing Business Continuity Management Attitudes in the Organizational Context. In N. Bajgoric (Ed.), *Always-On Enterprise Information Systems for Modern Organizations* (pp. 159–183). Hershey, PA: IGI Global. doi:10.4018/978-1-5225-3704-5.ch008

Steenkamp, A. L. (2018). Some Insights in Computer Science and Information Technology. In *Examining the Changing Role of Supervision in Doctoral Research Projects: Emerging Research and Opportunities* (pp. 113–133). Hershey, PA: IGI Global. doi:10.4018/978-1-5225-2610-0.ch005

Studdard, N., Dawson, M., Burton, S. L., Jackson, N., Leonard, B., Quisenberry, W., & Rahim, E. (2016). Nurturing Social Entrepreneurship and Building Social Entrepreneurial Self-Efficacy: Focusing on Primary and Secondary Schooling to Develop Future Social Entrepreneurs. In Z. Fields (Ed.), *Incorporating Business Models and Strategies into Social Entrepreneurship* (pp. 154–175). Hershey, PA: IGI Global. doi:10.4018/978-1-4666-8748-6.ch010

Sun, Z. (2016). A Framework for Developing Management Intelligent Systems. *International Journal of Systems and Service-Oriented Engineering*, *6*(1), 37–53. doi:10.4018/IJSSOE.2016010103

Swami, B., & Mphele, G. T. (2016). Problems Preventing Growth of Small Entrepreneurs: A Case Study of a Few Small Entrepreneurs in Botswana Sub-Urban Areas. In N. Baporikar (Ed.), *Handbook of Research on Entrepreneurship in the Contemporary Knowledge-Based Global Economy* (pp. 479–508). Hershey, PA: IGI Global. doi:10.4018/978-1-4666-8798-1.ch020

Tabach, A., & Croteau, A. (2017). Configurations of Information Technology Governance Practices and Business Unit Performance. *International Journal of IT/Business Alignment and Governance*, *8*(2), 1–27. doi:10.4018/IJITBAG.2017070101

Talaue, G. M., & Iqbal, T. (2017). Assessment of e-Business Mode of Selected Private Universities in the Philippines and Pakistan. *International Journal of Online Marketing*, *7*(4), 63–77. doi:10.4018/IJOM.2017100105

Tam, G. C. (2017). Project Manager Sustainability Competence. In *Managerial Strategies and Green Solutions for Project Sustainability* (pp. 178–207). Hershey, PA: IGI Global. doi:10.4018/978-1-5225-2371-0.ch008

Tambo, T. (2018). Fashion Retail Innovation: About Context, Antecedents, and Outcome in Technological Change Projects. In I. Management Association (Ed.), Fashion and Textiles: Breakthroughs in Research and Practice (pp. 233-260). Hershey, PA: IGI Global. doi:10.4018/978-1-5225-3432-7.ch010

Tambo, T., & Mikkelsen, O. E. (2016). Fashion Supply Chain Optimization: Linking Make-to-Order Purchasing and B2B E-Commerce. In S. Joshi & R. Joshi (Eds.), *Designing and Implementing Global Supply Chain Management* (pp. 1–21). Hershey, PA: IGI Global. doi:10.4018/978-1-4666-9720-1.ch001

Tandon, K. (2016). Innovative Andragogy: The Paradigm Shift to Heutagogy. In S. Tiwari & L. Nafees (Eds.), *Innovative Management Education Pedagogies for Preparing Next-Generation Leaders* (pp. 238–257). Hershey, PA: IGI Global. doi:10.4018/978-1-4666-9691-4.ch014

Tantau, A. D., & Frăţilă, L. C. (2018). Information and Management System for Renewable Energy Business. In *Entrepreneurship and Business Development in the Renewable Energy Sector* (pp. 200–244). Hershey, PA: IGI Global. doi:10.4018/978-1-5225-3625-3.ch006

Teixeira, N., Pardal, P. N., & Rafael, B. G. (2018). Internationalization, Financial Performance, and Organizational Challenges: A Success Case in Portugal. In L. Carvalho (Ed.), *Handbook of Research on Entrepreneurial Ecosystems and Social Dynamics in a Globalized World* (pp. 379–423). Hershey, PA: IGI Global. doi:10.4018/978-1-5225-3525-6.ch017

Trad, A., & Kalpić, D. (2016). The E-Business Transformation Framework for E-Commerce Architecture-Modeling Projects. In I. Lee (Ed.), *Encyclopedia of E-Commerce Development, Implementation, and Management* (pp. 733–753). Hershey, PA: IGI Global. doi:10.4018/978-1-4666-9787-4.ch052

Trad, A., & Kalpić, D. (2016). The E-Business Transformation Framework for E-Commerce Control and Monitoring Pattern. In I. Lee (Ed.), *Encyclopedia of E-Commerce Development, Implementation, and Management* (pp. 754–777). Hershey, PA: IGI Global. doi:10.4018/978-1-4666-9787-4.ch053

Trad, A., & Kalpić, D. (2018). The Business Transformation Framework, Agile Project and Change Management. In M. Khosrow-Pour, D.B.A. (Ed.), Encyclopedia of Information Science and Technology, Fourth Edition (pp. 620-635). Hershey, PA: IGI Global. doi:10.4018/978-1-5225-2255-3.ch054

Trad, A., & Kalpić, D. (2018). The Business Transformation and Enterprise Architecture Framework: The Financial Engineering E-Risk Management and E-Law Integration. In B. Sergi, F. Fidanoski, M. Ziolo, & V. Naumovski (Eds.), *Regaining Global Stability After the Financial Crisis* (pp. 46–65). Hershey, PA: IGI Global. doi:10.4018/978-1-5225-4026-7.ch003

Turulja, L., & Bajgoric, N. (2018). Business Continuity and Information Systems: A Systematic Literature Review. In N. Bajgoric (Ed.), *Always-On Enterprise Information Systems for Modern Organizations* (pp. 60–87). Hershey, PA: IGI Global. doi:10.4018/978-1-5225-3704-5.ch004

van Wessel, R. M., de Vries, H. J., & Ribbers, P. M. (2016). Business Benefits through Company IT Standardization. In K. Jakobs (Ed.), *Effective Standardization Management in Corporate Settings* (pp. 34–53). Hershey, PA: IGI Global. doi:10.4018/978-1-4666-9737-9.ch003

Vargas-Hernández, J. G. (2017). Professional Integrity in Business Management Education. In N. Baporikar (Ed.), *Management Education for Global Leadership* (pp. 70–89). Hershey, PA: IGI Global. doi:10.4018/978-1-5225-1013-0.ch004

Vasista, T. G., & AlAbdullatif, A. M. (2017). Role of Electronic Customer Relationship Management in Demand Chain Management: A Predictive Analytic Approach. *International Journal of Information Systems and Supply Chain Management, 10*(1), 53–67. doi:10.4018/IJISSCM.2017010104

Vergidis, K. (2016). Rediscovering Business Processes: Definitions, Patterns, and Modelling Approaches. In P. Papajorgji, F. Pinet, A. Guimarães, & J. Papathanasiou (Eds.), *Automated Enterprise Systems for Maximizing Business Performance* (pp. 97–122). Hershey, PA: IGI Global. doi:10.4018/978-1-4666-8841-4.ch007

Vieru, D., & Bourdeau, S. (2017). Survival in the Digital Era: A Digital Competence-Based Multi-Case Study in the Canadian SME Clothing Industry. *International Journal of Social and Organizational Dynamics in IT, 6*(1), 17–34. doi:10.4018/IJSODIT.2017010102

Vijayan, G., & Kamarulzaman, N. H. (2017). An Introduction to Sustainable Supply Chain Management and Business Implications. In M. Khan, M. Hussain, & M. Ajmal (Eds.), *Green Supply Chain Management for Sustainable Business Practice* (pp. 27–50). Hershey, PA: IGI Global. doi:10.4018/978-1-5225-0635-5.ch002

Vlachvei, A., & Notta, O. (2017). Firm Competitiveness: Theories, Evidence, and Measurement. In A. Vlachvei, O. Notta, K. Karantininis, & N. Tsounis (Eds.), *Factors Affecting Firm Competitiveness and Performance in the Modern Business World* (pp. 1–42). Hershey, PA: IGI Global. doi:10.4018/978-1-5225-0843-4.ch001

von Rosing, M., Fullington, N., & Walker, J. (2016). Using the Business Ontology and Enterprise Standards to Transform Three Leading Organizations. *International Journal of Conceptual Structures and Smart Applications, 4*(1), 71–99. doi:10.4018/IJCSSA.2016010104

von Rosing, M., & von Scheel, H. (2016). Using the Business Ontology to Develop Enterprise Standards. *International Journal of Conceptual Structures and Smart Applications, 4*(1), 48–70. doi:10.4018/IJCSSA.2016010103

Walczak, S. (2016). Artificial Neural Networks and other AI Applications for Business Management Decision Support. *International Journal of Sociotechnology and Knowledge Development, 8*(4), 1–20. doi:10.4018/IJSKD.2016100101

Wamba, S. F., Akter, S., Kang, H., Bhattacharya, M., & Upal, M. (2016). The Primer of Social Media Analytics. *Journal of Organizational and End User Computing, 28*(2), 1–12. doi:10.4018/JOEUC.2016040101

Wang, C., Schofield, M., Li, X., & Ou, X. (2017). Do Chinese Students in Public and Private Higher Education Institutes Perform at Different Level in One of the Leadership Skills: Critical Thinking?: An Exploratory Comparison. In V. Wang (Ed.), *Encyclopedia of Strategic Leadership and Management* (pp. 160–181). Hershey, PA: IGI Global. doi:10.4018/978-1-5225-1049-9.ch013

Wang, F., Raisinghani, M. S., Mora, M., & Wang, X. (2016). Strategic E-Business Management through a Balanced Scored Card Approach. In I. Lee (Ed.), *Encyclopedia of E-Commerce Development, Implementation, and Management* (pp. 361–386). Hershey, PA: IGI Global. doi:10.4018/978-1-4666-9787-4.ch027

Wang, J. (2017). Multi-Agent based Production Management Decision System Modelling for the Textile Enterprise. *Journal of Global Information Management, 25*(4), 1–15. doi:10.4018/JGIM.2017100101

Wiedemann, A., & Gewald, H. (2017). Examining Cross-Domain Alignment: The Correlation of Business Strategy, IT Management, and IT Business Value. *International Journal of IT/Business Alignment and Governance, 8*(1), 17–31. doi:10.4018/IJITBAG.2017010102

Wolf, R., & Thiel, M. (2018). Advancing Global Business Ethics in China: Reducing Poverty Through Human and Social Welfare. In S. Hipsher (Ed.), *Examining the Private Sector's Role in Wealth Creation and Poverty Reduction* (pp. 67–84). Hershey, PA: IGI Global. doi:10.4018/978-1-5225-3117-3.ch004

Wu, J., Ding, F., Xu, M., Mo, Z., & Jin, A. (2016). Investigating the Determinants of Decision-Making on Adoption of Public Cloud Computing in E-government. *Journal of Global Information Management, 24*(3), 71–89. doi:10.4018/JGIM.2016070104

Xu, L., & de Vrieze, P. (2016). Building Situational Applications for Virtual Enterprises. In I. Lee (Ed.), *Encyclopedia of E-Commerce Development, Implementation, and Management* (pp. 715–724). Hershey, PA: IGI Global. doi:10.4018/978-1-4666-9787-4.ch050

Yablonsky, S. (2018). Innovation Platforms: Data and Analytics Platforms. In *Multi-Sided Platforms (MSPs) and Sharing Strategies in the Digital Economy: Emerging Research and Opportunities* (pp. 72–95). Hershey, PA: IGI Global. doi:10.4018/978-1-5225-5457-8.ch003

Yusoff, A., Ahmad, N. H., & Halim, H. A. (2017). Agropreneurship among Gen Y in Malaysia: The Role of Academic Institutions. In N. Ahmad, T. Ramayah, H. Halim, & S. Rahman (Eds.), *Handbook of Research on Small and Medium Enterprises in Developing Countries* (pp. 23–47). Hershey, PA: IGI Global. doi:10.4018/978-1-5225-2165-5.ch002

Zanin, F., Comuzzi, E., & Costantini, A. (2018). The Effect of Business Strategy and Stock Market Listing on the Use of Risk Assessment Tools. In *Management Control Systems in Complex Settings: Emerging Research and Opportunities* (pp. 145–168). Hershey, PA: IGI Global. doi:10.4018/978-1-5225-3987-2.ch007

Zgheib, P. W. (2017). Corporate Innovation and Intrapreneurship in the Middle East. In P. Zgheib (Ed.), *Entrepreneurship and Business Innovation in the Middle East* (pp. 37–56). Hershey, PA: IGI Global. doi:10.4018/978-1-5225-2066-5.ch003

About the Contributors

André Riani Costa Perinotto has a Ph.D. in Communication Sciences (Mediatic Processes) - UNISINOS/RS-Brazil. Master's degree in Geography (Organization of Space) - UNESP/Rio Claro/SP-Brazil; Specialist in Teaching for Higher Education in Tourism and Hospitality by SENAC/SP-Brazil; and Graduated in Tourism by UNIMEP/SP-Brazil. Full Professor of the Bachelor's Degree in Tourism at the Universidade Federal do Delta do Parnaíba - UFDPar. Professor of the Professional Master's degree in Tourism Business Management at the UECE (State University of Ceará). Former Head of the Bachelor's Degree in Tourism of the UFDPar (Head of Department and Coordination of Course); Director of ABRATUR - International Academy for the Development of Tourism Research in Brazil.

Jakson Renner Rodrigues Soares is a Professor in the Department of Organization at University of Coruña (UDC) and Professor of the Master in Tourism Business Management at UECE (Brazil). Post-doctoral scholarship (jan.2018-mar.2022), granted by the Fundaçao Cearense de Apoio ao Desenvolvimento Científico e Tecnológico (FUNCAP) in collaboration with the Conselho Nacional de Apoio ao Desenvolvimento Científico e Tecnológico (CNPq). Leader of the GTES-Tourism, Economy and Sustainability Research Group (CNPq-UECE-Brazil).

* * *

Francisco Alexandre Barros has a Master in Tourism Business Management at Universidade Estadual do Ceará UECE - Brazil.

Mariana Cavalheiro is currently a postdoctoral researcher at the School of Tourism and Hospitality of the Fluminense Federal University (UFF), Rio de Janeiro, Brazil. She holds both PhD and Master of Science in Business Administration from Getulio Vargas Foundation (Rio de Janeiro, Brazil). Participated in the program "Certificate in Hospitality Management" held by Walt Disney World in partnership with UCF (USA). She was awarded in 2017, by the International Federation of

Information Technology and Travel & Tourism, with an IFITT ICT4D scholarship. Research interests lie in e- tourism, digital transformation, entrepreneurship, and nautical tourism.

Priscila Cembranel is a Professor at Universidade Sociedade Educacional Santa Catarina (UNISOCIESC-Brazil). Post-doctoral scholarship in Business Admonistration (Universidade Federal do Paraná).

Vicente de Paula Censi Borges Graduated in Tourism from the "Association of Teaching of Santa Catarina", Master in Administration - Strategic Management of Organizations from "University of the State of Santa Catarina" - UDESC and Doctor in Tourism, Leisure and Culture from the University of Coimbra - Portugal. He is currently a professor with exclusive dedication at the Universidade Federal do Delta do Parnaíba, member of the EITUR-Nucleus for Interdisciplinary Studies in Tourism (Brazil). He has experience in Tourism, Administration and Communication. The areas of interest of study are: Public Tourism Management, Gross Domestic Happiness and Strategic Planning.

Christian Ehrlich is Senior Lecturer in Organisational Behaviour at Oxford Brookes Business School. His research interests are around goal-setting and happiness. He developed the goal-striving reasons framework (Ehrlich, 2012; 2018; 2019; Ehrlich & Bipp, 2016). Based on this framework, Christian developed the "Happiness through Goal-Setting Training".

Ülkühan Bike Esen received a master's degree in Human Resources Management from Gazi University. In 2016, she completed her PhD at Ankara University, Department of Business Administration. During her doctorate, she worked as a research assistant at the same university. She taught leadership at Atılım University for two years as an instructor. She is currently working as an assistant professor at Çankırı Karatekin University. Her research areas include management and organization, organizational behavior, and human resources management.

Laura Mary Fernandes is a Professor of the Master in Tourism Business Management at UECE (Brazil). Visiting Professor of the Postgraduate Program in Geography at UECE (Brazil). http://lattes.cnpq.br/3941479209132567.

Victor Figueira is an Assistant Professor at the Polytechnic Institute of Beja (IPBeja) in the area of Tourism, Environment and Sustainable Development. PhD in Tourism by the University of Évora and is now currently linked to several projects, national and international concerning social, cultural and other related to social sustainability of rural communities, tourism and regional development and social responsibility in tourism. He is a member of IPBeja team who cooperates with the Tourism Observatory of Alentejo and also the Scientific Coordinator of the project for certification of the Alentejo and Ribatejo Regional Tourism destination. Presently, he is coordinator of the Tourism Lab at IPBeja's Technology and Management School. He is also member of the Governance, Competitiveness and Public Policy research unit (GOVCOPP), where he integrates the Tourism and Development group and also in Centre for Tourism Research, Development and Innovation (CiTUR).

Massimiliano Fissore is a PhD candidate in Innovation and Sustainability Management at the Department of Economics and Management of the University of Parma. He graduated cum laude in his Master's in Business Management, Marketing and Strategies at the University of Turin.

Carla Fraga is an associate professor in the Department of Tourism and Heritage of the School of Tourism of Federal University of the State of Rio de Janeiro (UNIRIO). She holds D.Sc. in Transport Engineering from the Transport Engineering Program of the Alberto Luiz Coimbra Institute for Graduate Studies and Research in Engineering, Rio de Janeiro Federal University (PET/COPPE/UFRJ). Studying in a specialization course in Neuroscience and Applied Psychology at Mackenzie Presbyterian University (2020 – in progress). She leads the Transport and Tourism Research Group (GPTT), which has been registered with the National Council for Scientific and Technological Development (CNPq). Her research interests are tourism, transportation and neurosciences. She is member of International Academy for Tourism Research Development in Brazil (ABRATUR), and of the Brazilian Association for Tourism Research and Post-Graduate Studies (ANPTUR).

Sabrina Latusi received her PhD in Business Administration from Parthenope University of Naples. She is an Associate Professor in Management at the University of Parma, Department of Economics and Management. She has published in national and international scientific journals on issues of place branding, place marketing, and consumer behaviour.

Aline Luz is a Master's student in Tourism at Universidade Federal Fluminense (UFF). Bachelor of Tourism at Universidade Federal Fluminense (UFF). Postgraduate fellow by the Euclides da Cunha Foundation at "Mapeamento e Promoção do Turismo Náutico em Niterói (Rio de Janeiro, Brazil)".

Osiris Marques is an Associate Professor at the School of Tourism and Hospitality, Fluminense Federal University (UFF), Rio de Janeiro, Brazil. Coordinator of the federal project Experiências do Brasil Rural. He holds a PhD in Economics at the Economics Institute of the Federal University of Rio de Janeiro/Brazil (UFRJ). He is the administrative director of the Brazilian Association for Tourism Research and Post-Graduate Studies (ANPTUR). His current research interests involve topics concerning behavioural economics applied to tourism and tourism economics. https://orcid.org/0000-0002-1902-9570.

Verônica Feder Mayer is an associate professor at the School of Tourism and Hospitality of the Fluminense Federal University (UFF), Rio de Janeiro, Brazil. She holds a PhD in Business Administration from The COPPEAD Graduate School of Business of The Federal University of Rio de Janeiro (UFRJ). Her current research projects involve topics concerning consumer behavior in tourism; behavioral economics applied to tourism; and well-being in tourism and hospitality. She is deputy chair of the International Academy for Tourism Research Development in Brazil (ABRATUR), and member of the Brazilian Association for Tourism Research and Post-Graduate Studies (ANPTUR).

Sashenka Milston is an Executive Team Coach. She also continues to lecture at Oxford Brookes Business School. She has taught at universities and colleges in Australia and the UK, and trained and coached teams and individuals internationally. Sashenka holds a PhD in Psychology.

Valério Rodrigues de Souza Neto is a Master candidate in Tourism at the Fluminense Federal University (UFF), Rio de Janeiro, Brazil, member of the research group Laboratory of Behavioural Studies in Tourism. He holds a Bachelor's degree in Tourism from the Federal University of Maranhão/Brazil (UFMA). His current research is in behavioural economics and nudge theory applied to tourism. He is also a member of the federal project Experiências do Brasil rural, a project that seeks to foster and support the development of tourism in rural areas through work focused on tourism-related products that are within tourist routes and belong to the agri-food chains of cheese, wine, beer and Amazon fruits. Valério is one of the founders of the Brazilian Network for Young Tourism Researchers (Branytour). https://orcid.org/0000-0003-4680-7697.

Letícia Silva is a Master's student in Tourism at Fluminense Federal University (UFF), Rio de Janeiro, Brazil. Bachelor in Tourism at Federal University of Maranhao (UFMA), Sao Luis, Brazil, with a sandwich degree in Tourism and Hospitality at Newcastle College, Newcastle Upon Tyne - United Kingdom. Her current research involves subjective well-being and museums.

Şükran Sirkintioğlu Yildirim is an assistant professor of Banking and Finance Department at Kastamonu University. She holds a PhD from Ankara University in department of Business Administration. Her main research interests are strategic management, innovation and structural equation modelling.

Bruno Sousa is a Professor in Polytechnic Institute of Cavado and Ave (IPCA, Portugal) Head of Master Program - Tourism Management - PhD Marketing and Strategy. He was Market Analist at Sonae Distribuição – Modelo e Continente, S.A. (2006 to 2009) and he was Marketing Assistant - Jornal O Jogo at Controlinveste (2005) - Teaching Award of the School of Economics and Management of the University of Minho 2015/2016 - Best Thesis in Tourism Award - ICIEMC 2015 - Management Graduation, University of Minho Award - Best performance (2006) - Merit Scholarship for Students in Public Higher Education Awards of Merit Scholarship by University of Minho in 2001/02 - 2002/03 - 2003/04 Rresearch centre: CiTUR and Applied Management Research Unit (UNIAG). He is author or co-author of several papers and her research interests include tourism management, marketing and strategy. Editorial board member of several peer reviewed scientific journals and ad-hoc reviewer of several peer-reviewed scientific journals. Member of the scientific committee of several national and international congresses and conferences.

Robert van der Veen is a Research Fellow and his research interests lie in tourism, hospitality, and services marketing. He specializes in quantitative research methods and structural equation modelling. His work has been published in journals such as the Annals of Tourism Research, Journal of Travel Research, and International Journal of Contemporary Hospitality Management.

Laurentina Vareiro is coordinating professor of Economics at the School of Management, Polytechnic Institute of Cávado and Ave (IPCA) Portugal, and research centre: Applied Management Research Unit (UNIAG), Portugal. She has a PhD in Economics (2008) at the University of Minho, Braga. She is the Quality Coordinator of the School of Management and she is author or co-author of several papers and her research interests include regional, rural and urban economics, tourism economics, rural and cultural tourism. Editorial board member of several peer reviewed scientific journals and ad-hoc reviewer of several peer-reviewed scientific journals, namely, Tourism Geographies, Tourism Management Perspectives, International Journal of Tourism Research, Asia Pacific Journal of Tourism Research, Anatolia, etc. Member of the scientific committee of several national and international congresses and conferences.

Index

A

B

C

D

E

Q

R

S

T

U

V

W

Recommended Reference Books

IGI Global's reference books can now be purchased from three unique pricing formats: Print Only, E-Book Only, or Print + E-Book.
Shipping fees may apply.

www.igi-global.com

ISBN: 978-1-7998-1048-3
EISBN: 978-1-7998-1049-0
© 2020; 605 pp.
List Price: US$ 285

ISBN: 978-1-5225-8933-4
EISBN: 978-1-5225-8934-1
© 2020; 667 pp.
List Price: US$ 295

ISBN: 978-1-7998-0357-7
EISBN: 978-1-7998-0359-1
© 2020; 451 pp.
List Price: US$ 235

ISBN: 978-1-7998-0070-5
EISBN: 978-1-7998-0071-2
© 2020; 144 pp.
List Price: US$ 175

ISBN: 978-1-5225-6980-0
EISBN: 978-1-5225-6981-7
© 2019; 339 pp.
List Price: US$ 235

ISBN: 978-1-5225-5390-8
EISBN: 978-1-5225-5391-5
© 2018; 125 pp.
List Price: US$ 165

Do you want to stay current on the latest research trends, product announcements, news, and special offers?

Join IGI Global's mailing list to receive customized recommendations, exclusive discounts, and more.
Sign up at: **www.igi-global.com/newsletters.**

Publisher of Peer-Reviewed, Timely, and Innovative Academic Research

IGI Global
PUBLISHER of TIMELY KNOWLEDGE

www.igi-global.com Sign up at www.igi-global.com/newsletters facebook.com/igiglobal twitter.com/igiglobal

Ensure Quality Research is Introduced to the Academic Community

Become an Evaluator for IGI Global Authored Book Projects

The overall success of an authored book project is dependent on quality and timely manuscript evaluations.

Applications and Inquiries may be sent to:
development@igi-global.com

Applicants must have a doctorate (or equivalent degree) as well as publishing, research, and reviewing experience. Authored Book Evaluators are appointed for one-year terms and are expected to complete at least three evaluations per term. Upon successful completion of this term, evaluators can be considered for an additional term.

If you have a colleague that may be interested in this opportunity, we encourage you to share this information with them.

IGI Global Author Services

Providing a high-quality, affordable, and expeditious service, IGI Global's Author Services enable authors to streamline their publishing process, increase chance of acceptance, and adhere to IGI Global's publication standards.

Benefits of Author Services:

- **Professional Service:** All our editors, designers, and translators are experts in their field with years of experience and professional certifications.
- **Quality Guarantee & Certificate:** Each order is returned with a quality guarantee and certificate of professional completion.
- **Timeliness:** All editorial orders have a guaranteed return timeframe of 3-5 business days and translation orders are guaranteed in 7-10 business days.
- **Affordable Pricing:** IGI Global Author Services are competitively priced compared to other industry service providers.
- **APC Reimbursement:** IGI Global authors publishing Open Access (OA) will be able to deduct the cost of editing and other IGI Global author services from their OA APC publishing fee.

Author Services Offered:

English Language Copy Editing
Professional, native English language copy editors improve your manuscript's grammar, spelling, punctuation, terminology, semantics, consistency, flow, formatting, and more.

Scientific & Scholarly Editing
A Ph.D. level review for qualities such as originality and significance, interest to researchers, level of methodology and analysis, coverage of literature, organization, quality of writing, and strengths and weaknesses.

Figure, Table, Chart & Equation Conversions
Work with IGI Global's graphic designers before submission to enhance and design all figures and charts to IGI Global's specific standards for clarity.

Translation
Providing 70 language options, including Simplified and Traditional Chinese, Spanish, Arabic, German, French, and more.

Hear What the Experts Are Saying About IGI Global's Author Services

*"Publishing with IGI Global has been **an amazing experience** for me for sharing my research. The **strong academic production** support ensures quality and timely completion."* – **Prof. Margaret Niess, Oregon State University, USA**

*"The service was **very fast, very thorough, and very helpful** in ensuring our chapter meets the criteria and requirements of the book's editors. I was **quite impressed and happy** with your service."* – **Prof. Tom Brinthaupt, Middle Tennessee State University, USA**

Learn More or Get Started Here:

For Questions, Contact IGI Global's Customer Service Team at cust@igi-global.com or 717-533-8845

IGI Global
PUBLISHER of TIMELY KNOWLEDGE
www.igi-global.com

IGI Global
PUBLISHER of TIMELY KNOWLEDGE
www.igi-global.com

Celebrating Over 30 Years of Scholarly Knowledge Creation & Dissemination

InfoSci®-Books

A Database of Nearly 6,000 Reference Books Containing Over 105,000+ Chapters Focusing on Emerging Research

InfoSci®-Books Database

The **InfoSci®-Books** is a database of nearly 6,000 IGI Global single and multi-volume reference books, handbooks of research, and encyclopedias, encompassing groundbreaking research from prominent experts worldwide that spans over 350+ topics in 11 core subject areas including business, computer science, education, science and engineering, social sciences, and more.

Open Access Fee Waiver (Read & Publish) Initiative

For any library that invests in IGI Global's InfoSci-Books and/or InfoSci-Journals (175+ scholarly journals) databases, IGI Global will match the library's investment with a fund of equal value to go toward **subsidizing the OA article processing charges (APCs) for their students, faculty, and staff** at that institution when their work is submitted and accepted under OA into an IGI Global journal.*

INFOSCI® PLATFORM FEATURES

- Unlimited Simultaneous Access
- No DRM
- No Set-Up or Maintenance Fees
- A Guarantee of No More Than a 5% Annual Increase for Subscriptions
- Full-Text HTML and PDF Viewing Options
- Downloadable MARC Records
- COUNTER 5 Compliant Reports
- Formatted Citations With Ability to Export to RefWorks and EasyBib
- No Embargo of Content (Research is Available Months in Advance of the Print Release)

**The fund will be offered on an annual basis and expire at the end of the subscription period. The fund would renew as the subscription is renewed for each year thereafter. The open access fees will be waived after the student, faculty, or staff's paper has been vetted and accepted into an IGI Global journal and the fund can only be used toward publishing OA in an IGI Global journal. Libraries in developing countries will have the match on their investment doubled.*

To Recommend or Request a Free Trial:
www.igi-global.com/infosci-books
eresources@igi-global.com • Toll Free: 1-866-342-6657 ext. 100 • Phone: 717-533-8845 x100

IGI Global
PUBLISHER of TIMELY KNOWLEDGE
www.igi-global.com

IGI Global
PUBLISHER of TIMELY KNOWLEDGE
www.igi-global.com

Publisher of Peer-Reviewed, Timely, and Innovative Academic Research Since 1988

IGI Global's Transformative Open Access (OA) Model: How to Turn Your University Library's Database Acquisitions Into a Source of OA Funding

Well in advance of Plan S, IGI Global unveiled their OA Fee Waiver (Read & Publish) Initiative. Under this initiative, librarians who invest in IGI Global's InfoSci-Books and/or InfoSci-Journals databases will be able to subsidize their patrons' OA article processing charges (APCs) when their work is submitted and accepted (after the peer review process) into an IGI Global journal.

How Does it Work?

Step 1: **Library Invests in the InfoSci-Databases:** A library perpetually purchases or subscribes to the InfoSci-Books, InfoSci-Journals, or discipline/subject databases.

Step 2: **IGI Global Matches the Library Investment with OA Subsidies Fund:** IGI Global provides a fund to go towards subsidizing the OA APCs for the library's patrons.

Step 3: **Patron of the Library is Accepted into IGI Global Journal (After Peer Review):** When a patron's paper is accepted into an IGI Global journal, they option to have their paper published under a traditional publishing model or as OA.

Step 4: **IGI Global Will Deduct APC Cost from OA Subsidies Fund:** If the author decides to publish under OA, the OA APC fee will be deducted from the OA subsidies fund.

Step 5: **Author's Work Becomes Freely Available:** The patron's work will be freely available under CC BY copyright license, enabling them to share it freely with the academic community.

Note: *This fund will be offered on an annual basis and will renew as the subscription is renewed for each year thereafter. IGI Global will manage the fund and award the APC waivers unless the librarian has a preference as to how the funds should be managed.*

Hear From the Experts on This Initiative:

"I'm very happy to have been able to make one of my recent research contributions ***freely available*** along with having access to the ***valuable resources*** found within IGI Global's InfoSci-Journals database."

– **Prof. Stuart Palmer**, Deakin University, Australia

"Receiving the support from IGI Global's OA Fee Waiver Initiative ***encourages me to continue my research work without any hesitation***."

– **Prof. Wenlong Liu**, College of Economics and Management at Nanjing University of Aeronautics & Astronautics, China

For More Information, Scan the QR Code or Contact:
IGI Global's Digital Resources Team at eresources@igi-global.com.

IGI Global
PUBLISHER of TIMELY KNOWLEDGE

Are You Ready to **Publish Your Research?**

IGI Global
PUBLISHER of TIMELY KNOWLEDGE

IGI Global offers book authorship and editorship opportunities across 11 subject areas, including business, computer science, education, science and engineering, social sciences, and more!

Benefits of Publishing with IGI Global:

- Free one-on-one editorial and promotional support.
- Expedited publishing timelines that can take your book from start to finish in less than one (1) year.
- Choose from a variety of formats, including: Edited and Authored References, Handbooks of Research, Encyclopedias, and Research Insights.
- Utilize IGI Global's eEditorial Discovery® submission system in support of conducting the submission and double-blind peer review process.
- IGI Global maintains a strict adherence to ethical practices due in part to our full membership with the Committee on Publication Ethics (COPE).
- Indexing potential in prestigious indices such as Scopus®, Web of Science™, PsycINFO®, and ERIC – Education Resources Information Center.
- Ability to connect your ORCID iD to your IGI Global publications.
- Earn honorariums and royalties on your full book publications as well as complimentary copies and exclusive discounts.

Join Your Colleagues from Prestigious Institutions, Including:

Australian National University
Massachusetts Institute of Technology
Johns Hopkins University
Harvard University
Columbia University in the City of New York

Learn More at: www.igi-global.com/publish

or by Contacting the Acquisitions Department at: acquisition@igi-global.com

Purchase Individual IGI Global InfoSci-OnDemand Book Chapters and Journal Articles

InfoSci®-OnDemand

Pay-Per-View Articles and Chapters from IGI Global

For More Information, Visit: www.igi-global.com/e-resources/infosci-ondemand

Browse through nearly 150,000+ articles/chapters to find specific research related to their current studies and projects that have been contributed by international researchers from prestigious institutions, including MIT, Harvard University, Columbia University, and many more.

Easily Identify, Acquire, and Utilize Published Peer-Reviewed Findings in Support of Your Current Research:

Accurate and Advanced Search: Utilize the advanced InfoSci-OnDemand search engine to identify research in your area of interest.

Affordably Acquire Research: Provides an affordable alternative to purchasing an entire reference book or scholarly journal.

Fast and Easy One-Click Shopping: Simply utilize the OnDemand "Buy Now" button on the webpage.

Instantly Download/Access Your Content: Receive an immediate link to download your InfoSci-OnDemand content to your personal computer and device.

Access Anywhere, Anytime, and on Any Device: Additionally, receive access to your OnDemand articles and chapters through a personal electronic library.

Benefit from the InfoSci Platform Features: Providing formatted citations, full-text PDF and HTML format, no embargo of content, no DRM, and more.

"It really provides **an excellent entry into the research literature of the field**. *It presents a manageable number of* **highly relevant sources** *on topics of interest to a wide range of researchers. The sources are* **scholarly, but also accessible** *to 'practitioners'."*

\- Ms. Lisa Stimatz, MLS, University of North Carolina at Chapel Hill, USA

Interested in Additional Savings and Acquiring Multiple Articles and Chapters Through InfoSci-OnDemand?

Subscribe to InfoSci-OnDemand Plus

Learn More

www.ingramcontent.com/pod-product-compliance
Lightning Source LLC
LaVergne TN
LVHW080557300926
849162LV00005B/979